One Hundred Years of Chemical Warfare: Research, Deployment, Consequences

Bretislav Friedrich · Dieter Hoffmann
Jürgen Renn · Florian Schmaltz
Martin Wolf
Editors

One Hundred Years of Chemical Warfare: Research, Deployment, Consequences

Springer Open

Editors
Bretislav Friedrich
Fritz Haber Institute of the Max Planck Society
Berlin
Germany

Dieter Hoffmann
Max Planck Institute for the History of Science
Berlin
Germany

Jürgen Renn
Max Planck Institute for the History of Science
Berlin
Germany

Florian Schmaltz
Max Planck Institute for the History of Science
Berlin
Germany

Martin Wolf
Fritz Haber Institute of the Max Planck Society
Berlin
Germany

ISBN 978-3-319-84712-2 ISBN 978-3-319-51664-6 (eBook)
DOI 10.1007/978-3-319-51664-6

Printed on acid-free paper

This Springer imprint is published by Springer Nature
The registered company is Springer International Publishing AG
The registered company address is: Gewerbestrasse 11, 6330 Cham, Switzerland

Foreword

This book is a collection of the contributions to the symposium organized at the Fritz Haber Institute in commemoration of the 100th anniversary of the gas attack at Ypres during the First World War. A centennial is normally a celebratory event, but certainly not so in this case: The centennial of the first large-scale gas attack in Ypres is an event that we commemorate as a dark hour in human history, an event where I as a chemist—as many chemists—feel ashamed, and it is also an event that gives us reason to consider the responsibility of scientists for their actions—or the lack thereof. If one reads the reports, even now, one hundred years later, one can only feel deeply affected by the suffering inflicted on the gas-attack victims and on their families.

The First World War—like almost any war—was a time during which numerous atrocities were committed in the name of patriotism. The use of gas to kill and incapacitate soldiers was one of the worst of these atrocities, and it made use of the science and technology developed by the chemical industry, which at that time flourished and prospered, especially in Germany. Science and technology have been used time and again in human history in order to improve weapons technology, but the chemical warfare in the First World War was arguably the first time at which the precision of the modern scientific methods was employed for making war more efficient and deadly, an approach that culminated about 30 years later in the Manhattan Project. Our view of events and actions changes with time, and what seemed right at the time it was done may appear utterly wrong when judged later from a distance. However, chemical warfare was not judged unequivocally as being right even during the First World War, and it was internationally banned by the Geneva Convention in 1925.

These introductory words I write in two different roles: first as a chemist, as a professional fellow of those who developed and deployed this cruel weapon, or, maybe more accurately, means of indiscriminate mass killing. I feel responsible for preventing anything like it from happening again. As a chemist, I hope that the brilliant minds in our science will turn toward research that will help mankind and not toward research to kill fellow humans. We have to teach this lesson to the younger generation, and I am convinced that the vast majority of chemists today feel the same.

At the venue of the symposium, the Harnack House, a stone's throw from the Fritz Haber Institute, it was not possible to discuss the chemical attack at Ypres without talking about Fritz Haber. Thus, I am writing these words also as vice president of the Max Planck Society, the successor of the Kaiser Wilhelm Society, which was the scientific home of Fritz Haber, one of the most prominent proponents of gas warfare. We are also the organization to which an institute bearing Fritz Haber's name belongs.

The institute was named after its founder in 1952—as an act of redress after the injustice committed by the Nazi regime against Fritz Haber. But as the plaque at the institute's building says, "the name of the institution is equally reminiscent of bright and dark sides of an eminent researcher in German history and is therefore a piece of living memory which should not be abandoned. The name is not solely intended as a tribute to Haber the scientist, but is a critical appreciation of an exemplary life in his time, which gives an impetus for reflection, for a differentiating assessment, and a memento for our own time."

This very institute was one of the organizers of this symposium, together with the Max Planck Institute for the History of Science, which clearly shows that we are also well aware of the dark sides in the life of its name-giver, and that we take responsibility by calling attention to these dark sides.

Most human beings and their deeds in the world are neither black nor white; they exhibit different shades of gray, possibly at different times in their lives. Fritz Haber was such a Janus-faced man. He developed the ammonia synthesis, a process without which the world would not be able to sustain the population of today; a process which is needed to feed the world. His publication on the ammonia synthesis, together with Robert le Rossignol, in 1913, is one of the true landmarks of science. I recommend it to each and every of my students as a masterpiece from the history of science. It is a single paper which nowadays would carry many scientists through their whole careers. It clarified the thermodynamics, described the innovative recycling concept, the high pressure technology needed, several classes of catalysts, and the kinetics of the process observed with these catalysts. On the other hand, this scientific hero turned his attention less than two years later to chemical warfare, like many other of his fellow chemists and physicists. With the same scientific approach, the same precision he had used in the research on the ammonia process, he developed the basics for the deployment of chlorine as an agent for chemical warfare. The same brilliant mind was then pursuing a goal that is nowadays clearly rejected as an aberration of science; the precision in the investigation of the most efficient ways of killing other human beings makes me shudder today.

However, while this view is probably unanimously shared today, it was not fully prevalent at the time, which is in the focus of this symposium. Scientists in the belligerent countries were working on gas warfare, and the question of whether this was ethically, morally, and legally permitted was highly disputed. Humankind progresses, hopefully, and we should take the lessons learned in order to turn the power of science toward the betterment of the human condition. Ypres stands as an admonition of where science can lead humans. Fortunately, most scientists involved in this effort during the First World War worked on weapons research only for a limited time. Except for this dark time, they made many discoveries that helped to lay the foundations of the technology our societies rely on today. The latter discoveries are the aspects of science that we should foster and highlight as exemplary for the younger generation, so that we never revert to the kind of research performed in the dark years of World War I.

The two Max Planck Institutes that organized this commemorative event and edited this book did it in the way that is appropriate for the Max Planck Society: in the form of a scientific symposium and a scientific publication, in order to elucidate the various aspects of chemical warfare introduced at the beginning of the last century: "Research, Deployment, Consequences," as the subtitle of the symposium reads. May this book bring new insights for its readers, but more importantly, may it serve as a remembrance of the many victims of chemical warfare, and may it remind us to never let science be corrupted in this manner again.

Ferdi Schüth
Vice President of the Max Planck Society

Contents

Introduction

Jürgen Renn

Taking the horrific events that took place at Ypres in 1915 as its point of departure, this volume traces the development of chemical weapons from their first use as weapons of mass destruction by German troops in Belgium to their deployment in Syria in the summer of 2013. The book has emerged from a conference commemorating the centenary of the events at Ypres, held at the Fritz Haber Institute of the Max Planck Society in Berlin. The contributions focus on the preconditions and immediate consequences of this war crime, but also cover, by way of examples, the subsequent history of chemical weapons, including their role in World War II, their global spread, and their recent deployment. The volume ends with a documentation of the commemoration ceremony closing the conference, comprising speeches of the Green Cross director Paul Walker, the Belgian ambassador Ghislain D'hoop, and the Nobel laureate Gerhard Ertl.

The first part of the volume deals with "Research on and Deployment of Chemical Weapons in World War I," as well as with the roles of the key actors involved. The dual-use characteristics of chemistry are, first and foremost, emblematically represented by the figure of Fritz Haber. In the history of science, he has played a double role. On the one hand, he is one of the most outstanding chemists of the twentieth century and even a benefactor of mankind. Through his development of ammonia synthesis, he is the most influential chemist as regards the history of humanity as a whole. Haber played a key role in negotiating the conditions and contracts for the large-scale industrial synthesis of ammonia required by the German Army for the production of munition to continue the war. His commitment to these negotiations led to the establishment of industrial capacities based on the industrial process that had been brought to maturity in 1912 by Carl Bosch and Alwin Mittasch, based on the scientific work of Haber. This dual-use process is still the present-day basis for the production of fertilizers, without which modern agriculture would not be able to feed the current world population that has grown

J. Renn (✉)
Chair of the Human Sciences Section of the Max Planck Society, Munich, Germany

B. Friedrich et al. (eds.), *One Hundred Years of Chemical Warfare: Research, Deployment, Consequences*, DOI 10.1007/978-3-319-51664-6_1

from around 1 billion in 1900 to over 7 billion today. On the other hand, Fritz Haber was also the pioneer of the deployment of science-based weapons of mass destruction.

Margit Szöllösi-Janze, author of the authoritative biography of Haber, investigates this dual role. In her contribution, she describes how the "expert culture" that gave rise to "big science" and "big industry" came about in Germany during World War I and also Fritz Haber's key role in this historic change. In particular, she reconstructs the conditions that led, on April 22, 1915, to the use of chlorine gas in Ypres following the initiative of Fritz Haber, who also provided the scientific guidance for its deployment. Motivating this action were the concerns of the German military about a possible "explosives shortage" that could ensue should the war drag on and access to supplies of natural nitrates be blocked by the allies. Given this concern, Haber's ammonia synthesis would prove to be very opportune.

In September of 1914, the military had already suggested that by-products from the manufacture of explosives could be used as chemical weapons. This solution also served industrial interests. The chief of staff, Erich von Falkenhayn, took up these suggestions and installed a commission that later included Haber. He was among those scientists and experts who offered their services to the military when the war broke out. Not only was Haber driven by the ambition to solve the problems of war in a technocratic way, that is, by means of science and technology, but he also sought to create a network connecting industry, academia, the military, and the politicians, thereby promoting the societal role of scientists. At the end of the war, around 1000 scientists were involved in the development of gas warfare in Germany, 150 alone from Haber's rapidly expanding Kaiser Wilhelm Institute. This represented a striking success that would have lasting consequences for the relation between science and the military.

The above history of the inception and implementation of chemical warfare in World War I Germany is described in the contribution by Bretislav Friedrich and Jeremiah James, who follow Haber's pathway from science to chemical warfare in greater detail and show how the Kaiser Wilhelm Institute for Physical Chemistry and Electrochemistry in Berlin-Dahlem became a center for the development of chemical weapons and of protective measures against them. They analyze, in particular, the role of Fritz Haber and his Kaiser Wilhelm Institute in the acceptance and use by the German military of chemical weapons as a means of resolving the greatest strategic challenge of World War I, namely the stalemate of trench warfare. The paper details the path from the Ni-Stoff and T-Stoff to the chlorine cloud and beyond, the transformation of Haber's Kaiser Wilhelm Institute after it fell under military command, as well as Haber's views on chemical warfare. The implications of this transformation of science into a military resource are investigated in further contributions to this volume.

But also the personal contexts of Haber's involvement in the war are illuminated. The contribution by Bretislav Friedrich and Dieter Hoffmann examines, on the basis of the available scholarly sources, the life of Fritz Haber's first wife, Clara, nee Immerwahr, including her suicide and its possible relation to her husband's involvement in chemical warfare. They also critically re-examine the origin of the "myth of Clara Immerwahr," according to which she was a top scientist, not unlike Marie Curie, and an outspoken pacifist, not unlike Bertha von Suttner.

The German deployment of chemical weapons of mass destruction had far-reaching repercussions for warfare. Olivier Lepick investigates the French reaction to the German poison gas attack and shows how this attack initiated a political and moral chain reaction in the course of which any reservations concerning the use of chemical weapons were abandoned by all sides involved. In France, the relevant decisions were taken by the military, largely independently of the politicians. In this way, chemical warfare led to one of the first arms races, with massive involvement of science and industry. Great efforts were undertaken, for instance, to rapidly overcome the French deficit with respect to the German chemical industry. The sites and structures of chemical industry still reflect this race even today.

Ulf Schmidt has reconstructed the pertinent developments on the British side. British scientists shared with their German colleagues the ambition to support the military with the development of an arsenal of chemical weapons. As in Germany, ethical concerns and the norms of international law were overruled in response to the perceived German radicalization of war and its disregard for any formerly established restrictions (*Entgrenzung*). Even Churchill strongly argued for the use of chemical weapons of mass destruction.

This situation is all the more surprising because during the nineteenth century, international law seemed to be taking huge steps forward and German international law was highly regarded throughout Europe. As the historian of law Miloš Vec argues in his contribution, the possibility of gas warfare was anticipated long before war broke out, but neither this nor the Haag Convention of 1907 had any impact on legal efforts to contain it. In any case, the Haag Convention left open loopholes that the contemporary actors used after the war to justify their decisions. It was argued, for instance, that the Haag Convention does not cover gas attacks originating from rigidly installed batteries rather than movable artillery or that "military necessity" could be used to justify violations of international laws.

Remarkably, the ethical and legal evaluation of gas warfare was severely hindered by several mechanisms. Among these mechanisms was not only the disregard for international law and the argument that war has its own logic, but also the military's aversion to overtly assume responsibility for the use of chemical weapons which clashed with the pretense to chivalry. Another mechanism was the relativization of chemical warfare in terms of its comparative assessment from a "sober" scientific perspective, as may be illustrated by Haber's infamous remark in a talk at the Reichstag: "Cyanide—there is no nicer way to die."[1]

Rather than simply documenting his inhumanity, as is often portrayed, it shows an attempt to shed responsibility by focusing on the scientific aspects, comparing

[1]"Die Einatmung der Blausäure belästigt in keiner Weise. Man kann nicht angenehmer sterben." Fritz Haber. 1924. *Zur Geschichte des Gaskrieges: Vortrag, gehalten vor dem parlamentarischen Untersuchungsausschuß des Deutschen Reichstages am 1. Oktober 1923*, p. 81. In Fritz Haber. *Fünf Vorträge aus den Jahren 1920–1923*. Berlin: Springer, 1924, p. 75–92.

the gruesome effects of ethyl bromo-acetate deployed by the French (already in August 1914) against German troops with those of hydrogen cyanide: While the toxicity of both gases was about the same, Haber argued, the effects of ethyl bromo-acetate caused torturous inhalation injuries in addition to death by asphyxiation. All of these maneuvers contributed to the repression of critical reflections and certainly played a role in the further expansion of chemical weapon capacities, which took place largely unnoticed by the public.

Another prominent theme in several contributions to this volume is how the junctions, networks, and structures resulting from World War I continued to shape the relation between science and war until World War II and far beyond. This path-dependency of economic, military, and scientific events is also at the center of the contribution by Jeffrey Johnson, which emphasizes the systemic character and institutionalization of the symbiosis between academic science and industry in the service of the military—in the course of and as a consequence of World War I. This symbiosis was also a consequence of feedback effects between German and Allied developments that led to what might be characterized as a veritable globalization of the "Haber complex" of science and the military.

Experimentation with the new warfare technology was followed by upgrades and successive improvements through ever more innovations and ever more widespread deployment of chemical weapons. Globalization and the dual-use character of chemical fertilizers, disinfectants, and drugs are mostly responsible for the persistent spread of chemical weapons of mass destruction in spite of the various attempts to ban them. These farther-reaching implications are more deeply explored in the second part of this book, dedicated to the "Contexts and Consequences of Chemical Weapons."

The sustained influence of the Haber complex was by no means obvious from the outset. The military historian Edward Spiers has argued that the "success" of the German gas attack and the use of chemical weapons in general were both highly controversial after the end of World War I. Certainly, chemical weapons were not decisive for the outcome of the war, partly because the integration of the new weapons into existing tactics and strategies was unclear. But as Spiers points out, the introduction of mustard gas in 1917 not only increased the number of gas casualties, but also promoted the use of chemical weapons during the later stages of the war. In any case, investments into the further development of chemical weapons seemed to be a worthwhile endeavor.

One thing was clear and is shown by several of the contributions to this volume: the horrific psychological effects of these weapons and their terrifying character. These effects are impressively illustrated and discussed in Doris Kaufmann's contribution on the gas war in European literature and art during the interwar period. Remarkably, there were no images that glorified gas warfare, while many images and literary accounts preserved the experience of the horror instilled by these weapons. They show the impersonality of war, the feeling of helplessness in gas attacks, and the shock of seeing one's comrades suffer. Doris Kauffmann also addresses the public battle over the interpretation and collective remembrance in the war's aftermath.

The immediate postwar period is also at the center of the contribution by Roy MacLeod, which analyzes the debate about chemical weapons in the USA and the questions of its willingness to deploy them as well as the means it had for protection against them. The paper traces, in particular, the sordid history of the refusal of the US political establishment to become party to the 1925 Geneva Protocol. The contribution also shows the extent to which this debate was part of a public discourse that in some respects resembles the later discussions about the protection—often naïve and futile—against nuclear weapons.

The discussion of the wider political and historical contexts should not let us forget the immense suffering induced by chemical weapons, even long after the end of World War I. In his contribution, Wolfgang Eckart has, on the basis of numerous historical documents, brought to light the misery that the war caused for the one thousand gas casualties and the one million wounded. Ultimately, there was no shelter from gas attacks and the injuries were unspeakable: blindness, suffocation, blistering of the skin, pulmonary edema, and, ultimately, an excruciatingly painful death. Many victims also suffered harrowing psychic damage that was treated by using electroshock therapy. If the victims managed to survive this, they were then sent back to the front. Those who remained were socially stigmatized. During the era of the Weimar Republic, the pensions of such victims were cut, and during the Nazi period, some soldiers involved in the gas war who had certifiable psychic damage became victims of "euthanasia" killings. Even in the early Federal Republic, the victims continued to be humiliated by Nazi doctors. Scientists and military personnel described their suffering in a sober and factual language—despite the fact that scientists like Haber, in particular due to work undertaken in the toxicological department of his institute, were intimately familiar with the suffering caused by gas.

The indifference of scientists to the suffering of gas victims turned into their outright instrumentalization by scientists under the Nazi regime, as is shown in the contribution by Florian Schmaltz that addresses the research on chemical weapons undertaken during this period. Schmaltz describes how soldiers and concentration camp inmates were forced to take part in inhumane experiments. His analysis makes it strikingly clear that these criminal experiments did not take place in a covert space but in the midst of a network of communications in which these experiments and their results were requested and evaluated by the military and other officials. To a large extent, the scientists who ran these experiments were able to continue their careers unhindered and suffered no consequences after World War II.

Two contributions are dedicated to the role of chemical warfare on the Pacific front during World War II. The paper by Walter Grunden is dedicated to a discussion of the remarkable phenomenon that the war between Japan and the allies did not lead to a mutual acceleration of gas attacks, as happened between Germany and France during World War I. This is all the more surprising because between 1937 and 1945, the Japanese military used chemical weapons on more than 2000 occasions, mainly in the war against China. The reason for the reluctance of the Japanese to deploy chemical weapon was fear of retaliation in kind, a counter-attack with the same weapons. The Japanese were very aware that the Americans were capable of retaliation, unlike the technologically inferior Chinese military.

As Jeanne Guillemin shows, even months before the end of the Pacific War on September 2, 1945, a commission on war crimes from the United Nations and the Chinese government had begun to take stock of Japanese war crimes. These included several instances of the use of chemical and biological weapons. Between 1945 and 1946, US officials also pursued an investigation into these war crimes with the participation of the United States Chemical Warfare Services, but with the aim of keeping this information secret for reasons of national security. One of the reasons for this secrecy was the fear that once the Americans publically accused the Japanese, they would themselves be liable to accusations of having used such weapons, as when they dropped atomic bombs on Japan. The Geneva Protocol of 1925 that unequivocally prohibits the use of chemical weapons was ratified by the USA only in 1972. A comprehensive consensus regarding the Geneva Protocol and the passing of the convention on chemical weapons emerged even later, after the end of the Cold War. This was also due to the fact that chemical weapons had in the meantime spread across many new states as, metaphorically speaking "the poor man's nuclear weapon."

The final part of the book is dedicated to current issues of "Dual Use, Storage and Disposal of Chemical Weapons Today." One of the key challenges is how to control the existence of chemical weapons under difficult conditions. The proliferation of chemical weapons continues due to global trade relations and the lack of export regulations from many national legislatures. More detailed inspections are necessary in order to guarantee compliance with international agreements, such as the convention on chemical weapons, which has been signed by 191 nations. Such inspections also require ever-larger scientific and technical efforts.

The production plants and storage spaces of chemical weapons from both world wars still profoundly shape Europe today, as the contribution by Johannes Preuss makes evident. The frontlines of World War I have not been cleared and only superficially buried. While there were no secret production plants for chemical weapons during World War I because they were produced by the traditional chemical industry, such plants were built and hidden in forests and at remote sites during the Nazi regime. A company founded by the Armament Office (Heereswaffenamt) under the camouflage name Verwertungsgesellschaft für Montanindustrie GmbH (collecting society for the coal and steel industry), founded in 1935, constructed more than 114 factories in which tens of thousands of tons of mustard gas, tabun, and other poisonous substances were produced that have sustained damage to and continue to impact on even today's environment. In Ammendorf, south of Halle an der Saale, for instance, more than 25,000 tons of mustard gas were produced. During the GDR period, the buildings on the site were demolished and simply covered by three meters of earth. Therefore, it can be expected that the territory is still very highly contaminated. In Falkenhagen, east of Berlin, chlorine trifluoride was produced using a technique developed by a collaborator from the Kaiser Wilhelm Institute for Physical Chemistry and Electrochemistry. After the war, many of these poisonous substances were only superficially removed. Large quantities of mustard gas have been filled and either brought to a mine in Italy or disposed of in the Baltic Sea. The decontamination of the production sites is feasible but would require considerable investment.

In his contribution, the biologist Matthew Meselson draws on his experience as advisor in the 1960s to the United States Arms Control and Disarmament Agency in Washington. He addresses the use of chemical weapons to destroy vegetation and in particular for deforestation and the destruction of crops during the Vietnam War, and also the debates surrounding these issues. The military ineffectiveness of this deployment soon became apparent, as enemy provision and supply lines were not seriously affected. It nevertheless took a considerable amount of time before a decision was taken to renounce their use. The circumstances under which President Nixon was presented with the relevant information turned out to be decisive. Despite the obvious uselessness and absurdity of the operation, the deployment of chemical weapons could only be stopped at the presidential level. Overcoming the compartmentalization of information in large institutions is evidently a crucial measure in increasing rational decision-making.

The account by Karin Mlodoch of the suffering of the Kurdish victims of the 1988 poison gas attack by the Bath Regime in Halabja confronts the reader with another aspect of the drama that still persists to this day. It traces the "indelible smell of apples" that has left indelible traces in the minds of the survivors of this attack. The persecution of around 6 million Kurdish people in Iraq began after 1978 and reached an apex during the Iraq–Iran war between 1980 and 1988 with around one million casualties. In 1987, Saddam Hussein's cousin Ali Hassan led a campaign against the rural areas inhabited by the Kurds and destroyed around one thousand villages. During this military campaign, chemical weapons such as cyanide, sarin, and VX were used, and not only in Halabja. German enterprises had delivered the expertise and technology for the production of these weapons; whether they had done this consciously or inadvertently remains controversial because of the dual-use issue. There was virtually no international aid for the thousands of victims. Many fled to Iran as the only location where medical help could be sought. Many survivors died in the following years, handicapped children were born, and orphans sought their identity. Today these victims struggle to have this poison gas attack recognized as genocide, rather than have it trivialized as the "collateral damage" of war.

In his contribution, the expert for disarmament Ralf Trapp reports on his experiences in the context of United Nations inspections dealing with the removal of Syrian poison gas stocks in the sequel of the deployment of sarin in 2013. This deployment differed from other uses because only small quantities were deployed with local effects. Nevertheless, an estimated 600 to 1300 people were killed. A further peculiarity of this case is that the removal of chemical weapons had to take place in the midst of a civil war. Contrary to the rules of the Organization for the Prohibition of Chemical Weapons (OPCW), the stocks were not destroyed on site but rather taken out of the country. For this purpose, a complex plant for the destruction of chemical weapons had to be installed on a ship where the highly toxic substances were decomposed into less toxic ones, which were then brought to Germany and Finland for final disposal. This case illustrates what some years ago was still inconceivable: A multilateral operation can work together to dispose of chemical weapons despite tensions and disagreements on essential questions of politics and security.

This hopeful view is confirmed by Paul Walker's magisterial review of a century of chemical warfare. Looking back at one hundred years of chemical warfare, the significance of a worldwide agreement on chemical weapons can hardly be overestimated. The 1993 Chemical Weapons Convention (CWC) is undoubtedly a key achievement, which in 2016 included 192 countries with 98% of the world's population. That this is more than a political framework becomes clear when considering that of the 72,525 metric tons of chemical agents declared to date in eight possessor states, over 66,000 metric tons—92%—have been destroyed in the last 25 years.

The volume closes with two authoritative statements. The ambassador of the Kingdom of Belgium, Ghislain D'hoop, points to the special role of diplomats and scientists "in making sure that the world fully understands the horrors of chemical warfare and unites in condemning its manufacturing, stockpiling, and use." Nobel Prize Winner Gerhard Ertl from the Fritz Haber Institute reminds us that the glory of Germany's rise after World War II is "tarnished by dark stains that have imbued later generations with sorrow and shame." These statements constitute a legacy to be taken to heart whenever we confront the challenges of the powerful potential of science, obliging us to take responsibility, also with a view to future generations, for its conscientious use.

The editors would like to thank all those who helped to organize the international conference on April 21–22, 2015, in particular Daria Haberland and Manuel Krüger (Fritz-Haber-Institut) and Norbert Domke (Harnack-Haus) and Kristina Schönfeld (Research Program for the History of the Max Planck Society). Invaluable help in securing many of the photos used in this volume was provided by Susanne Uebele (Archiv zur Geschichte der Max-Planck-Gesellschaft). The editors also thank Lindy Divarci (Max Planck Institute for the History of Science) for editing and coordinating the final manuscript, and Lea Marquart (Research Program for the History of the Max Planck Society) for preparing the papers in accordance with the publisher's standards. We also thank Sabine Lehr for overseeing the publication process at Springer.

Part I
Research on and Deployment of Chemical Weapons in World War I

The Scientist as Expert: Fritz Haber and German Chemical Warfare During the First World War and Beyond

Margit Szöllösi-Janze

Abstract In the course of the First World War, scientists who would in peacetime generate new knowledge assumed the role of experts, i.e., professionals who made extant knowledge accessible to non-scientist clients. The deepest conviction of Fritz Haber, the 1918 Chemistry Nobel laureate, was that problems faced by mankind could be solved by means of science and technology. Herein, Haber is interpreted as a personification of an early German expert culture. Acting as both mediator and organizer, Haber coaxed politicians, generals, industrial leaders, and scientists to join forces in developing new processes for the mass-production of war-relevant chemicals and in establishing large-scale industries for their manufacture. Among the chemicals produced were poison gases—the first weapons of mass extermination. Haber's leadership resulted in a conglomerate of enterprises similar to what we now call "big science". In close contact with "big industry", traditional science was transformed into a new type of applied research. With borderlines between the military and civilian use blurred, Fritz Haber's activities also represent an early example of what we now call "dual use". He initiated modern pest control by toxic substances, whereby he made use of a military product for civilian purposes, but went also the other way around: During the Weimar era, he used pest control as a disguise for illegal military research. Having emerged under the stress of war, scientific expertise would remain ambivalent—a permanent legacy of the First World War.

The first major poison gas attack, at Ypres, on April 22, 1915 is irrevocably linked with Fritz Haber, the 1918 Nobel laureate in chemistry. The developments that connect the place, time, and person, are paradigmatic. They had their origins in the late nineteenth century and came to full fruition in the Great War. They shaped new trends that would leave a deep imprint on the twentieth century. Rather than

M. Szöllösi-Janze (✉)
Ludwigs-Maximilians-Universität, Munich, Germany
e-mail: Margit.Szoelloesi-Janze@lrz.uni-muenchen.de

B. Friedrich et al. (eds.), *One Hundred Years of Chemical Warfare: Research, Deployment, Consequences*, DOI 10.1007/978-3-319-51664-6_2

describing at length Haber's life,[1] this contribution will focus on the impact of those trends on the ways in which the German physical chemist influenced future developments. More than any other person, Haber embodies the ambivalence of the modern scientist who has been praised as a benefactor of mankind and, at the same time, accused of being a war criminal. His scientific work transformed both food production and warfare. He was not just an eminent intellectual, but also belonged to the select group of experts who shaped in fundamental ways the functioning of modern societies in war.

With the outbreak of the First World War in 1914, the impact of Haber's research on warfare became increasingly apparent. We can differentiate two important strands of his activities, the first concerning the production of explosives. Cut off from its major supply of natural nitrates by the British blockade, Germany suffered a serious munitions crisis after only a few months of trench warfare. The Haber-Bosch process, for which Haber laid the scientific foundation, provided a large-scale supply of synthetic ammonia and thereby of nitric acid, its oxidation product. Haber was actively involved (technically, but also politically) in the development of the production facilities such as the huge chemical factory in Merseburg/Leuna (Szöllösi-Janze 2000a).

In the following, I will focus on the second strand of Haber's activities: as a physical chemist, Haber's main research focus was on the reactions of gases. This expertise had placed him at the centre of Germany's preparations for introducing an entirely new weapon—poison gas. I will condense my considerations into some major points.

1. The declarations of war in 1914 provided new spaces of warfare. Those spaces came about as a result of a massive use of science and technology which, in turn, were profoundly transformed by the war experience.

The battlefields of the First World War constitute a space extended into all three dimensions (Trischler 1996). Trench warfare created a new geometry of the battlefield that included new zones: no-man's-land in the crossfire of the artilleries, the widely branching system of trenches, wire fences, supply lines, the hinterland used for the necessary logistics. Extending warfare into the air and below the surface of the seas added the hitherto unknown experience of three-dimensional warfare, which developed its own dynamics. The experience of war was total: it took possession of all spatial dimensions.

Science and technology had a significant impact on these developments. In the three decades preceding the war's outbreak, railway systems had increased troop mobility. Telephone, telegraph, and radio improved communications. The military use of aircraft—not just balloons, but also the newly developed zeppelins and aeroplanes—as well as submarines added a new physical dimension to warfare.

[1]Some recent biographies on Haber are available in German and/or English: see Stoltzenberg (1994, 2004). Daniel Charles' more popular biography (Charles 2005) was published under different titles in the US and the UK with the same content. In what follows I will mainly refer to my own book (Szöllösi-Janze 1998, reprinted 2015).

None of these new weapons or technologies decided the course of the war. But they marked a turning point on the way to modern warfare. Not only did changes already underway speed up considerably but also completely new developments emerged because the total war forced the belligerents to find novel solutions to completely new challenges.

2. The integration of new spaces into warfare led to a formerly unknown convergence of the state, the military, the economy, and of science. In all belligerent nations, these developments were anything but visible at the beginning of the military confrontation.

As Jeffrey Allen Johnson underscored, in pre-war Germany, "the academic-industrial symbiosis still primarily consisted of its classical core", meaning personal ties between industrial chemists and their colleagues in academe, mutual research support and the "educational link", meaning the supply of trained manpower from the universities to a growing, science-based industry (Johnson 2000, 17–18). Johnson and MacLeod came to a similar assessment concerning the development of military technology. Areas such as munitions testing, military education, military medicine, and the technical disciplines within the military sector had long been established. However, as both authors also noted, "in no European country was there provision for the mobilization of scientific expertise, nor did anyone anticipate such mobilization to be necessary for a war that was expected to be short and fought with conventional military technologies" (Johnson and MacLeod 2002, 170).

Stagnation on the Western front meant that within a few months Germany faced multiple deepening crises, notably in areas such as access to raw materials and resources, munitions production, famine at the so-called "home front", and, last but not least, politics. Under the pressure of having to win this war, no matter what the circumstances and no matter what the cost, economy turned into wartime economy based on regulation, rationing, and technology. "War is a technological forcing house"—that's how Lutz Haber, an economic historian and Fritz Haber's son, called his chapter on the history of the chemical industry in the First World War (Haber 1971, 184, 208). The progressive integration of external technical and scientific expertise into decision-making transformed the style and methods of government. Advisory committees, personal councils and consolidated advisory task forces proliferated in a very short time. Such processes affected all belligerents, if at different times and in different ways. The national "styles of scientific thought" (Harwood 1993), which reflected the characteristics of each country's scientific cultures, went obviously hand in hand with national styles of scientific expertise.

3. In the course of the First World War, scientists and other academics adopted new roles as producers and re-producers of knowledge relevant to warfare.

Experts, and in particular their status, legitimacy, and control, have been the subject of heated sociological debates for decades (see Etzemüller 2009; Kohlrausch et al. 2010). Herein, I argue that the First World War contributed significantly to the emergence of the experts as a social stratum of intermediaries between the rulers and the ruled. In view of the rapidly evolving demands of modern warfare, only scientifically trained experts were able to maintain an overview over the extant knowledge in their fields. They were in a position to assess how science could contribute to the war effort by having the insights needed in industrial product design and manufacture. Through their advice, they placed themselves at the intersection between the military, the administration, and the industry. In this way, scientific experts started playing a highly significant intermediary role in society. They were anything but passive, but on the contrary actively helped to define possible solutions to all war-related military and social problems. Expert cultures mediating but also actively influencing government decision-making arose in all warring parties. This was particularly true for Germany. During the war, the traditional system of scientific research underwent rapid institutional change and functional differentiation. With state support, a whole system of highly centralized, closely linked research institutes, university seminars, and industrial laboratories had emerged.

4. Fritz Haber was among the first and most important scientists who offered their expert services to civil and military decision-makers.

Haber was by far not the only one to do so. Many scientists offered their help, and among those who did so, chemists played a prominent role. This demonstrates that the catchphrase about the reportedly "mandarin tradition" of autonomously researching German professors (Ringer 1969), originally coined in the humanities, is not appropriate for the sciences. Rather, it should make way for a more differentiated view of the transformation of the German academic community into a new type of scholarly self-understanding, which accelerated especially after 1914 (Johnson and MacLeod 2002, 176–177).

In hindsight, Haber's importance grew from his networking mind-set, which affected the ways in which he thought, communicated and acted. In support of modern warfare, he first had to establish the basic cooperation between the state, the military, the economy, and the scientific establishment. In April 1918, Haber quite consciously reflected on his role:

> Before the war, this relationship was incomplete. The general would live on the *bel étage* and would politely greet the scholar who lived in the same building, but there was no internal connection. For mediation, he would use the services of the industrialist who lived in that house as well (Haber 1918, 197).

Establishing this "internal connection" between the scholar, the general, and the industrialist was Haber's central aim during the war years.

5. Fritz Haber as a scientific expert had to simultaneously fulfil the triple role of mediator, organizer, and innovator.

The relative weight of the roles could vary depending on the task at hand (Szöllösi-Janze 2000b), but in the case of Haber's involvement in poison gas warfare, they played out as follows:

First, mediation[2]: Haber's research activities in physical chemistry led to the development of large-scale industrial solutions. From this long-term cooperation with BASF, he knew the different viewpoints and spoke the different jargons of both the industrialists and the scientists. During his time at the Kaiser-Wilhelm-Institute in Berlin, he also learned how to deal with political decision-makers. In wartime, Haber became something of a communications interface. He broke down communication barriers, translated the needs and aims of one party into the jargon of another, devised possible solutions drawn from his own scientific discipline, and applied these to large-scale crash programs. No institutions initially possessed such mediating abilities, which underscored the importance of individual experts.

Secondly, organization: as an organizer, the scientific expert goes beyond his role of just establishing communication. He aims to make communication and cooperation permanent, by establishing institutions and finding practical applications. In Germany, sharp borders between different academic fields existed. Haber succeeded in making the borders permeable: he would embody the interconnection among pure science, applied science, technical development, and a practical application. Although initially driven by wartime demands, such institutional cooperation was clearly intended to persist beyond the war's end.

Thirdly, innovation: this term describes Haber's contribution to the emergence of a modern type of scientific research. Later known as big science, it refers to a different way of organizing the research process. It grew out of the long-term cooperation among the state, the military, the economy, and the scientific establishment. However, politically networked, large-scale research had an immediate bearing on both the substance and the way of doing research.

6. The first use of poison gas at Ypres in April 1915 reflected Haber's early success as a mediator, organizer, and innovator.

As a simple volunteer advisor to the war ministry, Haber quickly understood the impact of the new spaces and dimensions of warfare. The layered system of trenches, dug outs, and command posts protected soldiers relatively well from conventional weapons. Chlorine gas, being heavier than air, would not just poison the surface of the battlefield but also sink into the structures built underground. Trenches and underground facilities would not be safe any longer. Haber also understood how to utilize the technological and scientific potential of large-scale chemical industry for military purposes and how to establish the necessary contacts. Of course he was aware that the German chemical industry produced and consumed toxic compounds for manufacturing intermediaries or for civilian goods. The chemical plants required only little adaptation to produce warfare agents: the carrier systems were common oxygen cylinders, distribution was well established, and budgeting was assured. The formerly export-oriented dye companies transformed themselves into producers of nitrates, explosives, and chemical weapons (see

[2]See Emil Fischer for another outstanding mediator among German chemists (Moy 1989).

Johnson 2000, 23). Haber was furthermore aware of the number of highly qualified German chemists. Their problem-solving skills easily translated from improving dyes to developing explosives or poison gas. An estimated 1,000 scientists supported Germany's poison gas efforts towards the end of the war, 150 of whom worked at Haber's own institute. The industrial laboratories employed most of them: Bayer alone had 200 chemists on its payrolls (Martinetz 1996, 30; MacLeod 1993).

Finally, Haber knew how to sell the potential contribution of science to warfare even to the mostly hesitant or dismissive military leaders. In doing so, he helped to permanently integrate science into warfare—even though its impact never reached the amplitude or acquired the strategic importance he had been hoping for. With warfare becoming more industrialized and technological, the officer corps underwent professionalization that resulted in growing numbers of officers with a technical education or scientific interest. Science and technology turned the German military, as Michael Geyer put it, into "a complex corporation for the highly efficient production of violence" (Geyer 1984, 99). Moreover, specialized military units began to emerge. Haber easily persuaded his military interlocutors that well-trained, specialized troops were needed to handle poison gas safely and reduce risk of its employment to their own troops. They were to become the so-called "Gas Pioneers". To this end, he enlisted physicists, chemists, biologists, engineers, and meteorologists who readily exchanged the boredom of the trenches for the excitement of becoming experts in a novel mode of warfare. They included future Nobel laureates and eminent scientists such as Otto Hahn, James Franck, Gustav Hertz, Hans Geiger, Wilhelm Westphal, and Erwin Madelung. Only future Nobel Prize winner Max Born refused all offers to take part in chemical warfare, instead preferring the less brilliant field of developing radio equipment for air planes (Born 1975, 235, 261).

7. The ways in which Fritz Haber organized his activities led him to adopt novel approaches to research: big science.

To use chemistry as a metaphor: Haber played the role of a catalyst in forcing the existing, at first rather reluctant elements—state authorities, military, industry, and science—to blend in a fierce chemical reaction that unleashed its own dynamics. This dynamics had an impact on research itself. Haber developed a complex, goal-oriented style of research that aimed for politically relevant results. Many well-resourced teams of scientists and technicians from different fields worked systematically and on long-term basis on a given project. Interdisciplinary research removed the formerly impermeable borders between different scientific fields in ways similar to how the erstwhile sacrosanct boundaries between pure science, applied science, technical development, and industrial mass production would disappear (see as an example Trischler 2001, 80–83). But we can put on our record that Haber's approach was in many respects a nucleus of modern big science. Its immediate impact was to lay to rest the idea of the Humboldtian German professor conducting autonomous research in utter freedom and splendid isolation. Whether this idea ever reflected a reality (see Paletschek 2001) does not matter here. In any case, the First World War pretty much buried the Humboldtian concept.

The poison gas project employed several interdisciplinary teams of scientists, engineers, lab technicians, and auxiliary employees. Each team bore responsibility for its specific sub-projects and, interestingly, enjoyed relative freedom of research within the overall program. In 1916, Haber's Kaiser-Wilhelm-Institute came under military command (for the following, see: Szöllösi-Janze 1998, 333, 438–439). By September 1917, its budget had increased 50-fold over peacetime levels. It comprised nine departments, six of which were researching offensive chemical warfare, including the projection of new warfare agents, the analysis of enemy agents, research in toxicology and pharmacology, and the generation of aerosols. Several teams specialized in the control of the risky large-scale production of warfare agents, gas munitions and gas mine launchers. The remaining three departments worked on gas defense and protection. Besides chemists, the teams included physicists, biochemists, pharmacologists, physicians, veterinarians, zoologists, botanists, and meteorologists, and were supported by engineers, explosives experts, medical officers, and technicians. Together with the auxiliary and temporary staff, the Kaiser-Wilhelm-Institute employed some 2,000 people.

The big science type structures of research in Haber's institute faced their share of criticism among colleagues: "I hope the lion does not lay its hand on our modest department", complained Lise Meitner who worked next door to Haber at the Kaiser-Wilhelm-Institute for Chemistry, "the Haber people treat us of course like conquered territory; they take whatever they want, not what they need" (in Charles 2005, 169).

8. Fritz Haber's commitment to the German war effort always implied plans to apply his wartime experiences to the future.

Haber always tried to transfer the results from poison gas research to future civilian uses. In a programmatic talk to officers of the War Ministry on November 11, 1918, he explicitly coined the phrase that his motivation was "to turn the means of extermination into sources of new prosperity". But he also realized that to apply his capacity for networked solutions to large-scale problems over a longer time, he had to maintain the all-important interaction between the state, military, economic and scientific communities (Haber 1924, 28–29). I will mention just one example.

Still during the war, Haber found immediate peacetime application of his poison-gas research in chemical pest control (for the following, see Szöllösi-Janze 2001). The German population suffered from famine as a consequence of the Allied blockade. Pests in countless mills and granaries aggravated the food situation further. With the help of his military personnel experienced in handling poisonous gases, Haber developed new methods for rooting out harmful pests. His teams organized systematic regional "gassing cycles" of mills and granaries and developed suitable operational techniques and systems to implement them all over the country. However, it was a typical ploy to consciously exploit the dual-use nature of the science and technology underlying gas warfare.

Already in the final phase of the war, Haber pursued the idea to continue military poison gas research under the pretence of civilian pest control. He was quite aware that the victorious Entente would prohibit any further military research, and he wanted to avoid anything that could be used as a pretext to close down his Institute as a whole. He pulled off an ingenious coup when he succeeded in transferring his institute's pharmacological department to the unsuspected Biological Reich Institute for Agriculture and Forestry (Biologische Reichsanstalt für Land- und Forstwirtschaft). There, it could hide under the cover of the laboratory for physiological zoology. Haber was able to raise generous anonymous funds to finance additional scientific and technical staff, new buildings, laboratory animals, and testing equipment. A deeper look into the sources reveals that he carried out a top-secret transfer of considerable funds from the German military, the Reichswehr, to the Biological Reichsanstalt. These funds covered the running costs of its physiological laboratory, which developed and tested poison gases not only for pest control, but also for military purposes. The long-term deal was initiated and arranged by Haber, whose role as mediator and organizer can hardly be overestimated (Szöllösi-Janze 1998, 452–467; 2001).

The dual use of poisonous gases for pest control, however, implied also an application which was absolutely beyond Haber's imagination. For it was within this far-reaching network of institutions, engaged covertly or overtly in research on toxic gases, that scientists developed processes to handle cyanides for pest control without the risk of harm to the technical staff. Among the substances they developed, there were Cyclon A and later the infamous Cyclon B, whose potential for dual use shows the tragic ambivalence of Haber's commitment. Cyclon B was a result of the conversion of military into civil poison gas research. Only some twenty years later, it was used against human beings as a means of mass extermination in the extermination camps. Haber was convinced that he could keep the interconversion of poison gas research under control, but in the Age of Extremes (Eric Hobsbawm), this was not possible.

9. "I was one of the mightiest men of Germany"—the technological imperative.

Many commentators have explained Haber's extensive involvement in the German war effort by pointing to his burning patriotism. He was indeed convinced that Germany had been pulled into the war against its will and was waging it for a just cause. Almost all other German scientists shared this view; they were, however, less intensely involved. It cannot have been mere patriotism, then, even though Haber also appeared to have felt a very "Prussian", state-oriented sense of duty and had a keen interest in the military. He also volunteered the services of his institute in support of the German cause, just like many other scientists. He showed initiative when approaching military leaders to offer his assistance. In line with his classical education, he saw a role model in Archimedes, who was said to have served "the progress of mankind in peace, but his home in wartime" (Haber 1920, 352).

In my view, his sense of power played a larger role for Haber than his patriotism. He was well aware of the power that the expert-scientist wielded as an intermediary between the ruling and the ruled. Especially during the first half of the war, the role of experts was informal—they connected with individuals rather than with institutions and stood outside formal bureaucratic structures. It was precisely this informality that they were able to use to their advantage. Scientific experts were flexible enough to take on tasks that cut across fields, including the early stages of policy advice (see Fisch and Rudloff 2004). Haber typified this transformation. As director of the Kaiser-Wilhelm-Institute for Physical Chemistry and Electrochemistry, he presented himself as a war volunteer who described his function simply as "adviser to the war ministry". He thus offered his scientific expertise and network of connections in an act of patriotic self-mobilization for the German war effort. Only later during the war, he became gradually integrated into the military-governmental apparatus. At the same time, he was perfectly aware of the fact that he was not only influential but also in control of a sector relevant to modern technological warfare. In hindsight, in August 1933, he reflected on his earlier power:

> I was one of the mightiest men in Germany. I was more than a great army commander, more than a captain of industry. I was the founder of industries; my work was essential for the economic and military expansion of Germany. All doors were open to me (Weizmann 1950, 437).

Haber's exercise of power went hand in hand with a technocratic mind-set— and a technocratic rhetoric. He was convinced that there was a scientific and technological solution to all societal problems. As a technocrat through and through, the demands of modern warfare challenged him intellectually. He was fascinated by the opportunities offered by modern science and technology to solve political, military, and economic problems. His notable ability for networking and strategic thinking served his remarkable creativity in addressing desperate situations. Typically, in one of his few remarks about his personal involvement in chemical warfare, Haber transpires as a gambler who had been provoked, with an almost physical sensation of risk, to play and win big in the game of high-tech warfare. In a letter to Carl Duisberg from February 1919, he wrote that he felt challenged to apply his own "scientific imagination" to future problems of warfare and find possible solutions at the forefront of scientific and technological progress. He portrayed conventional warfare dominated by artillery as a simple game of checkers that "turned into chess by poison gas warfare and the defence against it".[3]

So, is the scientific expert ultimately a mere technocrat fascinated by gambling at the large board of modern mass warfare?

10. As a key player in the high-tech combat of chemical warfare, Haber was aware of the underlying "human factor".

[3]Haber to Carl Duisberg, 26 February 1919. Abt. V, Rep. 13 (Haber Collection), no. 860, Archives of the Max Planck Society, Berlin.

It is no coincidence that the First World War accelerated the development towards the "scientification", the *Verwissenschaftlichung* that brought along with it the idea that the "human factor" is measurable. As a result, military leaders from all belligerent countries discovered the utility of the new discipline of psychology for their ways and means, such as intelligence tests in the US and the "psycho-technical" surveys of aircraft pilots in Germany (Raphael 1996, 174–176; Geuter 1984). Fritz Haber was highly conscious of the strong psychological dimension of chemical warfare. Like others, he used a specific gas warfare discourse. He rejected the suggestion that poison gas use was "unchivalrous" as initially argued by traditionally minded officers. On the contrary, he underlined that chemical weapons were more "humane" than conventional weapon technology, since their wide-spread use would shorten the war. This is, of course, a first-strike rhetoric. History did not bear out this argument, because weapon innovation set in motion an endless dynamics of increasingly lethal weapon technologies. It is well known that less than twenty-four hours after the German chlorine cloud attack at Ypres, the British commander in France and Belgium, Sir John French, sent a telegram to London:

> Urge that immediate steps be taken to supply similar means of most effective kind for use by our troops. Also essential that our troops should be immediately provided with means of counteracting effect of enemy gases which should be suitable for use when on the move (In Charles 2005, 164; Schmidt 2015, 26–28).

Haber's insight into the psychological dimension of chemical warfare went deeper yet. It was common knowledge—also among the Allied Forces—that poison gas war could unsettle the morale of the troops as well as on the home front (Schmidt 2015, 23). But Haber reflected on the impact of gas on the frontline soldier in a specific way. To him, the toxicity of chemical warfare agents was less relevant than the fact that the chemicals forced troops to wear respirators and use other protective devices. This demanded, as he wrote to Carl Duisberg, "better leadership and higher military ability".[4] The conviction that chemical warfare demanded a higher mind-set led to a curious expression of social Darwinism in Haber as well as in many other proponents of chemical warfare—just to mention Colonel Max Bauer, Haber's military protector, who used to ruminate on the "selection of the fittest" through poison gas warfare.[5] In this sense, Haber viewed chemical warfare primarily as a quest for psychological superiority. In modern scientific war, he wrote, the "psychological imponderables" are decisive.

> A strict selection divides the men capable of withstanding pressures thanks to this gas discipline and fulfilling their military duties from the inferior mass of soldiers who break up and leave their battle position (Haber 1924, 36, 39).

---

[4]Haber to Carl Duisberg, 26 February 1919. Abt. V, Rep. 13 (Haber Collection), no. 860, Archives of the Max Planck Society, Berlin.

[5]See Max Bauer's Memorandum (1918) in Brauch and Müller (1985, 81).

"Gas discipline" as a means to select the fitter soldiers from inferior ones—this is twentieth-century social Darwinism at its best.[6]

Fritz Haber, however, was also the product of the nineteenth century. His personal duty was to remain loyal to the state and to commit himself unconditionally to the German cause. Even if his personal conviction had been different, Haber would not have questioned the German agenda, including chemical warfare. Just like millions of others, he never asked himself who exactly had set that agenda. For his personal morality, he relied on the presumed morality of the state, which he never doubted. His son, Lutz Haber, later described his father as "a Prussian, with an uncritical acceptance of the State's wisdom, as interpreted by bureaucrats" (Haber 1986, 2). The British physical chemist J. E. Coates discerned one of Haber's most important characteristics in his wish "to be a great soldier, to obey and be obeyed [...] autocratic and ruthless in his will to victory" (Coates 1951, 146).

So Daniel Charles is quite right when he deems that Haber wasn't much preoccupied with the morality of his innovation because it arose from "a kind of technological imperative", which he viewed as "simply inevitable". Charles also correctly points out that Haber's vision was strictly limited to the battlefield. He never anticipated the possibility that future warlords would use poison gas or other weapons of mass destruction against civilian populations. "In this respect", Charles concludes, "Fritz Haber's imagination remained trapped in the nineteenth century" (Charles 2005, 174).

References

Born, Max. 1975. *Mein Leben. Die Erinnerungen des Nobelpreisträgers*. Munich: Nymphenburger Verlagsanstalt.

Brauch, Hans Günter, and Rolf-Dieter Müller (eds.). 1985. *Chemische Kriegführung—Chemische Abrüstung. Dokumente und Kommentare*. Berlin: Verlag Arno Spitz.

Charles, Daniel. 2005. *Master mind: The rise and fall of Fritz Haber, the Nobel laureate who launched the age of chemical warfare*. New York: Harper Collins; *Between genius and genocide: The tragedy of Fritz Haber, father of chemical warfare*. London: Jonathan Cape.

Coates, J. E. 1951. The Haber Memorial Lecture. Delivered before the Chemical Society on April 29th, 1937. In *Memorial Lectures Delivered before the Chemical Society 1933–1942*, vol. IV, ed. F. G. Donnan, 127–157. London: The Chemical Society.

Etzemüller, Thomas, ed. 2009. *Die Ordnung der Moderne. Social Engineering im 20. Jahrhundert*. Bielefeld: Transcript.

Fisch, Stefan, and Wilfried Rudloff (eds.). 2004. *Experten und Politik: Wissenschaftliche Politikberatung in geschichtlicher Perspektive*. Berlin: Duncker & Humblot.

Geuter, Ulfried. 1984. *Die Professionalisierung der deutschen Psychologie im Nationalsozialismus*. Frankfurt: Suhrkamp.

Geyer, Michael. 1984. *Deutsche Rüstungspolitik 1860–1980*. Frankfurt: Suhrkamp.

[6]However, the phrase of gas warfare as "a higher form of killing," which has been repeatedly attributed to Haber since Harris and Paxman's study of 1982, cannot be found in the sources (see in detail Schmidt 2015, 484–485).

Haber, Fritz. 1918. Das Verhältnis zwischen Heereswesen und exakten Naturwissenschaften. *Chemiker-Zeitung* 42/no. 49 (24 April): 197.

Haber, Fritz. 1920. Die chemische Industrie und der Krieg. *Die chemische Industrie* 43: 350–352.

Haber, Fritz. 1924. Die Chemie im Kriege (November 11, 1918). In *Fünf Vorträge aus den Jahren 1920–1923*, ed. Fritz Haber, 25–41. Berlin: Julius Springer.

Haber, Ludwig F. 1971. *The chemical industry 1900–1930: International growth and technological change*. Oxford: Clarendon Press.

Haber, Ludwig F. 1986. *The poisonous cloud. Chemical Warfare in the First World War*. Oxford: Oxford University Press.

Harwood, Jonathan. 1993. *Styles of scientific thought. The German genetics community 1900–1933*. Chicago, IL: University of Chicago Press.

Johnson, Jeffrey A. 2000. The academic-industrial symbiosis in German chemical research, 1905–1939. In *The German chemical industry in the twentieth century*, ed. John E. Lesch, 15–56. Dordrecht/Boston/London: Kluwer Academic Publisher.

Johnson, Jeffrey A., and Roy MacLeod. 2002. War work and scientific self-image. Pursuing comparative perspectives on German and allied scientists in the Great War. In *Wissenschaften und Wissenschaftspolitik. Bestandsaufnahmen zu Formationen, Brüchen und Kontinuitäten im Deutschland des 20. Jahrhunderts*, ed. Rüdiger vom Bruch and Brigitte Kaderas, 169–179. Stuttgart: Franz Steiner Verlag.

Kohlrausch, Martin, Katrin Steffen, and Stefan Wiederkehr, eds. 2010. *Expert cultures in Central Eastern Europe. The internationalization of knowledge and the transformation of Nation States since World War I*. Osnabrück: Fibre.

MacLeod, Roy. 1993. The chemists go to war: The mobilization of civilian chemists and the British war effort, 1914–1918. *Annals of Science* 50: 455–481.

Martinetz, Dieter. 1996. *Der Gaskrieg 1914/18. Entwicklung, Herstellung und Einsatz chemischer Kampfstoffe. Das Zusammenwirken von militärischer Führung, Wissenschaft und Industrie*. Bonn: Bernhard & Graefe.

Moy, Timothy. 1989. Emil Fischer as 'Chemical Mediator'. Science, industry, and government in World War One. *Ambix* 36: 109–120.

Paletschek, Sylvia. 2001. The invention of Humboldt and the impact of National Socialism: The German university idea in the first half of the twentieth century. In *Science in the Third Reich*, ed. Margit Szöllösi-Janze, 37–58. Oxford and New York: Berg Publishers.

Raphael, Lutz. 1996. Die Verwissenschaftlichung des Sozialen als methodische und konzeptionelle Herausforderung für eine Sozialgeschichte des 20. Jahrhunderts. *Geschichte und Gesellschaft* 22: 165–193.

Ringer, Fritz K. 1969. *Decline of the German mandarins: The German academic community, 1890–1933*. Cambridge, MA: Harvard University Press.

Schmidt, Ulf. 2015. *Secret science. A century of poison warfare and human experiments*. Oxford: Oxford University Press.

Stoltzenberg, Dietrich. 1994. *Fritz Haber: Chemiker, Nobelpreisträger, Deutscher, Jude*. Weinheim: VCH.

Stoltzenberg, Dietrich. 2004. *Fritz Haber: Chemist, Nobel Laureate, German, Jew*. Trans. Jenny Kein, Theodor Benfey, and Frances Kohler. Philadelphia, PA: Chemical Heritage Foundation.

Szöllösi-Janze, Margit. 1998 (reprint 2015). *Fritz Haber 1868–1934. Eine Biographie*. Munich: C. H. Beck.

Szöllösi-Janze, Margit. 2000a. Losing the war, but gaining ground: The German chemical industry during World War I. In *The German chemical industry in the twentieth century*, ed. John E. Lesch, 91–121. Dordrecht/Boston/London: Kluwer Academic Publisher.

Szöllösi-Janze, Margit. 2000b. Der Wissenschaftler als Experte. Kooperationsverhältnisse von Staat, Militär, Wirtschaft und Wissenschaft, 1914–1933. In *Geschichte der Kaiser-Wilhelm-Gesellschaft im Nationalsozialismus. Bestandsaufnahme und Perspektiven der Forschung*, vol. 1, ed. Doris Kaufmann, 146–64. Göttingen: Wallstein.

Szöllösi-Janze, Margit. 2001. Pesticides and war: The case of Fritz Haber. *European Review* 9 (1): 97–108.

Trischler, Helmuth. 1996. Die neue Räumlichkeit des Krieges: Wissenschaft und Technik im Ersten Weltkrieg. *Berichte zur Wissenschaftsgeschichte* 19 (2–3): 95–103.
Trischler, Helmuth. 2001. Aeronautical research under National Socialism: Big Science or small science? In *Science in the Third Reich*, ed. Margit Szöllösi-Janze, 79–110. Oxford and New York: Berg Publishers.
Weizmann, Chaim. 1950. *Trial and error. The autobiography of Chaim Weizmann*, 4th ed. London: Hamilton.

From Berlin-Dahlem to the Fronts of World War I: The Role of Fritz Haber and His Kaiser Wilhelm Institute in German Chemical Warfare

Bretislav Friedrich and Jeremiah James

Abstract There is little doubt that Fritz Haber (1868–1934) was the driving force behind the centrally directed development of chemical warfare in Germany, whose use during World War I violated international law and elicited both immediate and enduring moral criticism. The chlorine cloud attack at Ypres on 22 April 1915 amounted to the first use of a weapon of mass destruction and as such marks a turning point in world history. Following the "success" at Ypres, Haber, eager to employ science in resolving the greatest strategic challenge of the war—the stalemate of trench warfare—promptly transformed his Kaiser Wilhelm Institute for Physical Chemistry and Electrochemistry in Berlin-Dahlem into a center for the development of chemical weapons and of protective measures against them. This article traces in some detail the path from Berlin-Dahlem to the fronts of World War I, lays out the indispensible role of Fritz Haber in German chemical warfare and provides a summary of his views on chemical weapons, which he never renounced.

1 The Run-up to Ypres

The lingering idea of using chemicals to incapacitate enemy troops had been rekindled by the belligerents in World War I once trench warfare produced a strategic stalemate on the war's Western front. On the German side, it was likely Max Bauer of the Supreme Army Command (*Oberste Heeresleitung*, OHL), see Fig. 1, who proposed to OHL's Chief, Erich von Falkenhayn, already in September 1914, to consider the use of chemical weapons in trench combat (Haber 1924, 85). In response to Bauer's proposal—and in the face of the shock of the Battle of the Marne[1]—von Falkenhayn promptly established a committee comprised of scien-

[1]Ending in Allied victory, this week-long battle (September 5–12, 1914) set the stage for the immovable trench warfare of the next four years of WWI.

B. Friedrich (✉) · J. James
Fritz Haber Institute of the Max Planck Society, Faradayweg 4-6, 14195 Berlin, Germany
e-mail: brich@fhi-berlin.mpg.de

B. Friedrich et al. (eds.), *One Hundred Years of Chemical Warfare: Research, Deployment, Consequences*, DOI 10.1007/978-3-319-51664-6_3

tists, industrialists and military officers, to assess the suitability and availability of various chemicals as weapons (Szöllösi-Janze 1998, 321–332 and references cited therein). Among the committee members was Fritz Haber's scientific rival, the physical chemist Walther Nernst, as well as the chemist and industrialist Carl Duisberg—but not Haber himself. The brainchild of the fledgling chemical warfare committee was the ineffective sneezing powder (*Ni-Stoff*) that was used amid the hostilities at Neuve-Chapelle already during the same month that the committee was formed. In response to the *Ni-Stoff* fiasco, Gerhard Tappen of the OHL, see Fig. 1, turned for help to his brother Hans Tappen, a chemist trained by Emil Fischer. Hans Tappen proposed to fill artillery shells with the lachrymator xylyl bromide, hence dubbed *T-Stoff*, that was first tested in January 1915. It was promptly used at the Eastern front, near Lodz, in cold weather with little or no effect due to its low vapor pressure at low ambient temperatures. At that point, von Falkenhayn decided to take his gloves off, abandon the "smelly stuff" and make use of lethal chemicals (substances that "incapacitate permanently," as he put it) in order to break the stalemate of trench warfare (Martinetz 1996, 18). To this end, Fritz Haber, who had meanwhile become, along with Emil Fischer, a member of Falkenhayn's chemical warfare committee and was privy to the failure of *T-Stoff*, proposed the use of chlorine as a chemical weapon. Heavier than air and thus suitable for striking enemy troops inside their trenches, chlorine held the promise of not only killing enemy combatants but also incapacitating their "conventional" weapons by corrosion. Although Haber would have preferred delivery via a barrage of artillery shells, he bowed to the need of the military to save ammunition and proposed to discharge the lethal gas from a great number of cylinders in the form of a cloud.

An attempt to test chlorine as a chemical weapon at the proving ground in Wahn near Köln in January 1915 was aborted due to the dangerousness of the gas and a decision was made by the OHL to test a chlorine cloud directly in battle on the Western front. Although viewed with skepticism and mistrust by most of the military, including von Falkenhayn himself, the idea of deploying a chlorine cloud found support from generals Berthold von Deimling and Emil Ilse, who were in charge of German operations in Flanders and who set their eye on the Ypres Salient, which, according to the Schliefen Plan, lay on the German Army's route into France. Von Deimling: "War is self-defense that knows no rules" (Deimling 1930, 201).

Much of the high-ranking German military took a more scrupulous—or chivalrous—approach to chemical weapons and at first openly detested them, thereby furnishing a curious substitute for the adherence to the spirit, if not letter, of the Hague conventions from 1899 and 1907 that limited the use of poisonous substances in warfare.

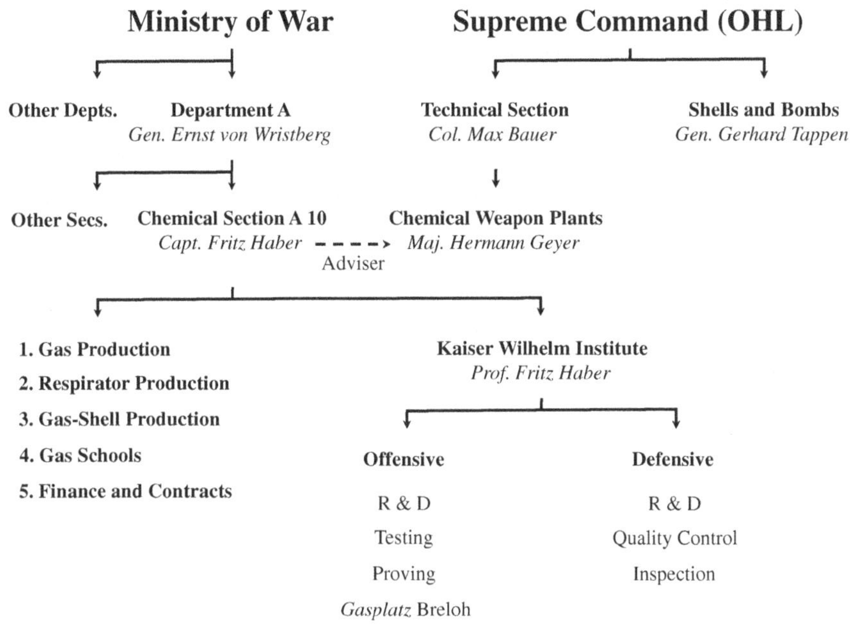

Fig. 1 Organization chart of the German chemical warfare from the end of 1916 on. After Haber 1986, 140

2 Ypres, 22 April, 1915, 1700 GMT

The commander of the gas units was Col. Peterson, with Fritz Haber co-opted as a member of his four-member staff.[2] The gas-warfare personnel, comprised of volunteers, was trained mainly in handling of gas cylinders, their transport, and protection. The steel cylinders were to be dug-in upright behind the battlements of the trenches and shielded with sand bags. Only the lead pipes were to peep out from the parapets, see Fig. 2. The training took place in Wahn during January 1915, but soon the chemical units, posing as "disinfection crews," started moving to their positions on the Western front, reaching Ypres in February 1915. In the meantime, on Haber's urging, many of the officer positions of the gas units were staffed with scientists—including meteorologists. These units would later become a part of the army corps of engineers (*Pionierregiment* or *-bataillon*). The movement of the chemical units and of their equipment to the front was accompanied by many mishaps, not least of them a serious injury suffered by Fritz Haber and Max Bauer on April 2 when they were caught in the midst of a small chlorine cloud released for testing purposes (Haber 1924, 88).

[2]The other members were adjutant Otto Lummitsch and Col. Ludwig Hermann (Szöllösi-Janze 1998, 327).

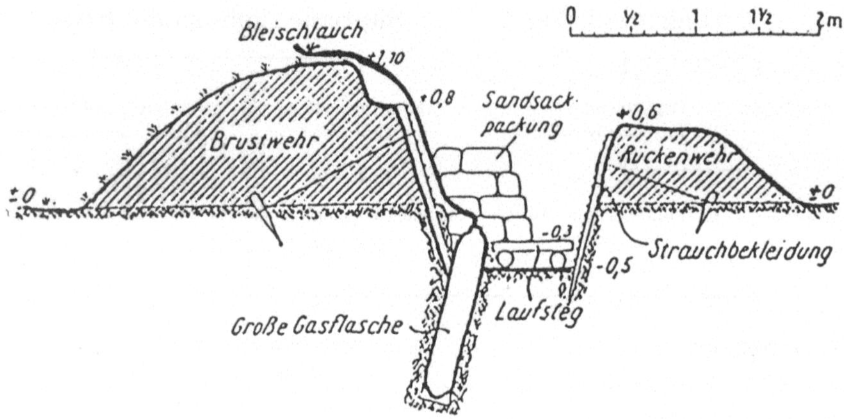

Fig. 2 Cross-sectional schematic view of a trench with a chlorine cylinder in position. After Martinetz 1996, VII

The first chlorine cloud attack on enemy positions (French and British) took place on April 22, 1915, at 1700 GMT (18:00 CET) in the perimeter of the Langemarck village near Ypres, when the prevalent wind finally turned in the northerly direction. Seven previous attempts had to be aborted because of unfavorable wind. The attack released 150 tons of chlorine gas from 1600 big and 4130 small cylinders placed at a distance of about 1 m from each other, covering about 6 km of the front. The chlorine gas concentration achieved was about 0.5% at a distance of 50–100 m from the cylinders, see Fig. 3. The first attack not only threw the Allied forces into a panic but reportedly injured about 5000 and killed about 1000. In addition, the Germans were able to capture 60 guns and a large swath of territory. To no avail, however, as the German military was unprepared to take advantage of the breakthrough:

> Unfortunately, the OHL had not prepared sufficient reserves because of doubts about the effectiveness of the new weapon. Otherwise it would have been possible to make a decisive advance (Lummitzsch 1955).

In spite of the "success" of the attack, it became clear that the predominant easterly winds made gas cloud attacks too unpredictable and unreliable a means upon which to base a new method of warfare. This led Haber to a renewed interest in poison gas grenades and shells, which were not so dependent upon rapidly changing meteorological conditions.

3 The Indispensable Fritz Haber

Fritz Haber, see Fig. 4, the founding director of the Kaiser Wilhelm Institute for Physical Chemistry and Electrochemistry, took part in the widespread enthusiasm that accompanied German mobilization and entry into the First World War in the

Fig. 3 Aerial view of the chlorine cloud released by the German troops on April 22, 1915 at Ypres. Photo: Martinetz 1996, VIII

summer of 1914 (Hoffmann 2014, 7–31) and registered for voluntary military service at the beginning of August. He gave voice to his euphoria in a letter to Svante Arrhenius in Stockholm, writing:

> This is a war in which our entire people is taking part with full sympathy and its utmost abilities. Those who don't bear arms work for the war, and everyone scrambles forward voluntarily for the slightest accomplishment. You know Germany all too well not to know that such a unanimous commitment to a cause is only possible amongst us when all are conscious that the good of the nation must be defended through a just struggle. You should give no credence to the absurd fiction, according to which we are conducting a war out of military interests … but we now see it as our ethical duty to take down our enemies with the use of all our strength and bring them to a peace that will make the return of such a war impossible for generations and give a solid foundation for the peaceful development of western Europe (Zott 1997, 77).

When Adolf von Harnack, as president of the Kaiser Wilhelm Society (KWG), called together all of the institute directors on August 12, 1914 to discuss possible consequences of the war for the work of the KWG, Haber was already occupied by military concerns and sent his deputy Gerhard Just, see Fig. 4, in his stead. Haber worked first as a scientific consultant in the Ministry of War for the Artillery Command and the Production Department, where his expertise in applied chemistry and ammonia synthesis were particularly valued. Representatives of German political establishment and industry had quickly realized that a limit on raw materials made a long war unsustainable for Germany. Nitrates were of particular concern, as Germany relied upon Chilean saltpeter to supply both its fertilizer needs and its production of explosives and propellants, but the British sea blockade threatened to cut off this source. The war would also lead to shortages of myriad other raw materials and create related bottlenecks in industrial production. Hence, German chemists faced the challenge of rationalizing use and production of these scarce materials or finding substitutes for them.

Haber not only followed his maxim "In peace for mankind, in war for the fatherland"[3] personally, but applied it to his entire institute and promptly redirected its resources toward projects relevant to the war. The conversion to military research projects proceeded surprisingly smoothly and without noticeable resistance. This raises the question whether the war euphoria alone eased the transition or whether something inherent to the research policy of the Kaiser Wilhelm Society, especially its chemical institutes, enabled such a conversion. As Johnson has argued:

> True to his nature, Fischer stamped the scientific program of the Kaiser Wilhelm Society with a dual character. On the one hand, it was aimed at the most fundamental problems of natural science; but on the other, it was intended to produce solutions to technological problems of the highest national interest, particularly with regard to providing domestically available synthetic or artificial substitutes for imported materials (Johnson 1990, 133).

The development of the catalytic process for ammonia synthesis was already one realization of the desire to manufacture domestic substitutes for key imported goods. It is also amongst the common tenets of the history of WWI that without the Haber-Bosch process, the German military would have run out of munitions in 1915. Similar intentions led Haber to his wartime partnership with the Raw Materials Department of the War Ministry under Walther Rathenau, which eventually led him to research chemical means for waging war. As Johnson pointedly summed up the progression: "the logic of Ersatz [substitute] led to the problems of munitions, and eventually to poison gas" (Johnson 1990, 133).

In the first months of the war, the Institute searched for ways to economize or provide substitutes for so-called "war materials"—substances required for the operation of firearms, artillery and other war machines; examples include toluene, glycerin and saltpeter. Gerhard Just made rapid progress in this field, in collaboration with Otto Sackur, see Fig. 4. Together they were able to demonstrate, through careful freezing and boiling point measurements, that a combination of xylene and certain water-soluble fractions of crude oil could replace toluene as an anti-freeze in engines. Their discovery meant a savings of roughly 400 tons of toluene per month that could then be used in the production of TNT and other explosives and munitions. In the autumn and winter of 1914, Haber and his colleagues also took part in the development of respiratory irritants and tear gases in connection with the already mentioned conservation efforts.

---

[3]Usually quoted in this abbreviated form. For a full quote in German, see Haber (1920). Haber was referring to the attitude of Archimedes.

Fig. 4 Distinguished German scientists involved in chemical warfare. Clockwise from *top left*: Fritz Haber, James Franck, Richard Willstätter, Gustav Hertz, Hans Geiger, Otto Hahn, Otto Sackur, and Gerhard Just. Photos: Archiv der Max-Planck-Gesellschaft, Berlin-Dahlem, Jahresbericht der Schlesischen Gesellschaft für Vaterländische Kultur 1915, 1: 35–37, private collections of the authors

In connection with their efforts to develop new and more effective explosives and propellants, Haber, Just and Sackur attempted to replace the irritant in the T-shells[4] with a substance that would act as both irritant and propellant. Someone thought of using cacodyl chloride, which Robert Bunsen had first synthesized in 1837, but which chemists had scarcely researched since, because it was such a powerful irritant and explosive. On 17 December 1914, during an experiment intended to tame cacodyl chloride, there was an explosion in the laboratory. Sackur was fatally wounded, Just lost his right hand. Haber had left the room shortly before the blast and remained physically unharmed. Nevertheless, he was unsettled by the death of his highly-talented colleague and steadfastly honored his memory (James et al. 2011, 27).

After the accident, research on cacodyl chloride at the Institute was halted, but the explosion also marked a turning point not wholly ascribable to its tragic consequence: the end of significant research on explosives at the Institute and the beginnings of poison gas research. Sometime in the first half of 1915, Haber redirected research at his institute toward the needs of gas warfare. Unfortunately, the available sources do not provide precise answers as to when or how this occurred.

[4]Shells filled with T-*Stoff*.

The men of the gas brigades who carried out the chlorine cloud attack at Ypres, amongst whom numbered many Dahlem and Berlin scientists, including James Franck, Otto Hahn, Gustav Hertz, and Hans Geiger, see Fig. 4, experienced an immediate improvement in status. Haber was even promoted by the Kaiser from the rank of staff sergeant to captain,[5] a potent recognition of the value of his efforts. This advance in rank appears to have further motivated Haber to the self-assigned task of promoting chemical warfare, first as chemical advisor to the Ministry of War, then, beginning in November 1915, as head of the "Central Office for Chemical Concerns" in the Artillery Division. He essentially abandoned scholarly research and concentrated upon the problems of chemical warfare. In the words of his biographer Dietrich Stoltzenberg:

> everything else in his life [faded] into the background. Wife and family now had almost no influence on his life. In fact, for him, family, friends, and acquaintances were just further sources of aid for his cause (Stoltzenberg 1994, 256).

4 Haber's Kaiser Wilhelm Institute Under Military Command

Like most academic institutes, Haber's was essentially abandoned with even Haber spending much of his time near the front directing preparations for gas attacks. Then, at the end of April 1915, the Supreme Army Command assigned Haber the task of developing defensive measures against gas attacks. Haber systematically redirected his institute toward the pursuit of this goal and built it up into a German center for gas warfare research. Over half of the expenditures of the Institute in 1915 were for military purposes, above all for "experiments on the development of gas warfare methods." As of February 1916, the Institute worked "almost exclusively" for the military, which covered all related costs, including standard operating expense; the Institute even took on an army administrator, Lieut. Alfred Glücksmann, amongst whose duties was managing the Institute's budget (Szöllösi-Janze 1998, 344).

As noted, the first task of Haber's institute was the exploration of defensive measures and the development of gas masks. To this end, a special department for "Chemistry of Gas Defense" was established under Hans Pick. Among the duties of this department was the development and testing of gas mask prototypes, in collaboration with production firms. At the conclusion of these tests, the preferred design was one proposed by Leopold Koppel's Auergesellschaft, which would produce the masks in great numbers.[6] Still, mask design was only one part of

[5]Non-Jewish members of the German professoriate typically received the rank of a major, Szöllösi-Janze (1998, 63–64, 267); Schmaltz (2014, 206).

[6]Leopold Koppel, a banker and enterpreneur, funded Haber's KWI.

effective gas defense; yet more important was the development of effective filters. Early in 1915 filter designers had to reckon with chlorine and phosgene attacks, but as new and ever more malicious poisons were introduced, including mixtures of compounds consciously designed to penetrate the masks, the filters needed to be continuously updated and retested. Responsibility for filter development initially fell to the neighboring KWI for Chemistry, in the person of Haber's colleague and friend Richard Willstätter, see Fig. 4. The first complete new gas masks, body and filter, were delivered to the troops in September 1915, and by the end of January 1916, the entire Western Front was equipped with masks, an enormous technical as well as logistical accomplishment.[7]

The initial fabrication of gas masks also marked the beginning of an unprecedented expansion of the Institute, which took place between the autumn of 1915 and the end of 1917. The Institute grew to include 10 departments and employ 1500 scientists and support staff, the latter composed overwhelmingly of women, see Fig. 5, all employed in the research, development and testing of gas warfare equipment. This number would rise to 2000 at the end of the war (Szöllösi-Janze 1998, 348). The expansion began with the departments of Reginald Herzog and Hans Pick, see Fig. 6 and Table 1. Herzog and his coworkers took over the supervision and testing of gas mask production, while Pick dedicated himself to the chemical aspects of gas defense. Then, in November 1915, Friedrich Kerschbaum, see Fig. 6 and Table 1, previously technical adjutant to Haber, established a department devoted to the study of enemy poisons and defensive measures, as well as the development of new poisons. The founding of this department and the near simultaneous appointment of Haber to head the Central Office for Chemical Concerns were the final, decisive steps in the commissioning of the KWI for Physical Chemistry and Electrochemistry for poison gas research by the military. The subsequent unprecedented expansion of the Institute would make it a prototypical example of "big science," not only with respect to its sheer size but also, and above all, with respect to the complexity and interdisciplinarity of its organizational structure and research methods (cf. Szöllösi-Janze and Trischler 1990). The institute also served as an early example of how quickly and smoothly resources for the establishment of optimal research conditions can be secured in a military or security context whereas, under normal circumstances, their attainment frequently involves near-endless, often unsuccessful struggle. As Johnson put it, "the Dahlem complex gradually assumed the character of a research center for tactical military science and technology" (Johnson 1990, 189). Or, looking forward with the words of Fritz Stern, the Haber Institute during the First World War became "a kind of forerunner of the Manhattan Project" (Stern 1999, 119).[8]

---

[7]We note that the testing of the gas masks was done, among others, by James Franck, Otto Hahn, and other scientists who let themselves be personally exposed to chemical warfare agents such as phosgen in the test chambers at Haber's KWI (Szöllösi-Janze 1998, 347–348).

[8]However, the Manhattan Project was by about two orders of magnitude larger in terms of human resources alone. See *Manhattan District History* Book I—General, Vol. 8—Personnel, dated 19 February 1946.

Similarly striking expansions of chemical warfare research occurred in the other warring nations, further encouraging German efforts. Although there was initially significant political debate in Britain as to whether German use of chemical weapons justified retaliation in kind, by May 1915 the British Ministry of Munitions had formed a chemical subcommittee, and in March 1916 the British opened a chemical weapons testing facility at Porton Down. France saw no similar political debate and had, in fact, already employed grenades filled with lethal chemicals early in the war, see below. In June 1915, France established a Directorship of Chemical War Materials under the Ministry of War and centralized chemical warfare research in laboratories near Paris (Lepick 1998). In so far as German efforts were exceptional, it was primarily in the degree to which research remained concentrated at the KWI for Physical Chemistry and Electrochemistry (Haber 1986; Martinetz 1996).

By October 1916, there were five departments dedicated to gas warfare research.[9] In addition to those already mentioned, Herbert Freundlich, see Fig. 6 and Table 1, headed a department dedicated to the supervision and testing of mask filters. That autumn, roughly one year into its expansion, the Institute employed a scientific staff of 77 chemists, pharmacists, and engineers, as well as a support staff numbering over 100. Up to that point, the Institute officially remained under the auspices of the Kaiser Wilhelm Society and the Koppel Foundation, which lodged muted complaints against the redirection of the Institute.

At the beginning of 1917, Haber's Office for Chemical Concerns was detached from the Artillery Division and refashioned into an independent department of the General Staff, Department A10, cf. Fig. 1. This department was then assigned control of Haber's institute, bringing the KWI for Physical Chemistry and Electrochemistry under military command. This reorganization was the result of a motion by the Ministry of War summarized by Haber in his letter to the minister of education that

> the whole operation of the institute be converted to a military one while keeping the name "KWI for Physical Chemistry and Electrochemistry" and that the staff currently working at the institute should be […] put on the staff of the military administration. The Ministry of War would […] as supervising authority still control the general direction of the institute on the condition that the position of Privy Counselor Capt. Haber […] as scientific director of the institute would remain unchanged, as would the positions of the others who had been members of the staff in peace time.[10]

This administrative move was accompanied by further enlargement of the Institute, primarily related to the intensification of efforts to develop new poisons. According to a report from Haber in September 1917, it was at this point that the Institute came to employ a staff of approximately 1500, with roughly 150 scientists and engineers, divided into 10 departments. Prominent scientific members included later Nobel laureate Heinrich Wieland, see Fig. 6 and Table 1, who worked on the production of new toxic gases, including the blistering agent mustard gas (LoSt,

[9]We draw on James (2011, 25–34).

[10]Fritz Haber, Letter to the Ministry of Education from 31 January 1916, Stoltzenberg (2004, 139).

Fig. 5 Out of the 1,500 employees of Haber's KWI after it became the German center of chemical warfare research and coordination, the majority were women. Photo: Archiv der Max-Planck-Gesellschaft, Berlin-Dahlem

named for Haber's coworkers Lommel and Steinkopf and also known as Yperite) and Ferdinand Flury, see Fig. 6 and Table 1, who tested newly-developed gases on live animals. Heinrich Wieland, who would receive the 1927 Chemistry Nobel Prize "for his investigations of the constitution of the bile acids and related substances," initiated his research into the relationship between steroids and the bile acids in 1912. During his 1917–1918 military service at Haber's Institute, for which he may have qualified due to his pre-war work on the fulminic acid, a primary explosive whose vapors are highly toxic, Wieland synthesized adamsite, in addition to his involvement in the synthesis of mustard gas. Wieland's penetration into the structure of the polymeric fulminic acids helped to lift the veil over what he called "an arid structural desert" under which he later discovered the constitution of bile acids as well (Karrer 1958, 342). To what extent Wieland was able to advance the line of research on steroids during his service at Haber's Institute remains unclear.

The two original Institute buildings could not contain all of the new departments and coworkers, and ever more barracks, kennels and other outbuildings sprang up on the Institute grounds, see Fig. 7. When even this space was consumed, rooms were requisitioned from the neighboring KWIs for Chemistry and for Biology, with the consent of their respective directors.[11] Toward the end of 1916, further growth

[11] And perhaps sometimes even without it. As Lise Meitner from the adjacent KWI for Chemistry noted: "Haber's people treated us like conquered territory; they didn't take what they needed but what they liked." (Stoltzenberg 1994, 255).

Fig. 6 Prominent heads of chemical warfare departments at Haber's KWI for Physical Chemistry and Electrochemistry when the institute was under military command. Clockwise from *top left*: Friedrich Kerschbaum, Ferdinand Flury, Heinrich Wieland, Reginald Oliver Herzog, Paul Friedländer, and Herbert Freundlich. Photos: Archiv der Max-Planck-Gesellschaft, Berlin-Dahlem and Martinetz 1996, II

Table 1 Departmental structure of Haber's KWI at the end of World War I, according to the 1921 "Hartley Report." The "defensive" departments are marked in red, the rest are the "offensive" departments (Haber 1986, 127)

Department	Assignment	Head
A	Development of respirators; Supervision of gas mask production	Reginald Oliver Herzog
B	Development of chemical weapons; Testing of enemy chemical weapons and protective gear	Friedrich Kerschbaum
C	Development of respirator drums and other protective appliances	Ludwig Hans Pick
D	Synthesis of new chemical warfare agents	Heinrich Wieland
E	Pharmacology and toxicology; Testing on animals	Ferdinand Flury
F	Supervision of the production of respirator drums	Herbert Freundlich
G	Supervision of the production of shells and fuses for gas munitions	Wilhelm Steinkopf
H	Trench mortars; Explosives	Otto Poppenberg
J	Supervision of the production of chemical warfare agents	Paul Friedländer
K	Particulate clouds – development of the "Gasbüchse"	Erich Regener

occasioned moves even further afield, and the Institute rented space in the Pharmaceutical Institute of the University, in a weaving school on Warschauer Platz, in the State Offices in Steglitz and on Königin-Luise-Straße in Dahlem (James et al. 2011, 32).

Compared to the extent of the research carried out on poison gases, the original documentation that remains is truly sparse. A single, relatively complete research report on chlorarsines remains from Johannes Jaenicke, at the time a member of Hans Pick's Department C, along with scattered reports from toxicity tests on animals and volunteers from the Institute staff (Haber 1986, 109). Extant correspondence between the main researchers is similarly sparse.

The post-war Allied forces reports on the Institute are somewhat more informative concerning the conduct of poison gas research, but even they are disappointingly superficial when it comes to questions of research procedures and practices, aside from quality testing protocols. Nevertheless, from these limited sources, the principal of them being the Hartley Report and its rendition in the book by Fritz Haber's son Ludwig (Lutz) Haber,[12] it is possible to sketch the following outline of chemical weapons research at Haber's institute.

The ongoing development of gas masks and filters, in increasingly close cooperation with industry, took place in Departments A (Herzog) and C (Pick). The work of these departments relied upon a steady exchange of knowledge between laboratory researchers and battlefield informants. Later in the war, prototypes from these departments would be tested against new poisons from Department B (Kerschbaum) (Hartley 1925, 39–42). Members of Kerschbaum's department strove to find and identify substances with potential for use in gas cloud and shell attacks. Their work consisted of a mixture of literature research to identify substances with an optimal combination of noxiousness, low boiling point and high vapor pressure, and experiments on animals and volunteers from the Institute staff to confirm the irritating or toxic effects of these substances (Hartley 1925, 45). Also, Kerschbaum along with Haber supervised the poison-gas production and filling facility ("*Gasplatz*") at Breloh, see Fig. 8. Department D (Wieland) focused specifically on deleterious arsenic and sulfur compounds, for instance mustard gas, and performed primarily laboratory research, including attempts to synthesize new substances with effects analogous to known poisons and irritants. Research on the physiological effects of various poisons, including careful study of their relative toxicity, occurred in Department E (Flury) and relied upon extensive animal experimentation. It was also Flury and his collaborators who promoted use of the so-called "Haber Constant," the product of the concentration and the exposure time required to cause death. This constant aided early efforts to define limits on

[12]Lutz Haber's personal interest in the topic of chemical warfare was fueled not just by his family lineage but also by his acquaintance and friendship with Harold Hartley, whose confidant—and in a sense heir of his extensive collection of materials connected with chemical warfare in WWI—Lutz Haber had become. Sir Harold Hartley was Fritz Haber's counterpart at the British War Office during WWI who, after the war, was in charge of inspecting German research and production facilities related to chemical warfare, and banned by the Versailles Treaty.

Fig. 7 Haber's KWI for Physical Chemistry and Electrochemistry with surrounding barracks, circa 1917. Photo: Archiv der Max Planck Gesellschaft

hazardous substances in the civilian sphere. Department J (Friedländer) was responsible for testing the quality of chemical weapons produced by industry, for which it employed predominantly classical analytical methods rather than measurements of physical constants (Hartley 1925, 50–52). Only in Department K, under Erich Regener, did techniques from physical chemistry play a central role. Regener's group used ultramicroscopes to study the small particles that constitute powders and smokes and their ability to penetrate existing gas mask filters.

Post-war assessments of the scientific value of this research by Allied representatives and later historians have been almost universally negative. In his own remarks on the subject during the 1920s, Haber emphasized the effectiveness and humaneness of chemical weapons, but, nonetheless, explained to Allied agents that all of the important toxic substances used in the war had already been synthesized and studied before 1914 and that "no systematic progress had been made" in their research. In the same vein, Richard Willstätter reported to the Allies that he did not consider the synthesis research pursued at Haber's institute particularly serious.

Apart from developing additional chemical agents at his Kaiser Wilhelm Institute (such as phosgene and mustard gas), Haber introduced the procedure of "Bunteschiessen" (variegated shelling), which consisted of first deploying "Maskenbrecher"—irritants based on organic arsenides (called Clark I and II) that penetrated all available filters and forced those under attack to remove their gas masks—and subsequently shelling with poisons such as phosgene or mustard gas (Haber 1986, 114–116).

Fig. 8 Inspection of the gas-testing and storage facility in Breloh near Münster on April 12, 1918. Head of the General War Department A (*Allgemeines Kriegsdepartment A*), Col. Ernst von Wrisberg (4th from *left*); Commander of the Breloh facility, Lieutenant-Colonel Ernst von Wangenheim (4th from *right*); Commander of the Field-Munitions Institute, Capt. Dr. Ludwig Hermann (3rd from *left*); Head of Department B at Haber's Institute, Dr. Friedrich Kerschbaum (1st from *left*). Capt. Fritz Haber (2nd from *left*). Photo: Archiv der Max Planck Gesellschaft

5 Haber's Views on Chemical Warfare

Haber advertised the first use of a chemical weapon as an important milestone in the "art of war"—and saw its psychological effect as key:

> All modern weapons, although seemingly aimed at causing the death of the adversary, in reality owe their success to the vigor with which they temporarily shatter the adversary's psychological strength (Haber 1924, 36):.

Haber also pointed out that key to success in chemical warfare are "intellectual imponderables" of the troops.

> A strict selection separates the troops that are capable of maintaining gas-discipline and who fulfill their combat task from the martially inferior mass of those who crumble and abandon their posts (ibid., 39).

He also emphasized that the variability of the effects of chemical weapons presents ever new demands on the "moral resistance" of the troops, as opposed to artillery shelling that is always the same and people get eventually used to it (ibid., 37).

Thus Haber viewed chemical weapons as a strategic means to break the stalemate of trench warfare by forcing the adversary to surrender, shorten the war, and thereby preclude the slaughter of millions by artillery and machine gun fire. In this sense he also referred to chemical weapons as "humane." He was not alone to do so, see below.

Haber's enthusiasm for chemical weapons had an important caveat, described by Haber's biographer Margit Szöllösi-Janze in this way:

> Was [Haber] a fanatic of gas-warfare? One must not overlook that he saw the significance of the gas weapon in its stupefying character and assessed the inherent dynamics of its use differently than the military. In his opinion chemical weapons could be effective and lead to a shortening of the war only if the war came to an end before the adversary would be able to develop suitable protective measures and even more dangerous chemical agents. For this reason, [Haber] voiced his opposition at a meeting in 1917 with [General Erich] Ludendorff [deputy of von Falkenhayn's successor as Chief of OHL, Paul von Hindenburg] to the use of mustard gas [that represented an escalation of the chemical war], which struck him sensible only if Germany could take advantage of its head start and win the war within a year (Szöllösi-Janze 1998, 332).

And indeed, the French began production of mustard gas in July 1918. Here's how Harold Hartley described the path of the Allies to mustard gas:

> [Hartley] was awakened early on that 13 July [1917] and was informed that the Germans had fired a new type of shell [at Ypres] that made a "plop"-like sound when it burst [...] The next day [Hartley] had located some unexploded shells with yellow cross markings. They were defused, taken to [the General Headquarters], opened, the contents analyzed [...] and the findings compared with the entry in *Beilstein*. By 16 July [the British] knew what the stuff was, and later analyses added little to this knowledge (Haber 1986, 192).

Haber's correspondence reveals his fascination with the new tactical possibilities opened by gas warfare and the room for "scientific imagination" that such technological means for conducting war had offered (Szöllösi-Janze 1998, 327). He used (and perhaps coined) the metaphor that conventional warfare was like playing checkers whereas warfare enhanced by chemical weapons was like playing chess. Haber never regretted his involvement in chemical warfare.

As for legal issues with chemical warfare, Haber put the blame for any transgressions against international law squarely on Erich von Falkenhayn. He did so in a testimony—which took the form of lectures—delivered to an investigative committee of the Reichstag in 1920–1923. However, in this testimony, Haber did not shy away from playing a legalistic shell game when he argued that German gas attacks were carried out either without the use of shells (like the chlorine cloud attacks) or with shells loaded, in addition to poison gas, with explosives (whereas the Hague conventions prohibited the use of shells or grenades filled solely with poisonous substances).

Haber also correctly claimed that chemical weapons were first used in WWI by the French—already in August 1914—when they fired rifle grenades filled with the highly toxic ethyl bromoacetate. Although ineffective because of the low concentrations achieved, the intended purpose was, according to Fritz Haber, the same as that of the German chlorine cloud: to force the enemy out of his trench positions by exposing him to an asphyxiating agent (Haber 1924, 87). This view was validated by the German parliamentary committee, which concluded:

> Neither the German nor the French governments, nor as far as is known, any other power participating in the war or a neutral one raised any objections against the modes of action in the gas war. From this it can be concluded that both sides viewed the Hague Conventions of 29 July 1899 and 18 October 1907 as obsolete and by silent agreement regarded them as annulled. Even when accepting this assumption, it remains a fact that the first obvious transgression of an international agreement was on the French side, whereas Germany only followed and thereby merely took a countermeasure as accepted in international law (Stoltzenberg 1994, 152).

6 The Legacy of Ypres

The universal abhorrence of chemical weapons as manifestly inhumane is surprisingly recent and so is their classification as weapons of mass destruction. While the latter is a concept of the nuclear age, the former is not... At the time of their use in the First World War, the perverse-sounding notion that chemical weapons were in fact humane had been a part of the vocabulary of munitions and war experts of the Central Powers and the Entente alike, including, for instance, that of the U.S. Assistant Secretary of War and Director of Munitions, Benedict Crowell:

> The methods of manufacturing toxic gases, the use of such gases, and the tactics connected with their use were new developments of this war; yet during the year 1918 from 20 to 30 per cent of all American battle casualties were due to gas, showing that toxic gas is one of the most powerful implements of war. The records show, however, that when armies were supplied with masks and other defensive appliances, only about 3 or 4 per cent of the gas casualties were fatal. This indicates that gas can be made not only one of the most effective implements of war, but one of the most humane (Crowell 1919, 396).

The few who forewarned Haber and the German military leadership that the German use of chemical weapons would lead to a quick retaliation by the Entente powers and a widespread use of chemical weapons were ignored. And indeed, the Entente introduced its own potent chemical arsenal within a few months of the first German chlorine cloud attack at Ypres, see also Table 2.

Table 2 Production of chemical weapons by country, in metric tons (Martinetz 1996, 120)

Country	Chlorine	Phosgen	Di-phosgen	Mustard gas	Chlor-picrin	HCN	Total
Germany	58,100	18,100	11,600	7,600	4,100	-	99,500
Britain	20,800	1,400	-	500	8,000	400	31,100
France	12,500	15,700	-	2,000	500	7,700	38,400
U.S.A.	2,400	1,400	-	900	2,500	-	7,200
Total	93,800	36,600	11,600	11,000	15,100	8,100	176,200

Artillery shells filled with chemical agents grew from a negligible proportion in 1915 to about 50% of the German, 35% of the French, 25% of the British, and 20% of the American ammunition expenditure by the Armistice (Spiers 2016). Providing little advantage to either of the equally equipped belligerents, chemical weapons greatly increased the already unspeakable suffering of the troops on both sides of both the Western and Eastern fronts. The British historian Edward Spiers recently characterized the WWI chemical weapons as "weapons of harassment" (ibid.). According to Augustin Prentiss's count (Prentiss 1937, 649), a total of about 90,000 soldiers were killed and 1.3 million injured by chemical weapons in WWI. What put finally an end to the war was the economic collapse of Germany (Mommsen 2011).

Albert Einstein's pacifist view contrasted sharply with that of his friend Haber. As he would put it later: "Warfare cannot be humanized. It can only be abolished" (Rowe and Schulmann 2007, 224). Strangely enough, there is no record of Einstein's criticism of Haber's WWI efforts, although Einstein occupied an office at Haber's institute at the time and must have been aware of what was going on. Gruesome as they were, chemical weapons have been banned only since 1997 (Organisation for the Prohibition of Chemical Weapons). Much of the military death toll in WWI (estimated to be at least 10 million troops) was, however, due to high explosives produced by the chemical industries of the warring nations. Hence the characterization of WWI as the "chemists' war," although chemical warfare surely added much weight to it. We note that the development and acquisition of the Haber-Bosch technology by Germany just in time for the Great War was key to sustaining her war effort: without it, the embargoed supplies of Chilean saltpeter would have run out within months and WWI would have indeed been as brief as anticipated by the German military planners, except that it would have ended not in Germany's speedy victory but rather her abrupt defeat.

Haber was without question the driving force behind the centrally-directed development of chemical warfare in Germany, whose use during WWI violated international law and elicited both immediate and enduring moral criticism, and has thereby inadvertently come to personify the moral flexibility of scientific research (Stern 2011; Dunikowska and Turko 2011). His efforts during WWI illustrate how quickly the fine line between the tolerable and the unacceptable can be crossed, in this case with fatal consequences.

As Haber's biographer Dietrich Stoltzenberg aptly noted: "It is easy to condemn [Haber]; it is much harder to make a sound judgment on him" (Stoltzenberg 1994, 153).

Acknowledgements We thank Dieter Hoffmann and Thomas Steinhauser (both Max Planck Institute for the History of Science) for most helpful discussions concerning chemical warfare research at Haber's institute and Robert Nye (Oregon State University) for his feedback on the manuscript.

References

Crowell, Benedict. 1919. *America's munitions 1917–1918: Report of Benedict Crowell, the Assistant Secretary of War, Director of Munitions*. Washington: Government Printing Office.

Dunikowska, Magda, and Ludwik Turko. 2011. Fritz Haber: The damned scientist. *Angewandte Chemie International Edition* 50: 10050–10062.

Haber, Fritz. 1920. Die chemische Industrie und der Krieg. *Die chemische Industrie* 43: 350–352.

Haber, Fritz. 1924. *Fünf Vorträge aus den Jahren 1920–1923*. Berlin: Springer.

Haber, Lutz F. 1986. *The poisonous cloud. Chemical warfare in the First World War*. Oxford: Clarendon Press.

Hartley, Harold. 1925. *Report on the German chemical warfare organization and policy, 1915–1918* (PRO/WO/33/1072).

Hoffmann, Dieter. 2014. … im Frieden der Menschheit, im Kriege dem Vaterland. Universität und Wissenschaft im Ersten Weltkrieg. In *Die Berliner Universität im Ersten Weltkrieg*, ed. Gabriele Metzger. Berlin: H. Henemann.

James, Jeremiah, Thomas Steinhauser, Dieter Hoffmann, and Bretislav Friedrich. 2011. *One Hundred Years at the Intersection of Chemistry and Physics. The Fritz Haber Institute of the Max Planck Society 1911–2011*. Berlin: De Gruyter.

Johnson, Jeffrey. 1990. *The Kaiser's Chemists. Science and Modernization in Imperial Germany*. Chapel Hill, NC: UNC Press.

Karrer, Paul. 1958. Heinrich Wieland 1877–1957. *Biographical Memoirs of Fellows of the Royal Society* 4: 340–352.

Lepick, Olivier. 1998. *La Grande Guerre Chimique, 1914–1918*. Paris: Presses Universitaires de France.

Lummitzsch, Otto. 1955. *Erinnerungen*. Haber Collection Va Rep. 5, 1480, Archiv der Max Planck Gesellschaft.

Martinetz, Dieter. 1996. *Der Gaskrieg 1914/18. Entwicklung, Herstellung und Einsatz chemischer Kampfstoffe. Das Zusammenwirken von militärischer Führung, Wissenschaft und Industrie*. Bad Neuenahr—Ahrweiler: Bernhard & Graefe Verlag.

Mommsen, Wolfgang. 2011. *Die Urkatastrophe Deutschlands. Der Erste Weltkrieg 1914–1918*. Stuttgart: Klett-Cotta.

Organisation for the Prohibition of Chemical Weapons. *Chemical Weapons Convention*. https://www.opcw.org/chemical-weapons-convention. Accessed 27 Sept 2016.

Prentiss, Augustin. 1937. *Chemicals in War: A treatise on chemical warfare*. New York, NY: McGraw-Hill.

Rowe, David E., and Robert Schulmann. 2007. *Einstein on politics*. Princeton, NJ: Princeton University Press.

Schmaltz, Florian. 2014. Chemie als Waffe: Fritz Haber und Richard Willstätter im Ersten Weltkrieg. In *Krieg! Juden zwischen den Fronten 1914–1918*, eds. U. Heikaus and J. B. Köhne (eds.). Berlin: Hentrich & Hentrich.

Spiers, Edward. 2016. The Gas War, 1915–1918: If not a war-winner, hardly a failure. In *One Hundred Years of Chemical Warfare: Research, Deployment, Consequences*, ed. B. Friedrich, D. Hoffmann, F. Schmaltz, J. Renn, and M. Wolf. Heidelberg: Springer.

Stern, Fritz. 1999. *Einstein's German world*. Princeton, NJ: Princeton University Press.

Stern, Fritz. 2011. Fritz Haber: Flawed greatness of person and country. *Angewandte Chemie Inetrnational Edition* 51: 50–56.

Stoltzenberg, Dietrich. 1994. *Fritz Haber—Chemiker, Nobelpreisträger, Deutscher, Jude*. Weinheim: Wiley-VCH.

Stoltzenberg, Dietrich. 2004. *Fritz Haber. Chemist, Nobel Laureate, German, Jew*. Philadelphia, PA: Chemical Heritage Press.

Szöllösi-Janze, Margit. 1998. *Fritz Haber 1868–1934. Eine Biographie*. München: C.H. Beck.

Szöllösi-Janze, Margit, and Helmuth Trischler (eds.). 1990. *Großforschung in Deutschland*. Frankfurt am Main: Campus.

Deimling, von Berthold. 1930. *Aus der alten in die neue Zeit*. Ullstein: Lebenserinnerungen. Berlin.

Zott, Regine (ed.). 1997. *Fritz Haber in seiner Korrespondenz mit Wilhelm Ostwald sowie in Briefen an Svante Arrhenius*. Berlin: ERS-Verlag.

Additional Open Access Information

Clara Immerwahr: A Life in the Shadow of Fritz Haber

Bretislav Friedrich and Dieter Hoffmann

Abstract We examine the life of Clara Haber, nee Immerwahr (1870–1915), including her tragic suicide and its possible relation to the involvement of her husband, Fritz Haber, in chemical warfare. Clara earned a doctorate in chemistry from the University of Breslau, in 1900, as the first woman ever, and married the physical chemist Fritz Haber within a year of her graduation. With no employment available for female scientists, Clara freelanced as an instructor in the continued education of women, mainly housewives, while struggling not to become a housewife herself. Her duties as the designated head of a posh household hardly brought fulfillment to her life. The outbreak of WWI further exacerbated the situation, as Fritz Haber applied himself in extraordinary ways to aid the German war effort, which included his initiative to develop chemical weapons. The night that he celebrated the "success" of the first chlorine cloud attack and his promotion to the rank of captain, Clara committed suicide. However, we found little evidence to support express claims that Clara was an outspoken pacifist who took her life because of her disapproval of her husband's engagement in chemical warfare. We examine the origin of this "myth of Clara Immerwahr" that took root in the 1990s from the perspective offered by the available scholarly sources, including those that have only recently come to light.

1 Prolog

On April 23, 1909, Clara Haber wrote to her PhD adviser and confidant, Richard Abegg, the following lines:

B. Friedrich (✉)
Fritz Haber Institute of the Max Planck Society, Faradayweg 4-6, 14195 Berlin, Germany
e-mail: brich@fhi-berlin.mpg.de

D. Hoffmann
Max Planck Institute for the History of Science, Boltzmannstraße 22, 14195 Berlin, Germany
e-mail: dh@mpiwg-berlin.mpg.de

B. Friedrich et al. (eds.), *One Hundred Years of Chemical Warfare: Research, Deployment, Consequences*, DOI 10.1007/978-3-319-51664-6_4

> What Fritz [Haber] has gained during these last eight years, I have lost, and what's left of me, fills me with the deepest dissatisfaction (Haber 1909).

This sobering summary of an eight-year marriage with Fritz Haber may serve as a key document about Clara's life and fate, not least in regard to her suicide six years later. Over the last twenty five years, Clara's suicide has been widely regarded not only as a personal tragedy and a result of a marital drama but, especially after the publication of her biography by Gerit von Leitner (Leitner 1993), as a consequence of Fritz Haber's involvement in chemical warfare in general and the first chlorine cloud attack at Ypres on 22 April 1915 in particular. More than that, it has been seen as a signal of a "feminine, life-preserving science" that opposes a patriarchal science, keen on securing power and on the exploitation of natural resources. Herein, we show that, based on the available biographical materials about Clara Haber, nee Immerwahr, this interpretation of her suicide is lopsided, lacking proper consideration of the complexity of Clara's personality and the circumstances of her life and time.

2 Clara Immerwahr's Background

Clara Immerwahr[1] was born on 21 June 1870 at the estate of Polkendorf near Breslau, where her father, a PhD chemist, withdrew after the failure of his chemical start-up company. Apart from becoming a highly successful agronomist in Polkendorf and its surroundings, he co-owned a flourishing specialty store in Breslau dealing in luxury fabrics and carpets. The family maintained an apartment in Breslau where the Immerwahrs would stay during their frequent visits to the city. And Clara would live there during her studies in Breslau.

Breslau, characterized by Goethe as a "noisy, dirty and stinking" town (Goethe 1949, 378), transformed itself during the second half of the nineteenth century into a prosperous metropolis teeming with business and industrial enterprise. This was accompanied by an enormous increase in population, which doubled during the 30 years since 1875, reaching 471,000 in 1905 (Rahden 2008, 32). At the same time, Breslau developed into a major center of science and culture with a large educated middle class. There was the Schlesische Friedrich-Wilhelms-Universität, founded in 1811 (whose forerunner was the Leopoldina, founded in 1702), a number of colleges, as well as an opera house, several orchestras, and a city theater —all of them of national significance (Davies and Moorhouse 2002).

The era of economic and cultural prosperity that the city enjoyed coincided with the childhood and youth of Clara Immerwahr, whose family belonged to Breslau's well-to-do Jewish middle class. After Berlin and Frankfurt, the Jewish community of Breslau was the third largest, at over twenty thousand Jewish residents (Rahden

[1]Herein, we draw on the biography of Clara Immerwahr as detailed in Szöllösi-Janze (1998, 124–131).

2008, 32), and its synagogue, consecrated in 1872, even the second largest in Germany (Scheuermann 1994). Breslau's Jewish community was academically oriented and represented the city's "intellectual aristocracy," to which also the Immerwahrs belonged (Noack 1959). However, they were assimilated Jews, who participated in communal cultural life but would only rarely, if at all, go to the Synagogue. Jewish religion, customs and practices played essentially no role in the family life. The political attitudes of the Immerwahr family were liberal, which however entailed the cultivation and demonstration of a certain degree of Prussian-German national awareness and patriotism, especially after the German unification of 1871 (Clark 2007). Prussian was also the simple lifestyle of the family, which was frugal—not because of need but out of principle. So despite the family's wealth, Clara and her three siblings were brought up in modesty.

Apart from the virtues of simplicity, frugality and modesty, a great value was attached to education—not just for the son and heir, but also for the three daughters. This was typical for the German Jewish middle class, as 40% of female students at the higher schools in Breslau were Jewish. As opposed to Switzerland or the Anglo-Saxon countries, German high schools (*Gymnasium*) were out of limits for women until the beginning of the twentieth century. The Grand-Duchy of Baden was the first state in Germany to institute, in 1900, admission of female students to universities. Before then, it was only possible for women to attend university by a special permission or as guest auditors (Johnson 1998).

Clara's path to education was shaped by these constraints. She started her studies at a Höhere Töchterschule (sometime translated as "women's college") in Breslau, which was supplemented during the summer months spent at the Polkendorf estate by instruction provided by a private tutor. Clara graduated from the Töchterschule in 1892 at age 22. The school was supposed to provide a basic education for young women that was compatible with their social status and to prepare them for their "natural purpose," that is, as companions of their husbands, as housewives, and as mothers. Nevertheless, Clara was up for more and after graduating from the Töchterschule she entered a teachers' seminary, which was the only type of institution that offered a higher professional education to women (Szöllösi-Janze 1998, 124). However, the graduates of the teachers' seminary only qualified to teach at girls' schools and remained ineligible to enter university and study, for example, science, which is what Clara wanted to do. So in order to qualify for the university, Clara had to take intensive private lessons and pass an exam equivalent to the *Abitur*. This exam was administered by a special committee set up at a *Realgymnasium* in Breslau and Clara passed it successfully at Easter 1896, when she was 26 years old.

Subsequently, Clara began her studies at the University of Breslau, however only as a guest auditor, since in Prussia women would only become legally admissible as university students as late as 1908. Prior to this, starting in 1895, women were only allowed to attend lectures as guests, and even that was contingent upon the support of the professor and faculty and permission from the Ministry; the last required a certificate of good conduct, character references, and so on. It is difficult today to imagine what it meant to women to break into the male domain of higher learning and what kind of discrimination and humiliation was connected with it. Talk of "intellectual Amazons" was not uncommon.

The attitude of Max Planck, who accepted Lise Meitner as an assistant in 1912 and was helpful in promoting her career even earlier, declared in 1895, in response to a poll, that

> Nature herself prescribed a role for women as mothers and housewives (Planck 1897, 256).

Thus according to the spirit of the time, Clara Immerwahr, with her wish to become a chemist, violated a law of nature. After her successful *Abitur* exam, Clara applied to the university curator's office for permission to attend lectures in experimental physics as a guest. And she had to proceed in a similarly awkward manner with all the other lecture courses that she wished to take.

From early on, Clara developed a keen interest in the then new field of physical chemistry.[2] Richard Abegg, one of this new field's pioneers and a friend of Fritz Haber's, played a key role in fostering Clara's interest in physical chemistry while paying little heed to Clara's guest auditor status. It was also Abegg who supervised Clara's PhD thesis—a part of the graduation requirement in chemistry—and who wrote a joint paper with Clara in 1899. The joint paper, published in 1900 (see next section and references therein), must have been perceived by the young female chemist as proof of her success and as an accolade. The following year, Clara submitted her dissertation and applied to be admitted to the *Rigorosum* final, which entailed exams in chemistry, physics, mineralogy, and philosophy. She passed the exams during the Fall and defended her thesis at the university's main auditorium on December 22, 1900.

Clara graduated with *magna cum laude* and her graduation was mentioned in the daily press, as Clara was the first woman on whom the University of Breslau conferred a doctoral degree. The left panel of Fig. 1 shows her photo during her university studies.

Richard Abegg (Fig. 2) assumed in 1899 an academic position at the Chemistry Institute of the University of Breslau, which belonged to the most prestigious in Germany. In 1909 Abegg became Ordinarius at the newly founded Technical University in Breslau. However, he would not live long enough to see through the construction of the new laboratory for physical chemistry at the Technical University, which was slated to be his own (Nernst 1913). Abegg was an early fan of balloon flying—and founded and presided over the Breslau ballooning club. He died in a ballooning accident in 1910 at the age of 41. As Nernst colorfully narrated in his obituary notice (Nernst 1913), Abegg was extremely hard working,

---

[2]Let us note that physical chemistry came about with a purpose, namely to save chemistry from taxonomy—from becoming a collection of little disconnected facts bred mainly by organic chemists. Its founders shared the view that chemistry should seek the general rather than cherish the particular and that the way to achieve it was to adopt the methods of mathematics and physics, Friedrich (2016). The success of physical chemistry in providing a common ground for chemistry was celebrated by Ostwald in his proclamation that "[p]hysical chemistry is not just a branch on but the blossom of the tree of knowledge" Ostwald (1887).

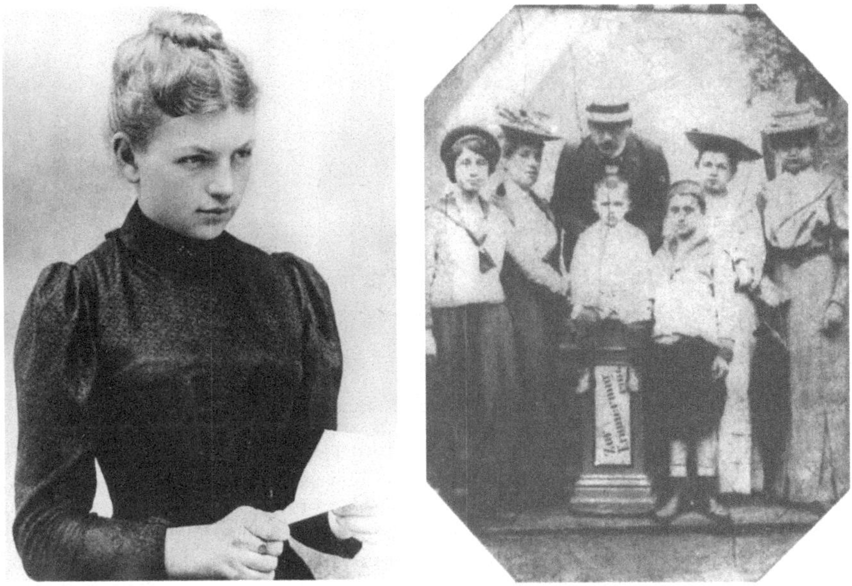

Fig. 1 *Left panel* Clara Immerwahr during her studies at the University of Breslau. *Right panel* Clara Haber, nee Immerwahr (2nd from *right*), on a family photo from 1906 with her son Hermann (seated in the center), her husband Fritz Haber (standing behind Hermann) and the landlady (2nd from *left*) of Habers' Karlsruhe apartment with her children; on the right is Habers' maid servant. Photos: Archiv der Max-Planck-Gesellschaft, Berlin-Dahlem

yet always had time for everybody. That must have surely made him an ideal academic teacher and adviser, particularly for Clara.

Otto Sackur (Fig. 2) was Clara's ten year junior *Kommilitone*, who studied chemistry at the University of Breslau, where, like Clara, he found an enlightened mentor in Richard Abegg. Sackur served on Clara's PhD committee as a referee.

As a *Privatdozent* at the University of Breslau, he was left after Abegg's death without an academic patron or a laboratory. It was during this period that Sackur launched his research at the intersection of thermodynamics and quantum theory. A reward in the form of a more senior appointment came at the end of 1913 when, thanks in part to mediation by Clara Haber, Sackur received a call to Haber's Kaiser-Wilhelm-Institut in Berlin. In 1914 he was promoted to the rank of a department head. After the outbreak of WWI he was enlisted in military research at Haber's institute, but continued on the side his experiments on the behavior of gases at low temperatures. In December of 1914, he was killed in a laboratory accident at his work bench, while trying to tame cacodyl chloride for use as an irritant and propellant (Badino and Friedrich 2013). He was just 34 years old.

While Abegg represented Clara's connection to science who, in addition, acted as her "cheerleader" and confidant in private matters, Otto Sackur was Clara's friend and *Kommilitone*. After Sackur's laboratory accident, Clara was among the first to attend to the injured. She proved capable of acting rationally in a situation

Fig. 2 *Left panel* Richard Abegg (*1869 Danzig; †1910 Tessin). Abegg graduated in chemistry from the Berlin University (1891) under August von Hofmann, received his *Habilitation* (1894) under Walther Nernst, was *Extraordinarius* at the University of Breslau (1899–1909), *Ordinarius* at the Technische Hochschule Breslau (1909), member of the Leopoldina (1900) and editor of the Zeitschrift für Elektrochemie (1901). Photo: Arrhenius, 1910. *Central panel* Otto Sackur (*1880 Breslau; †1914 Berlin). Sackur graduated in chemistry from the University of Breslau (1901) under Richard Abegg, received his *Habilitation* (1905) under Abegg, was department head at the Kaiser Wilhelm Institute for Physical Chemistry and Electrochemistry (1914). Photo: Jahresbericht der Schlesischen Gesellschaft für Vaterländische Kultur 1915, 1: 35–37. *Right panel* Fritz Haber (*1868 Breslau; †1934 Basel). Haber graduated in chemistry from the Berlin University (1891) under August von Hofmann, received his Habilitation (1894) under Hans Bunte, was *Extraordinarius* (1898) and *Ordinarius* (1906) at the Technische Hochschule Karlsruhe, founding Director of the KWI für Physikalische Chemie und Elektrochemie (1911–1933), *Honorarprofessor* (1912–1920) and *Ordinarius* (1920–1933) at the Berlin University, Member of the Prussian Academy (1914). He was awarded the Nobel Prize in Chemistry for 1918. Photo: Archiv der Max-Planck-Gesellschaft, Berlin-Dahlem

drastic to the extreme and to coordinate attempts to help the injured. However, Sackur died on the spot, before her eyes; Clara was crushed by Sackur's death. At the site of accident, Fritz Haber was just gasping for air in the arms of a coworker. He was shattered to the point that he stopped for good explosives research at his institute (James et al. 2011, 27).

3 The Scientific Work of Clara Immerwahr

Clara's scientific record consists of three research papers (Immerwahr and Abegg 1900; Immerwahr 1900a, 1901a), a supplement (Immerwahr 1900b) and an erratum (Immerwahr 1901b) to one of the research papers. Her first research paper is co-authored by her PhD adviser, Richard Abegg, the other two are solo. The second solo paper is an excerpt from Clara's PhD thesis. Clara's work concerned solution chemistry, one of the main preoccupations of physical chemistry at the time, and revolved about the connections among the conductivity, solubility, degree of dissociation, electrochemical potential and what was called electro-affinity.

The paper with Abegg, which expanded on the ideas of the 1899 Abegg-Bodländer paper that introduced the notion of electro-affinity as an organizing principle in chemistry (Abegg and Bodländer 1899), pretty much determined the topic and methodology of Clara's thesis paper. The thesis paper dealt in a more systematic way with the interplay between solubility of choice heavy metal salts and the electro-affinities of the constituent groups and atoms. Apart from providing tables of experimentally determined values of quantities such as equilibrium concentrations and relative electrode potentials, the paper aimed at assessing the issue of whether electro-affinities were additive quantities. The latter might be the reason for the relatively high number of citations (24) this paper has so far received.[3] However, one should keep in mind that quite a few of the publications citing Clara's paper are recent biographical articles about Clara rather than scientific papers.

Clara's second paper aimed to expand the solubility data base to include copper salts, using the ideas and methods developed by Walther Nernst, Wilhelm Ostwald and Friedrich Wilhelm Küster. The last was Clara's professor at the University of Breslau, who also deserves credit for arousing her interest in physical chemistry. He moved to the Bergakademie in Clausthal in 1899 and it was in Küster's Clausthal laboratory that Clara undertook the measurements reported in her second paper. As she noted, her data could be regarded as a corroboration of the Nernst-Ostwald-Küster theory.

Clara's PhD adviser Richard Abegg became well known for his work on valence that led to the octet rule. Clara's work on electro-affinity was somewhat related to this line of Abegg's research, but her contribution was not deemed significant enough to warrant Clara's inclusion in Svante Arrhenius's list of half a dozen or so of Abegg's former affiliates who had contributed to Abegg's research the most (Arrhenius 1910). To be sure, Sackur was not on that list either. However, Sackur made a name for himself in a research area that lay outside of Abegg's range of interests and published his key work only after Abegg's death (Badino and Friedrich 2013). It should also be noted that Clara's work, unlike Abegg's or Sackur's, did not seek to enrich the conceptual framework of physical chemistry in any way or to launch a new research direction.

Apart from her work as a researcher, Clara also gave public lectures, both in Breslau and later in Karlsruhe, on the broad topic of science in the household. Inspired by Lassar Cohn's popular book *Chemistry in daily life*,[4] Clara's lectures attracted audiences of up to one hundred women (Szöllösi-Janze 1998, 194).

---

[3]As of 7 September 2016.

[4]The book was translated into many European languages. Its author would become professor of chemistry at the University of Königsberg.

4 Clara's Husband: Fritz Haber

Apart from Abegg and Sackur, there was another pioneer of physical chemistry who entered Clara Immerwahr's life, namely Fritz Haber[5] (Fig. 2). Likewise a native of Breslau, Fritz likely met Clara at a dancing class (Szöllösi-Janze 1998, 124f). Little is known about this liaison, but Haber would later admit, at the occasion of his engagement with Clara in April 1901, that he was "in love with [his bride] as a [high school] student" and that during the intervening years he had "honestly but unsuccessfully" tried to forget her (Haber 1901). When the freshly minted Dr. Immerwahr appeared in April 1901 at the annual conference of the German Electrochemical Society in Freiburg—as the only female scientist—the affair between her and Haber was quickly rekindled.[6] As Haber would put it later in one of his letters (cf. Szöllösi-Janze 1998, 735, fn. 165), "we saw each other, we spoke and in the end Clara let herself be persuaded to give it a try with me." Clara would describe her motives for her acceptance of Fritz's advances in the already mentioned 1909 letter to her confidant Abegg:

> It has been my approach to life that it was only then worth living if one developed all one's abilities to the utmost and lived through everything that a human life can offer. And so I finally settled upon the idea of marriage [...] under the impulse that if I did not marry a decisive page in the book of my life and a string of my soul would lie idle. But the boost that I got from it was very short (Haber 1909).

As Margit Szöllösi-Janze, the biographer of both Fritz and Clara Haber, pointed out, their wedding, which took place already on August 3, 1901, marked the end of "the chapter 'chemical science' in Clara's book of life" which "must have been clear to the chemist" even without "any effects on the string of her soul" (Szöllösi-Janze 1998, 129).

Upon looking at the last decade of Clara's life, one has to agree. Although at the beginning she may have harbored the hope that she would be able to resume her scientific work at some point, she must have increasingly let go of such hopes as time went on. During the first years of her marriage, Clara appeared at lectures as well as in the laboratories of the Technische Hochschule in Karlsruhe, where her husband would soon become the founding director of its institute for physical chemistry and electrochemistry.

Moreover, it seems that at the time Fritz Haber would involve his wife in his research and share with her his scientific ideas, as suggested by the dedication of his 1905 classic textbook *Thermodynamics of technical gas-reactions*:

> To my dear wife Clara Haber, Ph.D., in gratitude for her silent co-operation (Haber 1908).

---

[5]Fritz Haber's authoritative biographies have been written by Szöllösi-Janze (1998) and Stoltzenberg (2004).

[6]In fact Haber asked Abegg to take Clara along to the Freiburg conference.

However, Clara was apparently not involved in doing any calculations for the book, as implied by the fact that this task fell to others (Haber 1908, x).

Nevertheless, that Clara's involvement in Haber's research entailed more than a silent co-operation transpires in her correspondence with Abegg, in which she reports about Haber's progress in writing the textbook, discusses academic appointments, and solicits advice about her own public talks. However, the dream of an equitable and reciprocal scientific marriage—such as that of Pierre and Marie Curie in Paris—did not come true.

The turning point likely occurred when their son Hermann was born in 1902[7] and/or when Haber became *Ordinarius* at Karlsruhe in 1906. Hermann was a sickly child, who claimed a lot of his mother's attention. Clara cared for the son lovingly while at the same time running a demanding household. At the beginning, the young family could not afford service staff and so Clara had to do a lot by herself. In a letter to Abegg written in 1901 from Karlsruhe, Clara declared that she would get back to the laboratory

> … once we become millionaires and will be able to afford servants. Because I cannot even think about giving up my [scientific work] (Haber 1901).

As we know, the Habers did get rich,[8] but nevertheless Clara would never return to the laboratory, despite Haber's positions as *Ordinarius* at Karlsruhe and later director of a Kaiser Wilhelm Institute. As the years went by, she would fall increasingly into the traditional role of a representative professorial wife, a housewife preoccupied with the well being of the family and a caring mother. This was aggravated by Haber's sharp-elbow mentality and his obsession with his work and career, which left little room for Clara's professional development and reduced her more and more to a mother/housewife. Clara broke down as a result and, as Szöllösi-Janze put it:

> the heyday that Haber had lived through in Karlsruhe was for his wife Clara her intellectual twilight (Szöllösi-Janze 1998, 353).

Clara saw it herself and committed her feelings to paper in her above mentioned letter to Abegg from 23 April 1909 where she said that

> [Fritz is a type of person] on the side of whom every other person who does not force his way even more recklessly at the other's expense than him, will perish. And that is the case with me (Haber 1909).

There were still six years left until Clara's voluntary exit from life on May 2, 1915. During this time Fritz Haber would enjoy further scientific and social ascent: in 1909, he laid the scientific foundations for the catalytic synthesis of ammonia from its elements ("bread from air") and in 1911 he became the founding director of the Kaiser Wilhelm Institute for Physical Chemistry and Electrochemistry in Berlin.

[7]This appears similar to the crisis of the marriage between Albert Einstein and Mileva Maric.

[8]Fritz Haber was awarded the patent rights for the ammonia synthesis in 1910, Szöllösi-Janze (1998).

Thereby Haber reached not only the Olympus of science in Germany but of science full stop. Clara could partake in the glory of it all—however not as a scientist but rather as a spouse of a scientist, a difference that the sensitive and earnest Clara surely must have reflected upon. The growing alienation of the couple was obvious to their contemporaries for whom

> the wearing down and the difficulties between the spouses were not of a petty kind but rather fundamental (Noack 1959, 301).

The strains and conflicts between Clara and Fritz further aggravated after the outbreak of WWI. In keeping with the maxim "In peace for mankind, in war for the fatherland,"[9] Fritz Haber applied himself in extraordinary ways to aid the German war effort (Stern 2011; Dunikowska and Turko 2011; Friedrich and James, this volume).

In part encouraged by the French use of tear gas (Haber 1924)—including its lethal variants—Haber took the initiative to employ chemistry in resolving the greatest strategic challenge of the war, namely the stalemate of trench warfare. Brought to glistening prominence by Germany's need to produce "gunpowder from air," Haber, backed by the profiteering chemical industry, was able to persuade his country's military leadership to stage a battlefield test of a chemical weapon—of "poison instead of air." This would earn him the epithet "father of chemical warfare."

The lethality of the April 22, 1915 chlorine cloud attack at Ypres lured the German military into adopting chemical warfare. Haber was promoted, by an imperial decree, to the rank of captain.

Haber celebrated the "success" at Ypres and his promotion at a gathering in his directorial mansion in Dahlem. The gathering took place in the evening of May 1, 1915. Afterwards, during the night from May 1 to May 2, Clara Haber committed suicide. She shot herself, with Haber's army pistol, in the garden of their mansion. Apparently, Haber, sedated by his daily allowance of sleeping pills, didn't hear the shots (there were two). Clara was found dying by their thirteen-year-old son Hermann.

5 Clara Haber's Suicide

Most of the materials related to Clara's suicide were generated nearly four decades later via interviews for the so-called Jaenicke Collection, named after Johannes Jaenicke (Hahn 1999), a Haber collaborator who planned to write Haber's biography and who headed the forerunner of the Archive of the Max Planck Society (Henning 1990). Mentions made in memoirs and personal correspondence of

[9] Usually quoted in this abbreviated form. For a full quote in German, cf. Haber (1920). Haber was referring to the attitude of Archimedes.

people who knew the Habers provide additional tidbits, albeit sometimes only between the lines. The coincidence of the suicide with the chlorine cloud attack at Ypres and Fritz Haber's key role in it gave rise to speculation and there were—as noted by Jaenicke—"numerous contradictory versions in circulation" (Jaenicke 1958). The Haber family treated the tragic event with utmost discretion, as a result of which there are no primary sources available, such as farewell letters, that would clarify the motive. Likewise, there are essentially no authentic contemporary testimonials available that shed light on the tragic event. Almost all of the extant testimonials are from the 1950s and 1960s, solicited and gathered by Johannes Jaenicke for his collection.[10] Twenty years earlier, in early 1940s America, Morris Goran, about whom little is known, except that he held a position at Roosevelt College in Chicago at some point, attempted to interview German émigré scientists about the German scientific establishment in general and about Fritz Haber in particular, however, with a mixed success. For instance James Franck, upon being contacted by Goran, characterized him as "the terrible guy in America [Goran], who wants to prove what a [great] man he is by writing about Fritz Haber" (Jaenicke 1958). In 1947, Goran published a rather hagiographic article about Fritz Haber (Goran 1947) and in 1967 a book *The Story of Fritz Haber* (Goran 1967), which contains a brief passage about Clara's suicide. In the passage, Goran stated that Clara was "vitally affected" (Goran 1967, 71) by her husband's involvement in WWI chemical warfare and committed suicide after a heated argument with Fritz about what she considered to be "a perversion of science" and "a sign of barbarism" (Goran 1967, 71). Goran gives no evidence or sources for either this scenario or these statements. Apparently, the much-quoted phrase about the perversion of science and barbarism, ascribed to Clara, is Goran's own. Apart from his political and moral categorization of Clara's suicide, Goran also points out for the first time that Clara was depressive and that

> chemical warfare was an avenue or excuse for the morbid worry she seemed to favor (Goran 1967, 71).

However, Goran doesn't give any references here either, which led Margit Szöllösi-Janze to the characterization of his book as one where "the line between a historically correct study and fiction is blurred" (Szöllösi-Janze 1998, 395).

In her Haber biography, Szöllösi-Janze has already critically evaluated the sources about Clara and her suicide that can be found in the Jaenicke Collection (Szöllösi-Janze 1998, 393–399), with the conclusion that the motives for Clara's suicide are as unclear as the available sources are ambiguous—and rare. However, the possible motives can be divided into two groups, listed on the left and right side of Table 1: on the left are sources suggesting that Clara's suicide had to do with Fritz Haber's engagement in the German war effort/chemical warfare and on the right with her unfulfilling personal (and absent professional) life and with depression.

[10]The individual items of the Jaenicke Collection have never been published.

Table 1 Possible motives for Clara Haber's suicide. Left: Sources suggesting that the motive had to do with disagreements about Haber's involvement in the German war effort. Right: Sources suggesting that Clara's suicide had to do with disagreements about personal matters and/or her depression

Disagreements about Fritz Haber's involvement in German war effort	Disagreements about personal matters; depression
Hans Krassa (1957); Archiv MPG[a]	Haber (1909); letter to Abegg[c]
Franck (1958); Archiv MPG[b]	Haber (1915); letter to Tamaru[d]
Goran (1967); source not given	Hahn (1915); letter to Otto Hahn[e]
Mendelssohn (1973); source not given	Meitner (1915); letter to Edith Hahn[e]
Von Leitner (1993); sources: either not given or Goran, Lütge, Lummitzsch tapped selectively	Lummitzsch (1955); Memoirs[f]
	Noack (1959); Archiv MPG[g]
	Goran (1967); source not given
	Lütge (1958); Archiv MPG[h]
	Haber (1970); Memoirs[i]
	Ebbinghaus (1993); review of von Leitner's book[j]

[a]Krassa (1957)
[b]Jaenicke (1958)
[c]Haber (1909)
[d]Clara (1915)
[e]Meitner (1915)
[f]Lummitzsch (1955)
[g]Noack (1959)
[h]Lütge (1958)
[i]Charlotte (1970)
[j]Ebbinghaus (1993)

Hermann Lütge, the former fine-mechanic at Haber's institute,[11] when asked by Jaenicke whether Clara committed suicide in response to "Haber's involvement in the abhorrent gas warfare" and whether Clara's "ethical asperity was a consequence of a hereditary depressive disposition," stated the following:

> No, this is not to be presumed. Frau Geheimrat [Clara] … was not in a state of mind to contemplate the abhorrence of chemical warfare … Yes, sometimes was Frau Haber gloomy, especially after being harshly rebuffed for mothering her husband too much (Lütge 1958, 260).

A similar answer was given to Jaenicke by Adelheid Noack, the niece of Clara's brother-in-law:

[11]Clara was the godmother of Hermann Lütge's son (private communication, Michael Lütge, 2016).

> There are various more or less pathetic accounts of her [Clara's] suicide, for instance that she had beseeched him [Fritz Haber] to abandon chemical warfare. These accounts are a nonsense (Noack 1959).

This is contrasted by James Franck's opinion, who stated in his conversation with Jaenicke that Clara was

> a good, talented person with distinct views, which often contradicted those of her husband ... she wanted to reform the world. That her husband was involved in chemical warfare had surely an effect in her suicide (Jaenicke 1958, 1449).

However, Franck added that Fritz Haber

> expended an immense effort to reconcile his and [Clara's] political and human views.

Another proponent of the view expressed by James Franck was the physical chemist Kurt Mendelssohn, who had worked before his emigration in 1933 in both Berlin and Breslau.[12] In his book *The World of Walther Nernst* he stated:

> ... there was a macabre sequel to his [Fritz Haber's] decision to develop poison gas. His wife, Dr. Clara Immerwahr, who was also a chemist, had pleaded with him [Fritz Haber] again and again not to work on gas warfare. His answer was that his first duty was to his country and that no argument, not even the entreaties of his wife, could shake his resolve. On the evening of Haber's departure for the front, Clara committed suicide (Mendelssohn 1973, 83).

An additional testimonial about a possible role of chemical warfare in Clara's suicide was delivered by her cousin Hans Krassa, according to which Clara visited Krassa's wife shortly before the suicide to confide to her about the "gruesome effects" of chemical warfare that she had witnessed, in particular the "testing on animals" (Krassa 1957, 1470). Krassa, however, added that other factors may had been at play as well. As far as Clara's disposition was concerned, Krassa stated that "the word gloom goes too far" and that "one certainly cannot speak of a hereditary depression."

That Clara was "exceedingly nervous," especially in the last years of her life, can be found in the testimonial by Otto Lummitzsch, the adjutant of the commander of the gas troops, who witnessed a visit by Fritz and Clara Haber at the proving ground in Wahn near Köln. He characterized Clara as

> a nervous lady, who was already then in sharp opposition to Geheimrat Haber's ventures to the Front along with the gas troops (Lummitzsch 1955).

This quote further attests to Clara's habit of "mothering" her husband.

Another aspect of Clara's personality transpires in the manner she behaved and dressed. According to James Franck,

> [Haber] liked to represent, whereas [Clara] exaggerated the simplicity of her manner and she dressed poorly – [perhaps] as a protest? (When I visited [the Habers] for the first time, the door was answered by a person whom I held for a cleaning woman. And I thought that it

[12]It is conceivable that James Franck was in fact the source of Mendelsohn's account.

would have been fitting if in such a fine household [as Habers'] the cleaning woman had dressed a little more nicely – but it was Frau Geheimrat [Clara] herself) (Franck 1958).

Out of the rest of the testimonials on the right of Table 1 we would like to bring to the fore additional tidbits provided by Adelheid Noack and by Hermann Lütge. In her conversation with Jaenicke, Noack also mentioned that Clara was "horrified by anything sensual," in keeping with the fact that she had quit the marital bedroom in 1902, never to return to it (Noack 1959). This fact as well as Noack's testimonial was corroborated by Haber's second wife, Charlotte Nathan, who had access to such intimate information more than anybody else (Haber 1970, 83 and 89). A real bombshell was dropped by Hermann Lütge, who testified that during the fateful night of May 1–2, 1915, Clara caught her husband *in flagranti* with Charlotte Nathan (Lütge 1958, 260). Charlotte worked as a manager of the then incipient club "Deutsche Gesellschaft 1914," where she and Haber got to know each other and was invited to the grand celebration of the "success" at Ypres in Habers' mansion (although Charlotte later contradicted it). The sociologist Angelika Ebbinghaus (Ebbinghaus 1993) as well as the historian Margit Szöllösi-Janze (Szöllösi-Janze 1998, 398) indicated that they tend to the view that Clara's discovery of her husband's affair may have been the actual trigger for her suicide.

Although provided by contemporaries, the above testimonials had been delivered with a delay of about 50 years, which makes them historiographically problematic. However, there are two recently surfaced documents that had been written within days of Clara's suicide and that answer some of the questions posed in connection with it: they are the letters (dated May 5, 1915) by Edith Hahn, the wife of the chemist Otto Hahn, to her husband and the letters (dated May 6 and 9, 1915) by Lise Meitner, Otto Hahn's collaborator and colleague at the Kaiser Wilhelm Institute for Chemistry,[13] to Edith Hahn. These letters, recently published by Eckart Henning (Henning 2016), the former director of the Max Planck Archive, confirm that Clara was mentally unstable. So Edith Hahn wrote:

Of course was the woman [Clara] ill, she'd been always strange – everybody was mocking her (Hahn 1915).

And Lise Meitner reports that

as of late [Clara] had always made an impression of being agitated (Meitner 1915).

The letters also agree that the reasons for Clara's act of desperation were to be found in her private life. Edith Hahn wrote to her husband that

he [Fritz Haber] [was] guilty. I have the feeling that she was [strongly] attached to him and that he treated her badly – or at least quite indifferently, and that she suffered more than we can imagine. Recently, she complained [to me] that he would never write to her [from the front], this came out inadvertently and was so sad that I lied to her that you write to me only seldom [as well] and [pointed out to her] that her husband has had even less time [than you

[13]The KWI's for Chemistry and for Physical Chemistry and Electrochemistry were located next to each other—and the on-campus Haber mansion.

> did]. Poor, poor woman. I've had always the feeling that he was fed up with her, which I could understand to a certain extent (Hahn 1915).

In line with this, Lise Meitner wrote that

> she [Clara] has recently made remarks to the effect that she was unhappy in her marriage. And that he [Fritz Haber] is not exactly an affectionate person. Anyway, it's a very sad story (Meitner 1915).

That the likely reasons for Clara's suicide were personal is supported by yet another contemporary document. At the turn of 1914/1915, an exchange of letters took place between Setsuro Tamaru, Haber's former Japanese collaborator, who had to leave Germany after the outbreak of the war, and Clara Haber. In his lengthy letter, written on Christmas Eve 1914, Tamaru complains about his personal situation as a guest in Theodore Richard's laboratory at Harvard, characterized by personal and scientific isolation; about being forced to leave Germany; and about receiving "not a single line whatsoever, no reply from Herrn Geheimrat [Fritz Haber]" (Tamaru 1914). Furthermore, Tamaru's six-page letter is concerned with the political and military situation during the first year of WWI and contains Tamaru's stance regarding war and peace:

> I am a pacifist of sorts and am always against war. A war doesn't decide anything, just breeds the next war (Tamaru 1914).

In her equally lengthy reply, Clara in no way reacts to Tamaru's stance and describes the "melancholy of our separation" and "your [Tamaru] being missed at the Christmas table" instead (Haber 1915). The silence of Fritz Haber and of others at the institute Clara explains by pointing out that

> … my husband is working 18-hour days, almost always in Berlin, I'm taking care of 57 poor children[14] and Hermann [the son] has been ill since November … Apart from that we are all adversely affected by the outrage and the dull pressure [of the war] that disable any impulse to do anything else than to help the country in the few remaining hours [of the day] (Haber 1915).

Clara also provides a brief report about Otto Sackur's "terrible accident" and the shock she suffered as a result and concludes by stating:

> To your political contentions, which were very interesting for us [to read], I will not respond; I am too ignorant in the matters of foreign affairs to be able to properly answer [your points]. You are certainly right in many respects, but have somewhat one-sided views on some points (Haber 1915).

Even if one takes into account that at the time international correspondence was subject to censorship, what transpires in Clara's letter is a woman made heavy-hearted by human suffering and the burdens of the war rather than a political activist or indeed a pacifist. This makes quite questionable the image of Clara, created in the 1990s—see below—according to which she was an outspoken

---

[14]Clara ran a makeshift kindergarten that made use of the premises of Haber's KWI. The fathers of the "57 poor children" were on the front and their mothers had to work in order to make ends meet.

pacifist. Likewise questionable is Clara's opposition to her husband's involvement in chemical warfare and thus the subsumed motive of her suicide as having been connected with it. So Dr. Kremmer, the principal of Hermann Haber's school, described in his condolence letter to Hermann upon the death of Fritz Haber how Hermann's "Frau Mutter" came to him

> to report on the success of the first gas attack at Ypres right after receiving a telegram from the front about it (Kremmer 1934).

And Hermann Lütge, in his testimonial, stated that

> the boss [Clara] was proud of the services provided by her husband (Lütge 1958).

Another controversy connected with Clara's suicide concerns Fritz Haber's behavior during the aftermath of the harrowing event. Haber's departure for the eastern Front the same day (May 2) was often portrayed as a reckless abandonment of his thirteen-year-old son Hermann and a sign of callousness and egotism. Even Szöllösi-Janze argues that the visibly shaken Haber may have regarded the Front as a place to which he could escape from the tragic reality at home. However, the above-quoted letter by Lise Meitner sheds new light on this aspect as well:

> As you know, Haber was supposed to leave in the morning, but stayed until the evening, when he was [finally] forced to depart. I'm told that he inquired at the [military] headquarters whether, out of consideration for the unfortunate event, he could postpone his departure, but his request was denied (Meitner 1915).

Although Lise Meitner qualified her statement by adding "Whether it's true, I of course don't know," the passage nevertheless suggests that Haber was not as unfeeling a chemical warrior who left his son in the lurch without a reason as had been previously conjectured.

6 The "Myth of Clara Immerwahr"

The scarcity and ambiguity of the historical record notwithstanding, during the 1990s a narrative took root according to which Clara Haber was supposedly a pacifist and decisive opponent of chemical warfare, in contrast to her husband Fritz Haber, who was chemical warfare's main proponent and Clara's oppressor to boot. It appears that this narrative was catapulted into the public sphere in Germany and beyond by Gerit von Leitner's book *Der Fall Clara Immerwahr. Leben für eine humane Wissenschaft*, published in 1993 (second edition 1994) as well as various dramatizations derived from it. In it, Clara is presented as an outspoken pacifist (not unlike the Czech-Austrian 1905 Nobel Peace Prize laureate Bertha von Suttner) and a star scientist (not unlike Marie Curie) who was destroyed—as both a person and a scientist—by her oppressive and opportunistic husband. The sources in von Leitner's book are either not given or tapped selectively, so as to provide a spotless image of Clara while portraying Fritz Haber as a kind of Dr. Evil. Von Leitner's account ignores other sources that suggest that the reasons for Clara's suicide may have had to do with her private life. These testimonials or opinions are listed on the right in Table 1.

The emphasis on Clara's 1909 letter to Richard Abegg (Haber 1909) is a case in point. Written on funeral stationary and opening with a tirade about her inability to locate a fountain pen (described—in pencil—on two pages out of twelve), Clara denounces her husband and details her unfulfilling life with him. The letter may have been triggered by jealousy, after Abegg, during his visit to Karlsruhe, congratulated Fritz Haber on his discovery of the catalytic synthesis of ammonia without mentioning Clara (Ebbinghaus 1993). Clara, however, had not been involved in research—her own or Haber's—since about 1901, as she had acknowledged in the same letter. The letter is special in that it is the only one written by Clara to Abegg (or anybody else for that matter) where she had lost her nerve and complained about Haber and their marriage.

Von Leitner's book (Leitner 1993) apparently struck a chord with the *Zeitgeist*, as it had been well—in some cases even euphorically—received not only in feminist and pacifist circles but also by a majority of German literary critics writing for Germany's leading newspapers and magazines. So, for instance, Volker Ullrich published in *Die Zeit* a review where he paid tribute to von Leitner's book as

> one of the best examples of a new, woman-inspired form of writing history, … a fascinating historical portrait … that reveals what was covered up and concealed for decades (Ullrich 1993).

Ullrich's review became emblematic for the reception of the book by other critics and its tenor can be found in many additional, roughly thirty reviews that we could identify, published in leading supra-regional newspapers such as *Frankfurter Allgemeine Zeitung*, *Frankfurter Rundschau*, *Die Welt*, *Süddeutsche Zeitung*, *Die Tageszeitung*, in regional periodicals such as *Sächsische Zeitung*, *Tagesspiegel*, *Westfalen Blatt*, *Main Echo*, *Emsdettener Tageblatt*, as well as in journals like *Emma*, *1999*, *Zeitschrift für Sozialgeschichte*, or on ARD radio stations. And the *New York Review of Books* chipped in as well (Perutz 1996). Time and again, the reviews had made a connection to contemporary events in the 1990s, including misuse of scientific research by the military and the Gulf War 1990–1991. Another issue discussed in the reviews, one that touches upon the core of von Leitner's book, is that of equality in scientific/academic marriages such as that of the Habers and the fostering of academic careers of female scientists. All that lent relevance to von Leitner's book vis-à-vis the political trends and debates of the 1990s and made it into a vehicle for furthering the opinions, ideals and *Wunschbilder* of the peace movement, feminism and antimilitarism. Clara's attempt to have a self-determined life as a woman, mother and scientist as well as her tragic suicide are interpreted as a "[beacon of a] feminine, life-preserving science" and juxtaposed with the male, patriarchal power-oriented science concerned with the exploitation of resources.

Volker Ullrich's review is a prime example of such an interpretation of von Leitner's book that had over time acquired an almost paradigmatic character. As apodictic appear Ullrich's statements according to which von Leitner tore down "the veil of falsified legend built [around Fritz Haber]." However, what had been overlooked is that, through the back door, another legend was being ushered in: the myth of Clara Immerwahr. According to this myth, Clara committed suicide in opposition to the gas warfare and as a desperate protest against the development of weapons of mass destruction by her husband, whose effort was contemptuous of human life. This interpretation is not only too monocausal and simplistic, but is difficult to support by the available historical sources, as already outlined above; in the best case, it can be viewed as a catchy hypothesis lacking supporting evidence. Incidentally, a criticism of this sort had been already leveled against von Leitner's book by several reviewers during the 1990s. For instance, the historian of science Ernst Peter Fischer writing in *Die Tageszeitung* (and also *Weltwoche*) denounced not only the stylistic and substantive shortcomings of the book, characterizing it as a "total failure [*total misslungen*]," but he also pointed out that because of the missing references it is unclear whether the book is a "reliable rendition [of historical facts]" and how one-sided its interpretations are (Fischer 1993a, b). A similar argument was presented in the review by the historian Jakob Vogel in the *Frankfurter Allgemeine Zeitung*, who noted that

> a personal fate whose meaning comes about through a true-to-life contrariness [is sacrificed to] political correctness (Vogel 1993).

The main deficiency of von Leitner's book was also commented upon by the sociologist Angelika Ebbinghaus in *1999. Zeitschrift für Sozialgeschichte des 20 und 21. Jahrhunderts*, where she pointed out that the documented fragments of Clara's life

> could have provided the basis for a novel. Such a novel could have rendered the historical truth without necessarily being literally true. A biography, however, must fulfill other criteria, namely whether reality at least resembled that what has been presented.

Although von Leitner chose the scholarly genre of biography rather than novel—she ditched the standards of scholarship in the process of writing her account, such as documenting her statements by critically evaluated references. In her account she often puts statements/opinions in the heroine's mouth or describes situations involving the characters of her book for which no record or evidence exist. For instance, she states that "Clara admired the courageous Bertha von Suttner" and even describes a scene in which Clara discusses women's rights with her husband and takes the side of von Suttner. A partial list of statements and quotations appearing in von Leitner's and other accounts of Clara Haber's relation to chemical warfare that are of unknown origin have been listed elsewhere (Friedrich and Hoffmann 2016).

Since neither Clara nor Fritz Haber left behind diaries or correspondence from which such opinions, conversations or situations could be reconstructed, these and other passages in von Leitner's book can only be regarded as an unscholarly mixture of fiction and historical fact. Of particular significance is von Leitner's contextualization of Clara's suicide, as this is presented as a decisive protest against the development and use of chemical weapons, as a signal "against the chemical mass destruction" (Leitner 1993) and as (Kokula 1988).[15]

> a signal for a new definition of natural sciences that had not been heard.

The evidence provided by historical sources is too thin for such a strong hypothesis, not to speak about von Leitner's handling of the historical record. Therefore, we cannot but agree with an earlier assessment by Szöllösi-Janze that:

> As regards the viability and validity of the sources, the record about the last months of Clara Immerwahr's life during the First World War consists chiefly of gaps rather than proven knowledge (Szöllösi-Janze 1998, 395).

Despite all these defects and their explicit critique in the press as well as in Szöllösi-Janze's authoritative Haber biography, the image of Clara Haber, nee Immerwahr as an outspoken pacifist and opponent of chemical warfare prevails in the public awareness until this day.

Herein, we plead for a more differentiated view based on the available historical record, according to which Clara Haber's suicide appears to have likely been the result of a "catastrophic failure" (to borrow an engineering term as a metaphor) brought about by a most unfortunate confluence of a host of circumstances that included, apart from her unfulfilling life, Haber's philandering, the tragic deaths of her close friends, Richard Abegg and Otto Sackur, as well as the death and destruction of the war itself, amplified by the perversions of chemical warfare.

7 Epilog

Our intention has been to make the above points without belittling in the least Clara's achievements and courage. Honoring Clara, for instance through the Clara Immerwahr Award of the Nobel-Prize winning organization International Physicians for the Prevention of Nuclear War (IPPNW) or the Clara Immerwahr Prize of the Berlin Excellence Cluster UniCat, is highly to the point and should not be questioned in any way. Haber's institute, named after its founding director in 1952 and incorporated into the Max Planck Society in 1953, had a memorial built for Clara in the garden of the institute in 2006 (Fig. 3).

However, we should refrain from projecting our contemporary ideas about women's rights activists or peace activists on Clara Haber in an ahistorical way.

[15]Kokula is the maiden name of Gerit von Leitner.

Fig. 3 *Left panel* Gravestone of Fritz and Clara Haber at the Hörnli Cemetery in Basel. In his testament, Haber expressed his wish to be buried alongside his first wife Clara—in Dahlem if possible, or elsewhere "if impossible or disagreeable." Haber's son Hermann became the will's executor. In accordance with this will, Clara's ashes were reburied beside Fritz Haber's in Basel. Photo: Archiv der Max-Planck-Gesellschaft, Berlin-Dahlem. *Right panel* Memorial for Clara Haber in the garden of the Fritz Haber Institute, installed in 2006. The photo by one of the authors (BF) shows the memorial at the centenary of Clara's suicide, on May 2, 2015

What she achieved in her time does not need to be embellished with exaggerations or even wishful thinking fashioned by present-day aspirations. Her achievements speak for themselves and should not be degraded or even compromised by mixing them up with fabrications and *Wunschbilder*.

Acknowledgements Our special thanks are due to Margit Szöllösi-Janze (Ludwig-Maximilians-Universität, München), whose authoritative biography *Fritz Haber: 1868–1934* (C.H. Beck, München 1998) that covers both Fritz and Clara Haber proved to be an invaluable source of scholarship on the subject. In addition, Margit Szöllösi-Janze kindly provided comments specific to this article. We also thank Eckart Henning (former director of the Archive of the Max Planck Society, Berlin) for introducing us to the 1915 letters by Edith Hahn and Lise Meitner that mention Clara's suicide and that he has only recently published. We are also grateful to Hideki Tamaru Oyama (Rikkyo University) for making available to us the letter by her grandfather Setsuro Tamaru to Clara from December 24, 1914. Last but not least, we thank Gerhard Ertl (Fritz Haber Institute), Hajo Freund (Fritz Haber Institute), and Matthew Meselson (Harvard University) for a critical reading of the manuscript.

References

Abegg, Richard, and Guido Bodländer. 1899. Die Elektroaffinität, ein neues Prinzip der chemischen Systematik. *Zeitschrift für Anorganische Chemie* 20: 453–499.
Arrhenius, Svante. 1910. Richard Abegg. *Zeitschrift für Elektrochemie* 16: 554–557.
Badino, Massimliano, and Bretislav Friedrich. 2013. Much polyphony but little harmony: Otto Sackur's groping for a quantum theory of gases. *Physics in Perspective* 15: 295–319.
Clark, Christopher. 2007. *Iron kingdom: The rise and downfall of Prussia, 1600–1947*. London: Penguin.
Davies, Norman, and Roger Moorhouse. 2002. *Die Blume Europas*. Breslau, Wroclaw, Vratislav. München: Droemer.
Dunikowska, Magda, and Ludwik Turko. 2011. Fritz Haber: The damned scientist. *Angewandte Chemie Int Ed* 50: 10050–10062.
Ebbinghaus, Angelika. 1993. 1999. *Zeitschrift für Sozialgeschichte des 20 und 21. Jahrhunderts*, Heft 4, 125–131.
Fischer, Ernst Peter. 1993a. Frau, Wissenschaftlerin, Pazifistin—und nach 80 Jahren noch unverstanden. *Die Weltwoche*, 19 August 1993, p. 23.
Fischer, Ernst Peter. 1993b. Professor Dr. Fritz. *Die Tageszeitung (taz)* v. 21.6.1993, S. 15.
Franck, James 1958. Note about Jaenicke's conversation with James Franck on 16–17 April 1958. Archiv der Max-Planck-Gesellschaft, Haber-Sammlung Va Abt., Rep. 5., Nr. 1449.
Friedrich, Bretislav. 2016. How did the tree of knowledge get its blossom? The rise of physical and theoretical chemistry, with an eye on Berlin and Leipzig. *Angewandte Chemie Int Ed* 55: 5378–5392.
Friedrich, Bretislav, and Dieter Hoffmann. 2016. Clara Haber, nee Immerwahr (1870–1915): Life, Work, and Legacy. *Zeitschrift für anorganische und allgemeine Chemie* 642: 437–448.
Goethe, Johann Wolfgang. 1949. *Goethe, the story of a man: Being the life of Johann Wolfgang Goethe as told in his own words and the words of his contemporaries*, vol. 1. New York: Farrar, Straus.
Goran, Morris. 1947. The present-day significance of Fritz Haber. *American Scientist* 35: 400–403.
Goran, Morris. 1967. *The story of Fritz Haber*. Norman: University of Oklahoma Press.
Haber, Charlotte. 1970. *Mein Leben mit Fritz Haber*. Düsseldorf: Econ-Verlag.
Haber, Clara. 1901. Clara Haber to Richard Abegg, 18 October 1901. Archiv der Max-Planck-Gesellschaft, Haber Sammlung Va Abt., Rep. 5., Nr. 812.
Haber, Clara. 1909. Clara Haber to Richard Abegg, 23 April 1909. Haber Collection. Archiv der Max-Planck-Gesellschaft, Haber-Sammlung Va Abt., Rep. 5., Nr. 812.
Haber, Clara. 1915. Clara Haber to Setsuro Tamaru, 15 January 1915. Oyama, Hideko. 2015. Setsuro Tamaru and Fritz Haber: Links between Japan and Germany in Science and Technology, Chemical Record 15: 535–549.
Haber, Fritz. 1901. Fritz Haber to Professor [unknown], 18 April 1901. Archiv der Max-Planck-Gesellschaft, Haber-Sammlung Va Abt., Rep. 5., Nr. 1874.
Haber, Fritz. 1908. *Thermodynamics of technical gas-reactions*. London: Longmans, Green & Co.
Haber, Fritz. 1920. *Die chemische Industrie* 43: 350–352.
Haber, Fritz. 1924. *Fünf Vorträge aus den Jahren 1920–1923*. Berlin: Springer.
Hahn, Edith. 1915. Edith Hahn to Otto Hahn, 5 May 1915. Archiv der Max-Planck-Gesellschaft, Va. Abt., Rep 55, Hahn-Sammlung, Nr. 96.
Hahn, Ralf. 1999. *Gold aus dem Meer. Die Forschungen des Nobelpreisträgers Fritz Haber in den Jahren 1922–1927*. Berlin, Diepholz: GNT.
Henning, Eckart. 1990. Die Haber-Sammlung im Archiv zur Geschichte der Max-Planck-Gesellschaft. *Berichte zur Wissenschaftsgeschichte* 13: 34–37.
Henning, Eckart. 2016. Freitod in Dahlem (1915): Unveröffentlichte Briefe von Edith Hahn und Lise Meitner über Dr. Clara Haber geb Immerwahr. *Zeitschrift für anorganische und allgemeine Chemie* 642: 432–436.

Immerwahr, Clara. 1900a. Potentiale von Kupferelektroden in Lösungen analytisch wichtiger Kupferniederschläge. *Zeitschrift für Anorganische Chemie* 24: 269–278.
Immerwahr, Clara. 1900b. Berichtigung zu meiner Arbeit 'Potentiale von Kupferelektroden in Lösungen analytisch wichtiger Kupferniederschläge'. *Zeitschrift für Anorganische Chemie* 24: 112.
Immerwahr, Clara. 1901a. Beiträge zur Kenntnis der Löslichkeit von Schwermetallniederschlägen auf elektrochemichem Wege. *Zeitschrift für Elektrochemie* 7: 477–483.
Immerwahr, Clara. 1901b. Berichtigung. *Zeitschrift für Elektrochemie* 7: 625.
Immerwahr, Clara, and Richard Abegg. 1900. Notiz über das Elektrochemische Verhalten des Fluorsilbers und des Fluors. *Zeitschrift für Physikalische Chemie, Stochiometrie und Verwandtschaftslehre* 32: 142–144.
Jaenicke, Johannes. 1958. J. Jaenicke to H. Lütge, Berlin 6 January 1958. Archiv der Max-Planck-Gesellschaft, Haber-Sammlung Va Abt., Rep. 5., Nr. 260.
James, Jeremiah, Thomas Steinhauser, Dieter Hoffmann, and Bretislav Friedrich. 2011. *One Hundred Years at the Intersection of Chemistry and Physics. The Fritz Haber Institute of the Max Planck Society 1911–2011*. Berlin: De Gruyter.
Johnson, Jeffrey. 1998. German women in chemistry, 1895–1925 (Part I). *NTM Zeitschrift für Geschichte der Wissenschaften, Technik und Medizin* 6: 1–21.
Kokula, Gerit. 1988. Gegen die Perversion. Vom verlorenem Kampf der Chemikerin Clara Immerwahr. *Tagesspiegel* v. 29.12.1991, p. IV (Supplement).
Krassa, Hans. 1957. Hans Krassa to Johannes Jaenicke, 2 November 1957, Archiv der Max-Planck-Gesellschaft, Haber-Sammlung Va Abt., Rep. 5., Nr. 1470.
Kremmer, [first name unknown]. 1934. Dr. Kremmer to Hermann Haber, 2 February 1934. Archiv der Max-Planck-Gesellschaft, Haber-Sammlung Va Abt., Rep. 5., Nr. 1222.
Leitner, von Gerit. 1993. *Der Fall Clara Immerwahr. Leben für eine humane Wissenschaft*, München: C.H. Beck.
Lummitzsch, Otto. 1955. *Erinnerungen*. Archiv der Max-Planck-Gesellschaft, Haber-Sammlung Va Abt., Rep. 5., Nr. 1480.
Lütge, Hermann. 1958. Hermann Lütge to Johannes Jaenicke, 9 and 17 January 1958, Archiv der Max-Planck-Gesellschaft, Haber-Sammlung Va Abt., Rep. 5., Nr. 260.
Meitner, Lise. 1915. Lise Meitner to Edith Hahn, 6 May 1915. Archiv der Max-Planck-Gesellschaft, Va. At, Rep. 55, Hahn-Sammlung, Nr. 94.
Mendelssohn, Kurt. 1973. *The world of Walther Nernst. The rise and fall of German Science.* London: MacMillan.
Nernst, Walther. 1913. Richard Abegg. *Chemische Berichte* 46: 619–628.
Noack, Adelheid. 1959. Recollection from 19 November 1959. Archiv der Max-Planck-Gesellschaft, Haber-Sammlung Va Abt., Rep. 5., Nr. 301.
Ostwald, Wilhelm. 1887. An die Leser. *Zeitschrift für Physikalische Chemie* 1: 1–4.
Perutz, M.F. 1996. The Cabinet of Dr. Haber. *New York Review of Books*, 20.6.1996, p. 31–36.
Planck, Max 1897. *Die Akademische Frau. Gutachten hervorragender Universitätsprofessoren, Frauenlehrer und Schriftsteller über die Befähigung der Frau zum wissenschaftlichen Studium und Berufe,* edited by Artur Kirchhoff, Berlin: Hugo Steinitz Verlag.
Scheuermann, Gerhard. 1994. *Das Breslauer Lexikon*. Dülmen: Laumann-Verlag.
Stern, Fritz. 2011. Fritz Haber: Flawed greatness of person and country. *Angewandte Chemie International Edition* 51: 50–56.
Stoltzenberg, Dietrich. 2004. *Fritz Haber. Chemist, Nobel Laureate, German, Jew*. Philadelphia: Chemical Heritage Press (in German already 1994, Weinheim: VCH).
Szöllösi-Janze, Margit. 1998. *Fritz Haber 1868–1934. Eine Biographie*. München: C.H. Beck.
Tamaru, Setsuro. 1914. Setsuro Tamaru to Clara Haber, 24 December 1914. Private collection of Hideko Tamaru Oyama: Rikkyo University, Tokyo.
Ullrich, Volker. 1993. Volker Ullrich. Die Zerstörung einer Frau, *Die Zeit* 4 (6): 1993.
Van Rahden, Til. 2008. *Jews and Other Germans: Civil Society, Religious Diversity, and Urban Politics in Breslau 1860–1925*. University of Wisconsin Press.
Vogel, Jakob. 1993. Die geistige Amazone. *Frankfurter Allgemeine Zeitung* 14 (9): 1993.

Addition to Open Access Information

Sections 2–7 reproduced with permission from Wiley-VCH Verlag GmbH & Co. KGaA. This material has been material published in Friedrich, Bretislav and Dieter Hoffmann, *Zeitschrift für anorganische und allgemeine Chemie* (2016) 642: 437–448. © The Authors. Published by Wiley-VCH Verlag GmbH & Co. KGaA, Weinheim as an open access article under the terms of the Creative Commons Attribution-NonCommercial-NoDerivs License, which permits use and distribution in any medium, provided the original work is properly cited, the use is non-commercial and no modifications or adaptations are made. Further permissions for Sects. 2–7 must be directly obtained from Wiley-VCH Verlag GmbH & Co. KGaA.

France's Political and Military Reaction in the Aftermath of the First German Chemical Offensive in April 1915: The Road to Retaliation in Kind

Olivier Lepick

Abstract Although France had been experimenting with chemical weapons when Germany launched its first lethal chemical offensive in spring 1915 in Langemark, the German initiative came as a huge tactical surprise to the country. Soon after the initial shock and the controversy that ensued on whether Germany had violated the laws of war that day, French authorities rapidly decided, without real political debate, to retaliate in kind. Although the country had to face heavy constraints, and due to a considerable scientific, industrial and financial effort, the French army was able to launch its first drifting cloud chemical attack on the battlefield only a few months after the German offensive. In the storm of the war and at this stage of conflict, when urgency was the only consideration and political influence far less than military, the French authorities did not realize that adopting chemical weapons in retaliation, ten months before Verdun, was one of the steps that would lead to the totalization of warfare and characterize the rest of the Great War.

1 Introduction

The sudden outburst of chemical warfare initiated by Germany in April 1915, which constituted a deliberate violation of the laws of war, dealt a real shock to French public opinion. Nevertheless, the decision was taken quite soon after to retaliate in kind. The French authorities launched a major industrial and scientific endeavor. And indeed, only a few months after the German surprise attack in Langemark, and despite many obstacles, the French army was ready to launch its first chemical offensive. The purpose of this paper is to describe the first days, weeks and months of the French response to the German chemical initiative, as well as the political and military context in which the decision to retaliate in kind was made by French authorities.

O. Lepick (✉)
Foundation for Strategic Research, Paris, France
e-mail: olivier@lepick.fr

B. Friedrich et al. (eds.), *One Hundred Years of Chemical Warfare: Research, Deployment, Consequences*, DOI 10.1007/978-3-319-51664-6_5

The day after the German chemical attack on April 22 in the region of Ypres, a lively polemic broke out between France and Germany, the two protagonists, each blaming the other for unleashing chemical hostilities. German arguments rested in part on the existence of a French note dated February 21, 1915, which gave instructions about the use of suffocating grenades and cartridges. The existence of these materials was confirmed by Ulrich Trumpener in a brilliant article published in 1975 (Trumpener 1975). This article proved, for the first time, and beyond any doubt, that even if the Germans were the first to introduce a lethal form of chemical warfare during the First World War, France had already used non-lethal chemical agents on the battlefield before April 22 (Haber 1986, 32–33; Lepick 1998, 53–66), and, along with Great Britain, the country was already preparing and planning potential military chemical initiatives at the time of the Langemark chemical surprise.

2 Retaliation in Kind: A Purely Military Decision

Looking back at the first days following Langemark, the reaction of the French military authorities to the German initiative was tremendously quick.[1] As soon as April 23, a military pharmacist, Major Charles Didier, who was near Langemark the day of the German chemical initiative, informed the French General Headquarters (GHQ) that the toxin used by the Germans was chlorine.[2] The same day, Ferdinand Foch, commander of the Northern Army Group, organized a meeting at his headquarters with a renowned French chemist in Cassel, André Kling. The very same day Kling, who was Director of the Paris City Laboratory, started to investigate the issue by visiting hospitals, meeting with victims of chemical weapons, and ordering many post-mortem examinations. As Didier's conclusions about chlorine were rapidly confirmed, Kling immediately began working to develop protective devices for frontline soldiers. On April 24, the first crude orders were issued to field commanders. Foch also asked the Ministry of Industry to provide some input about possible protective devices that could be used by the army, while Paul Louis Weiss, head of the Mines Directorate of the Ministry for Public Works (*Travaux Publics*), proposed to use the know-how of the mining industry regarding respiratory devices to help the GHQ provide means of defense to the armed forces.[3]

[1]It is puzzling to consider that, contrary to the British historiography, there are only very few studies dealing with the French response to the German chemical initiative following April 22, 1915. The only serious study of the history of French chemical warfare during the First World War is the minutes of a lecture that was given in March 1920 by Professor Charles Moureu from the Collège de France, which was later published as: Moureu (1920).

[2]Letter from Charles Didier to the General Commander Army Group Belgium, April 23, 1915, Service historique de l'armée de terre, Vincennes, SHAT/16N826.

[3]Report on the organization of war chemical materials by House Representative Alain Albert d'Aubigny, August 25, 1915, p. 1, Service historique de l'armée de terre, Vincennes, SHAT/16N826.

On April 25 the French Ministry of War decided to begin the production of a crude respiratory protection that was mainly composed of a tissue containing a sort of bag filled with cotton. The device was to be soaked with thiosulfide just before the poisonous cloud reached the first lines, and secured over the mouth in order to neutralize the chlorine before it reached the soldier's lungs. Two days later, the first shipments were leaving for Flanders. By mid-May, more than 500,000 of these protective masks had been distributed to the field.

It is clear that, in contrast to what happened in Great Britain, there were no real political debates at the highest echelons of the French government about whether or not to retaliate. All political authorities were convinced that the decision to retaliate in kind was a military necessity, and proclaimed as one united voice that this kind of warfare was simply abhorrent. French military authorities found absolutely no opposition to their willingness to retaliate. Less than 48 hours after the German offensive in Langemark, military and political leaders had agreed on the fact that the country was to respond to the German initiative as soon as possible.[4] Nevertheless, how can we explain the absence of political input concerning such an important decision, especially since the use of new weapons that could lead to political consequences was strictly controlled by the civil political authorities? Initially, at this point in the conflict and up to the dismissal of Alexandre Millerand as Minister of War in October 1915, the French government's policy was not to interfere in the conduct of the war. At no other point during this conflict had the French GHQ enjoyed such great room for maneuver, such independence from political influence. Furthermore, the French government deemed that the German attack had liberated France from its obligations accruing through its signature to the Hague Convention. From a governmental perspective, the decision to retaliate in kind was purely technical and military and fully in within Foch's ambit. And Foch's decision was taken a few hours only after the April 22 attack. On April 25, General Maxime Weygand captured the general opinion in the country by stating:

> The Germans took the initiative to use inhuman means of warfare that had been banned by international treaties. But for us, it was not about procedures but about preparing as fast as we could the means to protect ourselves and retaliate in kind to these attacks [...]. A new step toward total warfare has been taken by our enemies.[5]

On April 26, the Ministry of War asked all French chemical companies to report the amount of chemicals in their stocks which could be used to respond to the German initiative.[6]

---

[4]On this subject, the following books are highly valuable: Joffre (1932); Poincaré (1930, 173, 350).

[5]Les Allemands avaient pris l'initiative d'un moyen de lutte inhumain et condamné par les accords internationaux. Mais il ne s'agissait pas pour nous de procédures, il fallait sans retard trouver à la fois contre ces attaques, la protection et la riposte [...]. Un nouveau pas venait d'être fait pas nos adversaires dans la pratique de la guerre totale (Weygand 1953, 225). Translation by the author.

[6]Buat Archives, Service historique de l'armée de terre, Vincennes, SHAT/6N21; Dossier 18, Ministry of War, Notes on the measures adopted following the use of asphyxiating gases by the enemy on April 28, 1915.

Finally, on June 2, 1915 the French government published a press release for the attention of foreign countries. Starting with a vigorous denunciation of the awful techniques used by the German army, which violated all of the treaties signed by the Imperial government, the release closed with the statement that

> no government shall not respond to such barbarian initiatives without endangering its own troops. In this perspective, the French government intends, in the strict limits of its military needs, to use all necessary means that appear appropriate to stop the German military authorities from continuing to commit such horrible murders [...]. (Le Temps, 1915)

3 Between Eagerness and Constraints: Organizing the Chemical Response

On April 26, a note from the GHQ was transmitted to all armies. The memorandum summarized the outcome of the interrogation of German prisoners captured near Bixschoote two days earlier. This short text described the defensive measures to be taken in case of gas attacks, and was intended to comfort the troops by stating that "a tissue soaked with the liquid that was distributed on the front lines or even one soaked with water could easily protect the soldiers against these toxic gases."[7]

Very rapidly, the French authorities built the organization charged with directing the national chemical retaliation program.[8] As soon as April 28, a committee composed of both military and scientific representatives was installed under the command of Paul Louis Weiss. Days later, three different organizations were created:

- One commission headed by Kling, tasked with identifying the chemical agents used by the enemy,
- a second headed by Weiss in charge of offensive aspects,
- and a third dedicated to the production of chemical agents, which was headed by the Engineer Corps.

On April 30, the first live experiments with non-toxic fumes took place at the proving ground in Satory (15 km southwest of Paris). On May 4, the first attempt to produce a chlorine cloud was conducted, with limited results. As liquid chlorine was difficult to obtain in the country, other products had to be considered. On June 2, 1915, Weiss, as head of the *Commission des Etudes Chimiques de Guerre* proposed a tentative organization for the military chemical services. He proposed the creation of a completely new directorate attached to the Ministry of War. This proposition was not backed by Albert Thomas, under-secretary for artillery and

---

[7]Note to the Armies by General Pelle, *Les Armées Françaises dans la Grande Guerre*, Ministère des Affaires Étrangères, Paris, 1922–1939, vol. 2, Annexe N°1451, p. 1017.

[8]The organization of the French chemical warfare services is described very precisely in: Vinet (1919, 1377–1415).

munitions, who wanted to extend his department's responsibilities to chemical warfare. Nevertheless, a consensus was found, and on June 18 a Directorate of Chemical Material (DCM) was installed under Weiss's command.[9] The scope of this DCM's duties was limited to research and development.

In July 1915 a group of representatives requested the creation of a new independent directorate under the Ministry of War. In accordance with the proposal from the GHQ, a colonel named Paul Ozil was promoted to head up this now fully independent DCM, placed in charge of all aspects of chemical warfare for the Ministry of War. Its internal structure remained unchanged until the end of the conflict, consisting of the three different commissions created a few days earlier: R&D, materials, and production.[10] Military camps and proving grounds were soon dedicated to chemical trials, in Satory for the protection of individuals and for small gas emissions, in Fontainebleau for artillery tests, and in Vincennes for explosives and structural tests. One of the first possibilities explored by French researchers was aerial chemical bombing. On June 10, the GHQ suggested to the Ministry of War that "airplanes could be an interesting means to deliver chemical weapons, especially in counter-artillery operations in hidden areas."[11] The initiative was given to General Fernand-Alexandre Curmer, yet rapidly led to the conclusion that, due to the limited amount of toxins that an airplane could carry, the use of airplanes to deliver chemical attacks bore little potential. During the same period, the GHQ was trying to develop measures to improve defenses against drifting chlorine clouds. On May 28, a first order was delivered to all armies, containing instructions for protective measures, but also describing methods to disseminate the clouds by artillery fire—thus exposing the lack of any real solution to respond to this new technology with military means.[12]

4 Chemical War: Scientific War, Industrial War

At the time Germany initiated chemical warfare, France was not in a position to fight and respond to the attack. The Germans were aware of this situation. The French chemical industry and its production were meager, and so were its capacities to produce chlorine, bromine and sulfuric acid. The French military chemical organization had to invest an immense amount of work and resources in order to

[9]Report on the organization of war chemical materials by House Representative Alain Albert d'Aubigny, August 25, 1915, p. 1, Service historique de l'armée de terre, Vincennes, SHAT/16N826.

[10]Letter from the Commander in Chief to the Defense Minister, July 21, 1915, Service historique de l'armée de terre, Vincennes, SHAT/16N832.

[11]"Chemical Shells, 18 septembre 1914–28 septembre 1915," Note from GHQ, June 10, 1915, Service historique de l'armée de l'air, Vincennes, Cartons A54, Dossier 1.

[12]Instructions established by the North Army Corps concerning defense against toxic gases, May 28, 1915, Service historique de l'armée de terre, Vincennes, SHAT/6N7.

provide the French Army with compounds that could be militarized. French chemists rapidly realized that domestic chlorine production capacity was so low that months, or even years, would be needed before a counterstrike could be launched. The possibility was investigated nevertheless, and live tests with chlorine were conducted in May, June and July. Because of the absence of a real chemical industry, French researchers were obliged to turn to other compounds, and delivery methods other than drifting clouds, rather soon reaching the conclusion that artillery was probably the best means of delivery to deploy chemical agents. The first agent available was carbon tetrachloride, which was easy to synthesize. The first promising trials were conducted on the Vincennes Proving Grounds in May and July 1915. As soon as August, Joffre approved production of a first batch of 50,000 shells containing carbon tetrachloride.[13] The shell received the code name "*Obus N°1*"; more than 420,000 were produced and then used in September during the French offensives in Champagne.[14] Nevertheless they were rapidly abandoned due to their lack of efficiency and low toxicity.

This apparent conclusion about the efficiency of artillery shells to deliver chemical agents by French military researchers can probably explain the ineffective performance by the first units, called "*Compagnie Z*," formed to prepare and execute drifting cloud chemical attacks. During summer 1915, two units of 800 non-combat troops were created, and three more in June 1916. These units were to conduct all 51 of the drifting chemical cloud operations that the French Army would launch during the war, the first of which took place in mid-February 1916, more than 10 months after the first German chemical attack (Lepick 1998, 133–174). Many other compounds were under close study by French researchers. The first agent the French researchers seriously planned to militarize was phosgene, one of the very rare toxic chemicals that was produced in large quantities in France. A limited production of shells containing hydrocyanic acid was ordered in June 1915, but surprisingly enough, the French government decided to postpone their immediate use. Because these shells were undoubtedly lethal due to the high toxicity of hydrocyanic acid, their deployment would have represented a clear and obvious violation of the 1899 Hague Convention, which prohibited the use of toxic shells. These shells, the first real lethal ones shot during the Great War, were used only during the battle of Verdun in February 1916.[15]

France, whose chemical industrial capacity was very limited, undertook a massive industrial effort. In 1915, only Germany had a real chemical industry in terms of not only size, but know-how, agility and resources. Before the war France was completely dependent on German imports of liquid chlorine and bromine. These imports came to sudden halt, of course, as soon as war broke out.

[13]Memorandum on the use of poison gas in warfare by General Curmer, Service historique de l'armée de terre, October 1st, 1915, Vincennes, SHAT/16N839.

[14]Memorandum on the constitution of special units for the use of Z equipment, December 23, 1915, Service historique de l'armée de terre, Vincennes, SHAT/16N826.

[15]Secret message from the GHQ, February 18, 1916, Service historique de l'armée de terre, Vincennes, SHAT/16N707.

The challenge was for France to create adequate industrial capacity *ex nihilo*. The first demands for chlorine were covered by imports from Italy and mostly Great Britain. In early September, French authorities negotiated the acquisition of more than 50 tons of liquid chlorine per week from the Runcorn production facility,[16] the first batches of which arrived in France in October. The production capacities available for export in the UK were nevertheless far lower than those needed to supply a French military program. French authorities were forced to launch an ambitious industrial plan to build production facilities all over the country. As early as August 1915 a first industrial program was launched, planning the construction of six different chemical plants able to produce 30 tons of liquid chlorine a week. In early 1917, French liquid chlorine production capacity topped a level of 50 tons a day, a large share of which was produced in privately owned companies. For bromine, the problem was even worse, as Germany had a monopoly on bromine production in Europe. First supplies were imported from the United States as soon as June 1915. A solution emerged when a subterranean lake was discovered 600 km south of Tunis in Tunisia, at that time a French colonial possession, from which bromine could be extracted. A factory was built and began production in April 1916, yielding more than 900 tons of bromine during the war, plenty enough to fulfill the country's needs (Bloch 1926, 32).

Because of these industrial difficulties, the first French attack with drifting chlorine clouds did not take place until February 1916. The French conducted more than 50 such attacks, mostly during 1916. Some of them were massive, conducted on a front line more than 8 km long, and supported by more than 6,000 pressurized cylinders containing almost 300 tons of chlorine per operation.[17] The French rapidly abandoned this technique, for obvious reasons such as heavy meteorological constraints and early adoption of artillery as the main delivery system for chemical weapons, probably one of the most extraordinary offensive means of operation during World War I.

5 Retaliation in Kind: Towards Total War

The sudden outburst of chemical warfare initiated in April 1915 by Germany, which deliberately violated the laws of war, truly shocked the French public. Nevertheless, the decision was taken very rapidly by the French government to retaliate in kind, which entailed a huge industrial and scientific effort. And indeed, only a few months after Langemarck, the French army had surmounted many obstacles and was ready to launch their first chemical offensive. Great Britain was

---

[16]Letter from the 4th Directorate of the Ministry of Defence to the GHQ, October 17, 1915, Service historique de l'armée de terre, Vincennes, SHAT/16N827.

[17]General Curmer's archives. Service historique de l'armée de terre, Vincennes, SHAT/16N903.

able to launch the first allied chemical attack, on September 15, 1915 in Loos, followed by France in February 1916.

The scientific, industrial and financial effort was extraordinary for both countries, especially considering the starting point for France. But in the storm of the war, when urgency was the only law and political influence far less than military, the French authorities did not realize that by adopting chemical weapons in a mechanical way, as they did, they unconsciously took a first step leading to total war. Of course at that period of the conflict, gas was not yet a weapon of annihilation, but rather the weapon that promised local breakthroughs from the deadlock of trench warfare (Joffre 1932). This perspective was so crucial that political authorities stood aside. But, undoubtedly, ten long months before the battle of Verdun, the beginning of gas warfare was the first real step toward the totalization of the war.

References

Bloch, Paul. 1926. La guerre chimique. *Revue militaire française* 21: 19–47.

Haber, Ludwig F. 1986. *The poisonous cloud: Chemical warfare in the First World War*. Oxford: Clarendon.

Joffre, Joseph. 1932. *Les mémoires du maréchal Joffre*. Paris: Plon.

Lepick, Olivier. 1998. *La Grande Guerre chimique 1914–1918*. Paris: Presses Universitaires de France.

Les Armées Françaises dans la Grande Guerre, Ministère des Affaires Étrangères, Paris, 1922–1939.

Le Temps. Le gouvernement français répondra à l'emploi des gaz asphyxiants. June 2, 1915.

Moureu, Charles. 1920. *Chimie de guerre, les gaz de combat*. Paris: Librairie de l'enseignement technique.

Poincaré, Raymond. 1930. *Au service de la France: 1915, les tranchées*. Paris: Plon.

Trumpener, Ulrich. 1975. The road to Ypres: The beginning of gas warfare in World War I. *Journal of Modern History* 47: 460–480.

Vinet, Emile. 1919. La guerre des gaz et les travaux des services chimiques français. *Chimie & Industrie* 11–12: 1377–1415.

Weygand, Maxime. 1953. *Idéal vécu*. Paris: Flammarion.

Preparing for Poison Warfare: The Ethics and Politics of Britain's Chemical Weapons Program, 1915–1945

Ulf Schmidt

Abstract Allied political and military leaders have frequently been credited both with considerable foresight and with strategic and moral leadership for avoiding chemical warfare during the Second World War. Scholars have not, however, fully acknowledged how close Allied forces came to launching a full-scale chemical onslaught in various theatres of war. The paper offers a thorough reconstruction of Allied chemical warfare planning which takes a close look at the development of Britain's chemical weapons program since the First World War. The findings suggest that no "lack of preparedness," as it existed in the initial stages of the conflict in 1939/1940, would have deterred the Allies from launching chemical warfare if the military situation had required it. Allied forces were planning to launch retaliatory chemical warfare ever since they had been attacked with chlorine gas in 1915. Just War theorists at first opposed the use of this new weapon and campaigned for an internationally enforced legal ban. The paper argues, however, that post-war military and political exigencies forced the advocates of the Just War tradition to construct new arguments and principles which would make this type of war morally and militarily acceptable. The paper explores the ways in which military strategists, scientists, and government officials attempted to justify the development, possession, and use of chemical weapons, and contextualizes Britain's delicate balancing act between deterrence and disarmament in the interwar period.

1 Introduction

Allied political and military leaders have frequently been credited both with considerable foresight and with strategic and moral leadership for avoiding chemical warfare during the Second World War. Scholars have not, however, fully acknowledged how very close Allied forces came to launching a full-scale chemical

U. Schmidt (✉)
University of Kent, Canterbury, UK
e-mail: U.I.Schmidt@kent.ac.uk

B. Friedrich et al. (eds.), *One Hundred Years of Chemical Warfare: Research, Deployment, Consequences*, DOI 10.1007/978-3-319-51664-6_6

onslaught in the European and far eastern theaters of war.[1] Although Allied intelligence was aware of Japan's chemical warfare operations against China, which had commenced in 1937, the Allied military decided against retaliatory measures.[2] Chemical warfare would not only have violated international law and morality as it was known and understood at the time, but would have changed beyond recognition the image and conduct of modern warfare for generations to come. The fact that a potentially devastating event did not happen is seen as tangible evidence of the underlying morality and humanity of Western governments in defending modern civilization. A more thorough reconstruction of Allied chemical warfare planning, as is proposed here, one which incorporate and takes a close look at the development of Britain's chemical weapons program since the First World War, allows us to recognize that no "lack of preparedness," however serious it may have been in the initial stages of the conflict, would have deterred the Allies from launching chemical warfare if the military situation had required it. Allied forces were indeed planning for chemical warfare ever since they had been attacked with chlorine gas in 1915.[3]

From the moment chemical weapons appeared on the stage of armed conflict, Just War theorists opposed the use of this new weapon and campaigned for an internationally enforced legal ban. Chemical weapons, they argued, violated the requirement for non-combatant immunity because they indiscriminately killed and injured children, women, and the elderly. In the 1920s, however, military and political exigencies forced the advocates of the Just War tradition to construct new arguments and principles that would make this type of war morally and militarily acceptable. Responding to an international legal ban on poison gas, government experts began to condemn the inhumanity of armed conflict, while simultaneously accepting the need for this type of warfare in certain circumstances. There is therefore a need to examine the ways in which military strategists, scientists, diplomats, and government officials attempted to justify the development, possession, and use of chemical weapons through different means and methods of propaganda, and to contextualize Britain's delicate balancing act between deterrence and disarmament in the postwar period.

[1]For some of the scholarship on the history of biological and chemical warfare since the First World War see SIPRI (1971); Harris and Paxman (1982); Haber (1986); Richter (1994); Evans (2000); Balmer (2001); Hammond and Carter (2002); Schmaltz (2005, 2006a, b, c); Wheelis et al. (2006); Tucker (2006); Schmidt (2006, 2007a, b, 2013); Schmidt and Frewer (2007); Spiers (2010); Avery (2013). For a comprehensive analysis of chemical warfare research and human experiments during the twentieth and twenty-first centuries, see also Schmidt (2015). Sections of this chapter have been reproduced with permission by Palgrave published in Schmidt (2012).

[2]See Van Moon and Ellis (1989, 1996).

[3]Spiers, for instance, suggests that "lack of preparedness was a principal reason for non-use of chemical or biological weapons between the major belligerents in the Second World War" (Spiers 2010, 57). For the use of chlorine in 1915 see Cowell et al. (2007).

2 Ypres 1915

By the time the first major chemical warfare attack in modern history came to an end, the Allies had lost hundreds, if not thousands, of soldiers. Allied propaganda estimated that 5000 soldiers had been killed and 10,000 had been wounded, though these numbers are generally accepted to have been exaggerated (Szöllösi-Janze 1998, 318).[4] Whatever the exact casualty figures, witness accounts confirmed that Allied troops had been exposed to one of the first weapons of mass destruction, which killed men slowly and painfully from within rather than wounding them on the outside. Total panic had gripped thousands of seasoned soldiers and civilians who fled from the toxic fumes; the modern battlefield had become a site of unimaginable horror and untold human suffering.

Despite a four-mile hole in the Western Front, and an enemy army in disarray, the German military, having failed to anticipate the effects of the "new infernal invention," as some called it, and lacking the necessary reserves to break through Allied defenses, was unable to exploit their sudden strategic advantage (Buffetaut 2008). Among those disappointed by the German lack of planning was the head of the Kaiser Wilhelm Institute in Berlin, Fritz Haber, who was instrumental in developing German chemical warfare agents. If only the military authorities had launched a full-scale offensive, he complained, "instead of the experiment at Ypres, the Germans would have won" (Harris and Paxman 1982, 10). Haber saw gas warfare as a more "humane" weapon of war:

> The gas weapons are surely not more horrible than flying metal fragments, on the contrary, the percentage of deadly gas injuries is comparably smaller, there are no mutilations and nothing is known […] in terms of follow-up injuries (Haber 1924, 35; see 25–41).

In 1919, much to the shock of the civilized world, Haber was awarded the Nobel Prize for Chemistry.[5]

Rejected as immoral and illegal by many, the new weaponry was greatly feared by the soldiers on the battlefield. Gas warfare became as much a psychological as a physical weapon. Often the experience of being gassed led to real and imagined clinical symptoms for years to come. The possibility of being killed by asphyxiating gases triggered deep-seated emotional responses and occasional nervous breakdowns which psychiatrists classified as "gas neurosis"; in other cases, soldiers exposed to blistering agents were classed as suffering from "gas hysteria," since the substances could cause conjunctivitis and temporary blindness (Harrison 2010, 106–109). Eyewitnesses recalled that "gas shock was as frequent as shellshock" (Shephard 2000, 64).

---

[4]Piet Chielens from the Flanders Fields Museum, Ypres, Belgium, has recently suggested that the number of "casualties" of the German gas attack near the Belgian town of Ypres in April 1915 was significantly lower than previously assumed (Chielens 2014; also Corrigan 2003, 164–165).

[5]For Haber's biography see Szöllösi-Janze (1998); Stolzenberg (2004); Charles (2005); see also the account by his son Lutz F. Haber in Haber (1986).

No one had been prepared for this kind and scale of warfare; few had ever imagined that poison gas would be used; almost all were shocked that the world would never be the same, that armed conflict would forever be tainted by what many perceived to be an unmanly, dirty form of warfare. As early as 22 April 1915, Michel Toudy, a soldier of the Belgian Grenadiers tasked with strengthening front-line defenses in the immediate aftermath of the first gas attack, noted in his war diary: "Throughout the entire night French territorials arrive in our trenches coughing and saying that it is not permitted to attack aged family fathers with asphyxiating gas."[6] Many Allied servicemen believed at this point that Germany had violated international conventions governing the conduct of war, which in many ways it had—if not the letter of the law, then certainly its spirit.

The German use of steel cylinders for the delivery of poison gas was meant to ensure that a large area would be cleared for a ground offensive once the gas had dissipated but it also had another, more profound rationale. Since the end of the nineteenth century, international law had prohibited the use of poison gas. Fearing that the ongoing arms race with Germany could weaken the fledgling Russian economy, and further destabilize the regime through strikes and revolutionary activities, Tsar Nicholas II had initiated the First Peace Conference in The Hague to revise and ratify the declarations about the laws and customs of war that had been negotiated in 1874 in Brussels. In 1899, representatives of twenty-six countries, including Britain, France, Russia, and Germany, had signed The Hague Convention Respecting the Laws and Customs of War, which not only regulated the treatment of prisoners of war and the care of sick and wounded, but also banned certain types of warfare and the use of modern technology, including aerial bombardment, chemical warfare, and hollow point bullets. Article 23(a) specifically prohibited the employment of "poison or poisoned arms."[7] In a separately signed document, The Hague Declaration Concerning Asphyxiating Gases, the contracting states also pledged to outlaw the use of poison gas as a means of future warfare by "abstaining from the use of projectiles, the sole object of which is the diffusion of asphyxiating or deleterious gases" (Tucker 2006, 10f.). Attempting to ban weapons which did not yet exist, The Hague Declaration contained three major loopholes which the belligerents exploited during the First World War: the use of irritants, the employment of gas through means other than by using projectiles, and the use of gas-filled, yet shrapnel-causing bombs, were not covered by The Hague Declaration. Faced with a war of attrition, the German army was less concerned about the inherent legality or morality of gas warfare but more about semantics. Whereas the use of gas-filled *projectiles* was against international law, the German military considered the use of poison gas released from cylinders to be lawful. Days after Germany's first gas attack, the *Kölnische Zeitung* claimed that "the letting loose of smoke clouds,

---

[6]In Flanders Fields Museum, Documentation Centre, Toudy papers. I am grateful to Dominiek Dendooven for sharing this source with me. See also Dendooven (2005).

[7]Convention (II) with Respect to the Laws and Custom of War on Land and its Annex: Regulations Concerning the Laws and Customs of War on Land. The Hague, 29 July 1899.

which, in a gentle wind, move quite slowly towards the enemy, is not only permissible by international law, but is an extraordinarily mild method of war" (Harris and Paxman 1982, 5). The Allied powers, however, described it as an act of inhumanity that violated "all codes of civilized behaviour" (SIPRI 1971, 231). It certainly did not bode well that the German military had given the poisonous cloud the code-name "Disinfection," a cover to confuse Allied intelligence, surely, but also one which portrayed enemy soldiers and civilians as vermin to be exterminated (Buffetaut 2008, 20).

Twenty-four hours after Germany's first gas attack, Sir John French, the Commander of the British Expeditionary Force, inquired about the existing supply of respirators and requested from London that "immediate steps be taken in retaliation to supply similar means of the most effective kind for the use of our own troops."[8] In his reply, Lord Kitchener, the War Minister, called for caution: "The use of asphyxiating gases is, as you are aware, contrary to the rules and usages of war. Before we fall to the level of the degraded Germans I must submit the matter to the government."[9] To investigate the matter, Kitchener called upon two civilian scientists: John S. Haldane (1860–1936), a former reader in physiology at Oxford University who, as director of a research laboratory in Doncaster, had worked with the mining industry in developing respirators against the toxic effects of mine gases;[10] and Herbert B. Baker (1862–1935), a professor of chemistry at Imperial College. Both were dispatched to France to find out what kind of gas had been used and inspect the site of the first gas attack. At St Omer, close to the general headquarters in France, they managed to identify the gas that had been used as chlorine through the use of a school laboratory (Thorpe 1936, 525; also Foulkes 1934, 37). For all concerned, it was clear that "immediate defensive measures were required."[11] On their return to Britain, Haldane submitted a full report to Prime Minister Herbert Asquith, while Baker briefed Lord Kitchener about the situation; the latter told him to "do his damnedest" to ensure that Britain could soon retaliate (ibid.).

Despite these bold declarations of intent there was considerable uncertainty among members of the British government as to whether Germany had actually contravened the terms of The Hague Declaration. On 26 April, Asquith told King George V: "As the gases are apparently stored in and drawn from cylinders, and not "projectiles," the employment of them is not perhaps an infraction of the literal terms of The Hague Convention."[12] Given that Germany was widely perceived as having violated the spirit of the Declaration, however, and with pressure mounting

[8]Foulkes 1934 (19); also Carter (2000, 2), who misquotes French in this instance.

[9]Foulkes (1934, 19f); Harris and Paxman (1982, 5); Hobbs et al. (2007, 260). See also TNA, WO142/241, correspondence between Sir John French and Lord Kitchener, 23–24 April 1915.

[10]For John S. Haldane see Douglas (1936); Sturdy (1987); Goodman (2008); Sturdy (2011); see also Haldane (1925, 63) who recounts how his father was sent to France to identify the gas which the Germans had used.

[11]TNA, WO188/802, p. 1.

[12]Hobbs et al. (2007, 260); TNA, CAB 37/127/40, Asquith to George V, 26 April 1915.

on the War Office to retaliate, the government knew that such legal sophistry would have little truck with the British public.

Within days, after graphic accounts of gas casualties had been published by *The Times* and other newspapers, anti-German sentiment reached fever pitch (SIPRI 1971, 231ff.). On 29 April, *The Times* commented:

> The willful and systematic attempt to choke and poison our soldiers can have but one effect upon the British people and upon all the non-German people of the earth. It will deepen our indignation and our resolution, and it will fill all races with a new horror of the German name.[13]

On the same day, the *Daily Mirror* reported that the German military had again used "asphyxiating gases" contrary to The Hague Declaration.[14] In Germany, meanwhile, gas warfare was portrayed as a modern weapon that was not only lawful but humane, one that produced a "rapid end" rather than the misery resulting from turning the German trenches "into a terrible hell."[15] A week later, on 7 May, the sinking of the *Lusitania* off the coast of Ireland by a German U-boat, killing 1198 civilians on board, including American citizens, caused further international outrage and turned public opinion firmly against Germany (see also Spiers 2010, 40). By portraying Germany as an "inhuman enemy," and German soldiers as barbaric criminals, hell-bent on committing atrocities against civilians by means of poison gas and submarine warfare, and in flagrant violation of the rules of war, Allied officials managed to bring the United States into the conflict and justify, in the eyes of the public, Britain's retaliatory measures.

Germany's premeditated gas attack initiated a Europe-wide chemical arms race on an unprecedented scale, one in which there was "no time to worry about ethics" (Harris and Paxman 1982, 21). Even neutral Netherlands got involved in the production of hundreds of tons of poison gas (Van Bergen 2012). The German gas attack "both inspired and provoked the British into retaliating with illegal weaponry, thereby opening the door to a virtually unlimited chemical warfare" (in Hobbs et al. 2007, 261). After recovering from the initial shock, Britain and France wasted little time in establishing large-scale programs for the testing of toxic substances, and in preparing their armies for all-out technological warfare to be fought irrespective of any moral or legal boundaries. At the end of September 1915, British forces attempted, but largely failed, to use poison gas in a major offensive at Loos in Belgium (see Lloyd 2006). Despite months of preparation, the training of special gas brigades, the employment of chemical experts, and the shipment and positioning of thousands of gas-filled cylinders along the front line, military planners began to appreciate the enormous problems associated with chemical warfare. Gas warfare was highly unpredictable, scientifically complex, and dependent on prevalent weather and environmental conditions, and it quickly turned into a nightmare for military strategists. Whereas the human cost of the Battle of Loos was

---

[13]*The Times*, 29 April 1915, p. 9; also UoK, Porton Archive, A201, WWI CW Media Articles.

[14]*Daily Mirror*, 29 April 1915.

[15]*Frankfurter Zeitung*, 26 April 1915; quoted from SIPRI (1971, 232).

substantial, strategic gains were almost negligible. The British had captured some 3000 German prisoners of war. Yet with over 50,000 British casualties, and hundreds of troops gassed by their own side—after the toxic cloud had changed direction—together with 3 miles of ground taken and then lost again, the military agreed that the machinery of war needed to be modernized if Britain and her Empire were to sustain a prolonged military campaign. Moreover, by using the newly developed Stokes mortar, the *sole* purpose of which was the delivery of chemical projectiles into and behind enemy lines, Britain had become the first nation to contravene the literal terms of The Hague Convention, and thus international law (Hobbs et al. 2007, 260f; also Spiers 1986, 24).

3 Porton Down

At the end of 1915, officials in the Ministry of Munitions concluded that the modern war machine needed nothing less than a fully equipped, large-scale testing ground to keep abreast of rapid developments in science, technology, and medicine.[16] In September, the Trench Warfare Department duly instructed the Scientific Advisory Committee to find and requisition a suitable "ground for experimental purposes."[17] A few months later, in early 1916, some 2886 acres of land near the villages of Idmiston, Idmiston Down, and Porton, on the southern edge of Salisbury Plain in Wiltshire, formed the basis of what came to be known as Porton Down.[18]

Porton rapidly expanded to take over 6200 acres of largely woodland and farmland, accessible through a complex network of roads and a light railway that interlinked the administrative headquarters, army huts, workshops, laboratories, munitions depot, open-air testing station, and animal farm, a place teeming with service personnel and civilian scientists working under the leadership of Porton's commandant, Lieutenant Colonel Arthur W. Crossley, a Mancunian, who had made his career as a chemist at King's College London. During the war, Porton was divided into four departments: the Commandant, the Division Officer Royal Engineers, the Works Department, and the Experimental Department. While the Division Officer Royal Engineers was responsible for the general upkeep of the facility, the Works Department, line-managed by the Superintendent of Experimental Grounds, carried out the construction work through civilian laborers. By 1918, the Experimental Department was divided into six sections: the Chemical Laboratory, the Anti-Gas Department, the Physiology Laboratory, the Meteorological Station, the Experimental Battery RA, and the Experimental

[16]TNA, WO188/802, p. 1.

[17]TNA, WO142/264, Lt Col A.W. Crossley RE, 'The Royal Engineers Experimental Station, Porton' (1919); also TNA, WO188/802, p. 1; Carter (2000, 3).

[18]TNA, WO142/264, Lt Col A.W. Crossley RE, "The Royal Engineers Experimental Station, Porton" (1919), pp. 3f. Work at the Station is believed to have commenced on 7 March 1916 when the first officer from the Royal Engineers reported for permanent duty.

Company RE.[19] As sign that Porton was there to stay, certainly for the duration of the war, the organization soon saw the creation of more permanent laboratories, photographic and meteorological units, barracks, and welfare facilities. By 1918, Porton had become a large-scale research facility with over 900 members of staff, many of them officers, thirty-three women from the Queen Mary Army Auxiliary Corps, who were employed as typists, and some 500 civilian workmen who maintained the workshops and laboratories.[20] At first, much of the work was directed towards developing new weapons of mass destruction.

In September 1916, the first use of the Livens Projector, an ad hoc device consisting of a steel tube, about 3 feet in length and 8 inches wide, dug into the ground at an angle of 45° in batteries of twenty, and detonated remotely through an electrical charge, opened a new chapter in gas warfare. It was no longer necessary to rely on the right meteorological conditions: bombs containing 30 lb (15 kg) of chemical agents, generally phosgene (CG), could be fired directly into enemy lines, resulting in high numbers of casualties and deaths. The power of the new weapon lay in the number of projectiles that could be fired simultaneously, sometimes more than 1000 at a time.[21] In April 1917, at the Battle of Arras, the British used the Livens Projector for a full-scale, deadly attack. Although inexpensive and inaccurate, with a range limited to one mile, it was an effective but also terrifying weapon that served as a technological precursor to "multiple rocket launchers and… aircraft cluster bombs" (Harris and Paxman 1982, 22). Gas shells, on the other hand, used by Germany and France from 1916, required less preparation, offered greater targeting precision, and were able to be fired over longer distances. In Germany, the symbols on the shell cases represented the different chemical agents: a white cross stood for tear gas, a green cross for phosgene, and a yellow cross for mustard gas (HS).[22] With an estimated total of 66 million gas shells fired during the war, chemical warfare had turned into an ever-present threat for Allied and German forces.

In December 1915, the German military used phosgene for the first time, an almost colorless and odorless gas, eighteen times more toxic than chlorine, which, when inhaled, caused serious lung damage from excessive fluid accumulation, and death within a few hours.[23] Toxicologists called it an "inner drowning" of the lungs Klee (1997, 269). Retaliating in June 1916, the Allies employed phosgene with devastating effect during the battles of the Somme; by firing 4000 gas-filled shells

[19]TNA, WO188/802, pp. 3, 10f; for Crossley see TNA, WO142/264, Lt Col A.W. Crossley RE, "The Royal Engineers Experimental Station, Porton" (1919).

[20]TNA, WO188/802, p. 10; McCamley (2006, 97).

[21]TNA, WO188/802, p. 20.

[22]Mustard gas, or dichlorodiethyl sulphide, was code-named 'H' or 'HS'. Lewisite, or chlorovinyl dichloroarsine, was code-named 'L'. For the exact chemical names of the code-named warfare agents see Historical Survey (2006, 209).

[23]For the properties of phosgene as a chemical warfare agent see Marrs et al. (1996, 185–202).

simultaneously, and thus releasing a total of 54 tons of gas over the target area, the Allies wiped out hundreds, if not thousands, of German soldiers, horses, and wild animals.

The close proximity of Porton's laboratories to one another allowed scientists and service officers to conduct integrated research across disciplinary boundaries. Except for a few scientists who wanted to protect their academic independence, most researchers were given military ranks.[24] Physiologists, chemists, pathologists, meteorologists, and a range of technical and military experts all collaborated in designing and executing experiments, both outdoors on the test range and indoors in the laboratory, sampling station, or gas chamber; by sharing their research data, they managed to improve protective clothing, diagnostic tools, and forms of treatment. Sometimes, relevant expertise had to be brought in from the outside. Porton's first respirator and gas tests, for instance, were conducted by civilian rescue workers from Derbyshire, where mining accidents from gas explosions were not uncommon (McCamley 2006, 97). Teamwork was an essential ingredient of Porton's developing research culture. It provided scientists and military personnel with an incentive to join the establishment and work long, exhausting hours, late at night, or during weekends. Porton's collaborative "spirit and unity of purpose," as Crossley put it, strengthened their belief that they belonged to an exclusive group of professionals who were tasked by the government to develop chemical weapon technologies.[25]

New challenges brought about by modern chemical warfare also led to advances at Porton in defensive technologies for both soldiers and civilians, for example in the design and development of more efficient respirators (Sturdy 1998). Realizing that Allied respirators offered improved protection against certain gases, chlorine and phosgene especially, German scientists developed ever more lethal and incapacitating agents that attacked the body from the outside. Dichlorethyl sulphide, or mustard gas as it became known in Britain because of its distinct garlicky, mustard-like smell, attacked the skin, causing severe burns and blisters within a couple of hours.[26] If inhaled, mustard gas could cause serious inflammation of the lungs, followed by a slow and painful death from asphyxiation. In Germany, the agent was called "Lost" in recognition of the two scientists (Lommel and Steinkopf) who synthesized it, and in France it was called "Ypérite" in reference to Germany's first mustard gas attack in July 1917, when the military employed the agent to deadly effect in the area around, yet again, the heavily embattled Ypres. The onset of symptoms was delayed, and thousands of soldiers were unaware of having been exposed to a toxic agent, yet developed severe blisters on their hands and neck, and in armpits, groin, and buttocks. The blisters often became infected, leaving soldiers

---

[24]TNA, WO188/802, p. 3; see also Roughton (1949, 320).

[25]TNA, WO188/802, p. 8.

[26]Sulphur mustard (mustard gas) was first synthesized in the mid nineteenth century and developed as a chemical warfare agent during the First World War. For the chemical properties of sulphur mustard see Marrs et al. (1996, 139–173).

totally incapacitated and in need of medical treatment. Impregnated leather gloves and suits drenched in linseed oil provided some degree of protection, yet these could be worn only temporarily. British scientists quickly came to realize that the complex scientific problems linked to mustard gas, and the means to protect the human skin from it, needed to be studied in great detail after the end of hostilities. As the "King" or "Queen" of war gases (Tucker 2006, 19; Klee 1997, 269; Groehler 1989, 72), contaminating Allied troops and their equipment for prolonged periods of time, mustard gas stood at the forefront of Porton's research until the end of the Second World War.

4 Servants of the Realm

The generation of civilian scientists and service officers associated with Porton during the Great War had grown up in Victorian and Edwardian Britain, came from middle-class or more modest social backgrounds, studied at elite universities such as Cambridge, Oxford, or University College London, and occasionally married into the British establishment. Porton's origin as a defense establishment during the Great War was intricately connected with a generation of male researchers who were driven by a deep-seated desire for advancement and social prestige, an emerging "intellectual aristocracy" with strong social and professional bonds, determined to unlock the secrets of the world through science and experiment and thus realize their visionary ideas of modern society (see also Sturdy 2011). Those who believed in the power of science were men such as Lieutenant Colonel Charles Lovatt Evans (1884–1968) who, according to a friend, "possessed the great qualities of some of the most zealous and distinguished of the Victorians, who accomplished their life's work by an immense capacity for hard work and a burning zeal for achievement."[27] At Porton, Lovatt Evans' colleagues included the Cambridge physiologist Joseph Barcroft. Despite his Quaker upbringing, Barcroft felt the need to contribute to the war effort after Germany's premeditated gas attack.

In January 1917, prompted by the devastating effects of Germany's mustard gas attack, Porton established a permanent laboratory for physiological tests on humans at nearby Boscombe Down Farm.[28] The department seems to have been limited at first to a single hut, measuring 30 feet by 15 feet, which was converted into an office and physiological laboratory. To ensure close liaison with the Royal Army Medical Corps (RAMC), medical officers were attached to Porton.[29] Conditions were

[27] *BMJ*, Obituary, Charles Lovatt Evans, 3 (1968), 5619, pp. 684–685.

[28] The decision was taken by the Chemical Advisory Committee under Barcroft's leadership: TNA, WO142/264, Lt Col A.W. Crossley RE, "The Royal Engineers Experimental Station, Porton" (1919), pp. 11ff; also TNA, WO188/802, p. 7; Sturdy (1998, 70).

[29] During the First World War, Rudolph Peters worked as the medical officer under A.E. Kent, who was in charge of offensive chemical warfare on sections of the front controlled by the British First Army; TNA, WO188/802, p. 7.

austere, to say the least, with Peters not only living in the hut but taking his bath in a round tin on the floor of the post-mortem room where the animals were dissected (Thompson and Ogston 1983, 499). Other accounts mention a better-equipped outfit: an old brick building, laboratories, offices, and even a gas chamber for various animal experiments (Garner 2003, 138). Whether staff deliberately played down the existing working conditions to highlight their scientific achievements is difficult to tell in retrospect, yet what seems to be certain is that facilities were relatively simple, even by the standards of the day. Under Barcroft's leadership, and in collaboration with British universities, for example with Cambridge's Chemistry Department, itself engaged in the preparation of toxic agents under the leadership of Sir William Jackson Pope (1870–1939), these men set out to investigate the effects of chlorine, phosgene, adamsite (DM), an arsenical irritant, and mustard gas in experimental gas chambers and to analyze the results in improvised laboratories.[30]

Research had at first concentrated on assessing toxic agents for their ability to kill within forty-eight hours, though experts soon discovered the "casualty producing effects" of certain gases. Chemical warfare, they realized, was not so much about killing people but about incapacitating them for the duration of combat activity. Toxicity trials had revealed that the length of exposure to certain substances, and their concentration, were key in determining the degree to which agents were harmless or dangerous. By the end of the war, Barcroft's team had examined the toxicity and the possible remedies for 160 substances, including mustard gas and lewisite, which became known among Allied propagandists as the "dew of death," a description that overemphasized its actual killing potential.[31]

Although Porton encountered difficulties in retaining some of the civilian scientists after the war, with men such as Barcroft, Lovatt Evans, and Starling returning to their university positions, often as FRS, almost all of them continued to conduct research which was informed by their work on chemical warfare.[32] In the interwar period, and thereafter, the "Old Portonians" formed a closely knit group of experimental scientists who continued to have close links to the British defense community at Porton Down.[33] This was the generation of military men and civilian researchers for whom the experience of the Great War, and of tens of thousands of gassed soldiers, marked a watershed in their determination to prepare the country

---

[30]TNA, WO188/802, p. 15; see also Gibson (1941, 321f.); for Pope see also IWM, Photographic Collection, Portrait W. J. Pope (1870–1939).

[31]For a history of lewisite see Vilensky (2005).

[32]For Lovatt Evans' subsequent professional career see WL, PP/CLE/A.9, Jodrell Chair; PP/CLE/A.11, Service at Porton Experimental Station, Wiltshire, during the Second World War. Miscellaneous Correspondence re. Secondment and Service. For Barcroft's postwar career as a Reader in Physiology at Cambridge University see Roughton (1949). During the early 1920 s, Peters worked with F.G. Hopkins, Malcolm Dixon, and J.B.S. Haldane at the Balfour Laboratory at Cambridge University; see Peters (1959).

[33]Lt (later Maj) J.A. Sadd continued work as a senior civilian scientist at Porton until the 1950 s. Lt Col W.A. Salt, Lt Col A.E. Kent, Capt S.J. Steadman, and Lt A.C. Peacock all worked at Porton Down after the Great War.

for a future war, protect the army and civilians from the anticipated fallout, and supply the military with the means and methods to retaliate. On the eve of the Second World War, many of those who had fought in the previous war were ready to recommence research on chemical warfare to protect the United Kingdom and her allies.

5 Crisis of Legitimacy

Following the Armistice in November 1918, the victorious powers envisaged the creation of a demilitarized and largely peaceful world, free from violence and weapons of mass destruction. Undermined by feelings of revenge and demands for reparations, their vision got off to a difficult start. Under the Versailles Treaty, notorious for its humiliating terms, the Allies not only annexed territory, disarmed the German army, and extracted material resources from a traumatized, politically divided society that was barely coming to terms with military defeat, but also forced the government to admit sole responsibility for the war (Kershaw 1998, 136). To destroy any future chemical warfare capability, Germany was strictly prohibited, under Article 171, from using, producing, or importing chemical agents, including "asphyxiating, poisonous or other gases and all analogous liquids."[34]

Although far from homogenous, public opinion became a powerful force in shaping the international community's protracted disarmament negotiations.[35] In 1918, representatives of the British medical profession called for a ban on chemical warfare in *The Times*, describing it as an "unclean," uncontrollable, and malignant weapon of war which ought to be abolished.[36] Elsewhere, doctors and nurses employed by the armed forces protested against their involvement in this type of warfare. Some have argued that the interwar debate simply resulted from a "clash" between the wartime practicalities of using chemical weapons and the experienced or perceived horrors among "victims and observers alike," but this overlooks quite specific economic, political, and scientific factors as well as cultural traditions that shaped the discourse at a national level (see Van Bergen 2012). In the United States, where the chemical industries, like their British counterparts, launched a major publicity offensive, chemical warfare became a matter of domestic politics. Chemical warfare meant big business at a time of great economic uncertainty and guaranteed the employment of thousands of officers lecturing in US chemical

[34]"The use of asphyxiating, poisonous or other gases and all analogous liquids, materials or devices being prohibited, their manufacture and importation are strictly forbidden in Germany" (Treaty of Versailles, Article 171). Tucker (2006, 21); Hobbs et al. (2007, 278f.). Britain's proposal during the Versailles Treaty negotiations for full disclosure of Germany's wartime manufacturing processes was seen as an attempt at economic espionage, and rejected by the United States; see also SIPRI (1971, 235f.).

[35]See SIPRI (1971, 231–267), Chap. 3: "Popular Attitudes towards CBW", 1919–1939.

[36]*The Times*, 29 November 1918, p. 6.

warfare (CW) training facilities.[37] In a bid to improve chemical warfare preparedness and secure the postwar continuity of the Chemical Warfare Service (CWS), founded in mid 1918, stakeholders and major suppliers from the building, mining, and engineering trades, who advocated a more isolationist policy, became involved in a campaign to frustrate international disarmament negotiations. The proposed abolition of the CWS, in particular, threatened the existence of small, specialized companies supplying the US chemical warfare industry, which needed to adjust to peacetime conditions, for example through the sale of tear gas to law enforcement agencies. While Edgewood Arsenal highlighted the "relative humaneness" of toxic agents compared to high explosives in specially designed publications, engineering firms promoted their latest airtight steel tanks. Elsewhere, the producers of metal ores advertised their ability to deliver "gas by the ton."[38]

Yet the campaign also fuelled public anxieties against possible airborne attacks with toxic agents, and strengthened the resolve of organizations such as the British Association for the Advancement of Science and the International Committee of the Red Cross (ICRC) to protest against the use of poison gas in 1918 and call for an absolute ban on chemical warfare (SIPRI 1971, 239–242; Hobbs et al. 2007, 280; Van Bergen 2012). Having been criticized by the belligerents for abandoning its principle of impartiality, the ICRC subsequently took a more "neutral" position and "waged war on gas warfare" by campaigning for the improvement of defensive capabilities in the late 1920s in order to make the use of chemical warfare agents unworkable; there was even an ICRC-funded prize for innovative developments in the field of chemical defense. Anti-militarist groups and pacifists, however, became increasingly hostile towards the ICRC for viewing chemical weapons as an inevitable reality of future wars. In the aftermath of the Second World War, and in light of the Holocaust, the ICRC's stated policy of impartiality and non-interference became the subject of heated controversy, which has continued ever since.

Another major organization involved in shaping public opinion on the subject of chemical and biological warfare was the League of Nations. In the 1920s, it played a leading role in negotiating international agreements for the limitation and reduction of chemical weapons, and in prohibiting their use in future wars. Founded in 1920, the League of Nations was firmly committed to comprehensive disarmament, weapons control, and conflict resolution through international cooperation. Yet political setbacks during the League's formative years placed the United States in a powerful negotiating position. Held in Washington DC from November 1921 to February 1922, the Conference on the Limitation of Armament, organized by the United States to establish a new security framework in the Pacific area, sought, inter alia, a legally binding resolution for the prohibition of chemical weapons. During the negotiations, because of behind-the-scenes tensions between experts and

[37]CHF, Archive Collection, GB98.09, Williams—Miles Reprint Collection, William Williams, Notebook: US Gas School, 1918.

[38]CHF, The Edgewood Arsenal, Special Edition of *Chemical Warfare*, vol. 1, no. 5 (March 1919); for the origins of the CWS see Brophy and Fisher (1959); Ede (2011).

politicians about the real and imagined power of chemical agents and the ability to control them, careful management was required to preserve a united front (SIPRI 1971, 242ff.). Article 5 of the Washington Agreement prohibited the "use in war of asphyxiating, poisonous or other gases, and all analogous liquids, or materials, of devices," such use having been "justly condemned by the general opinion of the civilized world."[39] Despite reservations by Britain and France, which prevented the resolution from coming into force, the Washington Agreement marked an important milestone that galvanized public opinion and political power to work towards an international chemical weapons ban.[40]

Clearly affected by the international climate, Porton Down suffered a crisis of legitimacy after it transpired that Britain's chemical warfare program no longer enjoyed unconditional political and public support.[41] At first, almost all research activities ceased. Parliamentary questions were now being raised about Porton's annual cost to the taxpayer.[42] Reflecting public concerns about a substantially weakened economy, the MP Hugh Morrison queried in 1920 whether the government would not be well advised to "have it [Porton] closed down." In his cautious reply, which avoided revealing that the total cost of the establishment had been around £90,000 in 1919–1920,[43] Winston Churchill told the House of Commons that the government aimed to keep the experimental facility open "until the attitude of the League of Nations to chemical warfare is defined."[44] In March 1922, prompted by the Washington Agreement, the government came under renewed pressure, but insisted that it "would be failing in its duty if it failed to take all possible steps which might be necessary to protect the Forces of the Crown and the inhabitants of the country against gas attacks in time of war."[45]

---

[39]Article 5 of the Washington Agreement stated: "The use in war of asphyxiating, poisonous or other gases, and all analogous liquids, or materials, of devices, having been justly condemned by the general opinion of the civilized world, and a prohibition of such use having been declared in treaties to which a majority of the civilized powers are parties; now to the end that this prohibition shall be universally accepted as a part of international law, binding alike the conscience and practice of nations, the signatory powers declare their assent to such prohibition, agree to be bound thereby between themselves, and invite all other civilized nations to adhere thereto." See https://ihl-databases.icrc.org/applic/ihl/ihl-search.nsf/content.xsp.

[40]Whereas the British representative pointed out that it is "impossible to prevent a nation bent upon Chemical Warfare from making preparations in peacetime, no matter what the rules of war may be," the French Government reserved the right "to act in accordance with the circumstances" if an enemy refused to give a guarantee not "to use poison gas"; TNA, WO188/802, p. 48; Tucker (2006, 21); Hobbs et al. (2007, 280).

[41]Subordinated to the Chemical Warfare Department, Porton shared responsibility for chemical warfare research with a number of supervisory committees and organizations, including the Chemical Warfare Committee and university research facilities.

[42]Hansard, HC Debate, vol. 122 c60 W, Experimental Ground, Porton, 1 December 1919.

[43]For the total cost of Porton Down between 1919 and 1924 see Hansard, HC Debate, vol. 181 c1108, Chemical Warfare Research Department, 10 March 1925.

[44]Hansard, HC Debate, vol. 130 c1063, Chemical Experimental Ground, 15 June 1920.

[45]Hansard, HC Debate, vol. 152 c984 W, Asphyxiating Gas (Washington Treaty), 27 March 1922.

Unbeknown to the public, the Cabinet had accepted the recommendations of the Holland Committee in May 1920 to expedite chemical warfare research and reorganize Porton Down. Made up of experienced military and civilian experts, the committee had concluded that the "safety of the Empire" could not be left to chance: "A nation which is unprepared for gas warfare lays itself open to sudden and irretrievable disaster."[46] Separating defensive from offensive research was seen to be impossible, because one could not be understood without the other. Recommended changes to the organization involved a reconstituted Chemical Warfare Committee, the attachment of experts to the Director of Military Intelligence, the consolidation of "research, design and supply" under the control of the Ministry of Supply, and improved liaison between Porton's scientists and the armed services, a subject which had caused some considerable controversy during the war.[47] It was recommended that Porton's staff should, in future, be composed "partly of soldiers and partly of men of science," the latter to be of "high standing" and "independent of outside inspection and criticism."[48] To attract scientists of the highest caliber, and because staff sacrificed parts of their careers and occasionally risked their own health in the pursuit of knowledge, the authorities were asked to offer substantial inducements in the form of salaries, security of tenure, pensions, and the right to publish.[49] Largely oblivious to stringent cuts to the military budget during a period of economic austerity, the committee weighed Porton's "considerable" running costs on the basis of national security considerations. It also believed that Porton's discoveries were likely to have scientific and commercial value that would transform the organization into a "very valuable national asset."

At the same time military interference with Porton's activities needed to be kept to a minimum, provided the General Staff could "indicate the general lines" which appeared to be the most promising. The tension between the ability to conduct independent research, free from external pressures, and the practical demands by the military to defend the country against potential chemical warfare attacks, together with the need for a credible retaliatory capability, have characterized Porton ever since.

Close liaison between Porton's scientists and expert networks elsewhere in Britain and overseas, essential in maintaining a first-class research facility, was to be assured through the Chemical Warfare Committee, which was broadly

---

[46]UoK, Porton Archive, A207, "Report of the Committee on Chemical Warfare Organization", 7 July 1919, p. 5; also TNA, WO188/802, p. 30.

[47]Subordinating field trials to the military requirements of war had, according to some, slowed down, if not inhibited, Britain's chemical warfare program during the Great War; UoK, Porton Archive, A207, 'Report of the Committee on Chemical Warfare Organization', 7 July 1919, pp. 3ff; TNA, WO188/802, pp. 13, 17.

[48]UoK, Porton Archive, A207, "Report of the Committee on Chemical Warfare Organization", 7 July 1919, p. 5.

[49]UoK, Porton Archive, A207, "Report of the Committee on Chemical Warfare Organization", 7 July 1919, p. 5; TNA, WO188/802, pp. 30f.

representative of the wider scientific, military, and business community.[50] To ensure the coordinated production of toxic agents, including those for testing purposes at Porton, the committee recommended the creation of a state-controlled factory for chemical warfare products at Sutton Oak, near St Helens in Lancashire, which later became the Chemical Defence Research Establishment (see Carter and Pearson 1996, 60f; Carter 2000, 52f.) A representative of Porton liaised with members of the committee about planned field trials. It was this coordinated approach to chemical warfare through an external body of experts and stakeholders that other nations, the United States and Canada especially, began to emulate.

6 Collaboration

On a bilateral level, Britain and the United States joined forces in developing offensive and defensive chemical warfare capabilities that required the sharing of information and resources. In 1918, after the American Expeditionary Force (AEF) sustained disproportionately high numbers of chemical warfare casualties due to an inadequate level of preparedness, the US Army attached liaison officers to Porton to keep abreast of Britain's advances in chemical warfare work, a tradition which continued thereafter.[51] Britain's scientists, on the other hand, developed close links with their counterparts at Edgewood Arsenal, near Baltimore, Maryland, which became the United States' headquarters for chemical warfare research and development (Chemical Corps Association 1948, 14ff.). Given the exclusivity of the field, together with the need to preserve the utmost secrecy, research networks which had been established during and after the First World War were central in creating a long-term system of bilateral, and later tripartite, cooperation on chemical warfare between the Allied powers.

Still unresolved questions about the legitimacy of chemical warfare, together with the widespread condemnation of toxic agents as a means of warfare, turned intelligence sharing between Britain and the United States into a sensitive issue requiring a clear understanding about confidentiality arrangements and levels of secrecy. By assigning the highest security classification to chemical and biological warfare matters, and by avoiding the publication of details that could inform other

[50]In 1920, members of the Chemical Warfare Committee included: Joseph Barcroft (professor of physiology, Cambridge University), A.W. Crossley (former Commandant Porton and Director of the Cotton Industry Research Association), C.G. Douglas (physiologist, Oxford), Harold Hartley (chemist, Oxford), H. Levinstein (representative of Levinstein Limited, chemical manufacturers), Sir William Pope (professor of chemistry, Cambridge University), Jocelyn F. Thorpe (professor of organic chemistry at Imperial College and representative of the Department of Scientific and Industrial Research), A.M. Tyndall (professor of physics, Bristol University); see TNA, WO188/802, pp. 32ff.

[51]TNA, WO188/802, p. 11.

countries about the nature and extent of the work undertaken, Britain attempted to ensure that its expanding chemical warfare program would not become public knowledge.[52]

In addition, Britain was outsourcing certain types of research, some of it offensive in nature, to British-controlled laboratories overseas to deflect public attention from its expanding chemical warfare program.[53] Subsidiary research facilities in India and Australia, established in the 1920s, allowed British scientists to investigate the effect of chemical warfare agents under specific climatic conditions and among different population groups. Between 1921 and 1924, one of Porton's officers, Lieutenant Colonel W.A. Salt, ran the Military Chemical Laboratory in Dehra Dun in India, which conducted high-altitude and smoke trials to test different types of respirators suitable for bearded Sikhs.[54] Porton's service personnel and physiological staff also served as instructors to the Indian Chemical Warfare School in Begaum, a center of the armed forces for the British Raj. In 1929, the British authorities set up a Chemical Warfare Research Establishment in Rawalpindi, in the Punjab, staffed by scientists and officers from Porton, who engaged in smoke trials for the protection of bridges and other strategic sites.[55] To forge better relations, Indian representatives were invited to Britain for an appreciation of the power of chemical warfare. Around the same time, Britain established closer links with the Australian Chemical Warfare Board to study the effects of tropical and subtropical conditions on chemical warfare, attached Australian, Canadian, and South African representatives to Porton, and organized chemical warfare courses in the Dominions. In some cases, Porton helped Allied governments to deal with civil unrest by providing defensive technologies and chemical agents; in 1930, for example, Porton supplied South Africa with specially developed bombs filled with tear gas which the government employed against opposition groups. Most of Porton's activities overseas were strictly classified, not only to protect existing expertise but also to preserve Britain's political credibility in ongoing disarmament talks. At an international level, though, and in public, the subject of chemical warfare was openly discussed.[56]

---

[52]UoK, Porton Archive, A205, Porton Experiments 1920 s, Atkisson to Chief of CWS, Washington, 15 July 1924.

[53]TNA, WO188/802, pp. 86–90.

[54]TNA, WO188/802, pp. 86ff.

[55]TNA, WO188/802, p. 88; see also Evans (2007).

[56]A pamphlet published by the League of Nations noted that "everywhere except Germany, experiments in Chemical Warfare openly proceed […]. It will not necessarily inflict more pain than high explosive, but will tend to aggravate the burden of war upon the civilian population"; TNA, WO188/802, p. 49.

7 The Geneva Protocol

The "Protocol for the Prohibition of the Use in War of Asphyxiating, Poisonous or other Gases and of Bacteriological Methods of Warfare," known as the Geneva Protocol, and modeled on Article 5 of the Washington Agreement, outlawed the employment of chemical and biological weapons. However, it failed to establish an international verification and enforcement system, and exposed deep-seated disagreements about disarmament. The United States was opposed to prohibiting the use of tear gas in war on the grounds that it was also used by police forces against civilians in peacetime, especially as a weapon for riot control, and they refused to ratify the Protocol until 1975. The French and the British were likewise reluctant to go ahead and ratify, and this further limited the scope of the Protocol to a "no-first-use" agreement (Hobbs et al. 2007, 286f.).[57]

Questions relating to chemical and bacteriological weapons routinely surfaced in the discussions of the Preparatory Commission for the World Disarmament Conference[58] that opened under the chairmanship of the former British Foreign Secretary and Labour politician Arthur Henderson (1863–1935) in Geneva in February 1932. Preliminary meetings had highlighted the relative ease with which a chemical industry could be adapted to the production of toxic agents, and their potential delivery from the air. Secret intelligence further suggested that almost all countries that had signed up to the Geneva Protocol were pursuing an offensive chemical weapons capability.[59] It therefore came as little surprise that the negotiations were beset by disagreements over what constituted "offensive" and "defensive" weapons and by Germany's belligerent posturing. Britain's high-profile role during the negotiations left senior officials back in London distinctly nervous about granting permission for human experiments involving chemical warfare agents. By the time Britain proposed a draft convention at the World Disarmament Conference in March 1933, two months after Hitler's accession as Reich Chancellor, it had become clear that Europe, if not the world, was faced with an extraordinarily brutal military dictatorship which had no intention of settling international disputes by peaceful means. The talks collapsed after Germany withdrew firstly from the Geneva World Disarmament Conference and then, in October 1933, from the League of Nations. Breathing tests with toxic substances remained prohibited until the outbreak of war in 1939 changed the ethics of human experimentation at Porton.

Whereas some have hailed the Geneva Protocol as "the high-water mark of the hostility of public opinion towards CW," others have stressed the role of the international community in reasserting its authority after the contravention of The

[57]See also Hobbs et al. (2007, 255–295). For the use of tear gas in dealing with civil disturbances in the United States see SIPRI (1971, 270).

[58]The title of the conference was "Conference for the Reduction and Limitation of Armaments."

[59]TNA, WO188/802, p. 49; see also the debate in parliament about Britain's "offensive" and "defensive" chemical warfare capability; Hansard, HC Debate, vol. 272 cc811–12, Chemical Warfare, 30 November 1932.

Hague Declaration during the First World War (SIPRI 1971, 247; Hobbs et al. 2007, 287ff.). Irrespective of whether chemical weapons had been "politicized" before, during, or after the war, whether politicians had responded to public opinion, or whether chemical weapons themselves were inhumane and immoral, the Geneva Protocol established a new international law which not only prohibited the use of chemical and biological weapons, but which, perhaps more importantly, most nations perceived to be obligatory.

8 Foreboding

Far from being a "sudden outburst" of idealism, the Geneva Protocol was the League's "attempt to meet a grave and increasing practical danger, viz., the insecurity of European peace and, resulting therefrom, the rise of a new competition in armaments" (Hobbs et al. 2007, 288f.).[60] In the context of emerging European dictatorships, this constituted a realistic assessment. Since the early 1920s, the German *Reichswehr* and the Soviet Red Army had been involved in clandestine military operations that included weapons development and arms trade. Although the manufacture of chemical weapons was banned under the Versailles Treaty, and outlawed by national legislation, Germany's chemical industry and the military were organizing shipments of poison gas from Soviet Russia. The accidental release of phosgene from a storage tank in Hamburg in 1928 alerted the international community to the fact that Germany was flouting the Versailles Treaty. By the early 1930s, Germany's rearmament program had reached such alarming proportions that another war in Europe seemed a realistic possibility, especially in the context of a Hitler-led regime.

In the context of the Europe-wide rearmament programs that preceded the outbreak of the Second World War, issues relating to chemical warfare became absorbed into debates about national security. Almost all European governments, including the Soviet Union and Britain, employed the threat of chemical weapons as a way of accelerating the introduction of comprehensive civil defense measures. Whereas Soviet citizens received anti-gas drills in simulated gas attacks on Leningrad and Kiev in 1928, the British public was exposed to exaggerated reports about the power of chemical weapons. One estimate predicted that all men, women, and children in Central London would be killed if a large poison gas bomb were dropped onto Piccadilly Circus; another estimated the death of all Londoners if 40 tons of newly developed toxic agents were released. Italy's widely reported, but at first vehemently denied, use of chemical weapons in Ethiopia in 1935 and 1936, which involved the alleged use of mustard gas bombs against civilians and hospital patients, led to demands for sanctions by the League of Nations and increased the

[60]See also TNA, PRO30/69/1273, "Some Questions on the Geneva Protocol" (1925).

value of chemical weapons as a propaganda tool among anti-fascist groups.[61] In the Middle East, the British government pursued a dual strategy of attempting to broker a political settlement in conjunction with providing practical support; Porton's experimental officer was dispatched to Egypt, Aden, the Sudan, and Palestine to advise military officials about defensive chemical warfare technologies.[62] The outbreak of the Spanish Civil War in 1936 led to a fierce propaganda war over chemical warfare, with insurgents and government forces each alleging the enemy's use of poison gas. In the context of Britain's appeasement policy, official support was limited to public condemnation and the supply of respirators to aid the Spanish government. Although the Spanish military had shown few moral qualms in employing chemical weapons against Moroccans in the mid 1920s, reports confirming the use of chemical agents during the Spanish Civil War never materialized, apart from one incident involving the alleged use of tear gas (SIPRI 1971, 258ff.; Balfour 2002, 123–156). At the same time, intelligence from Germany and the Soviet Union suggested increased chemical and biological warfare activities.[63]

Retaliatory preparations were likewise set in train. In 1936, in addition to existing facilities which produced 20 tons of mustard gas per week, the Committee of Imperial Defence ordered the development of a pilot plant with an estimated output of 50 tons per week for the production of a new chemical warfare agent, code-named HT, better known as Runcol.[64] Less than a year later, over 5 million respirators were reported to be in storage for a national emergency. Additional storage facilities for defensive equipment were set up in Canada and South Africa.[65] Following the notorious Munich Agreement of September 1938, and Hitler's invasion of what was left of Czechoslovakia in March 1939, the government distributed over 30 million respirators. Two years into the war, Britain had at its disposal a total of almost 4 million children's respirators and anti-gas helmets alone (Grayzel 2012, 250; Harris and Paxman 1982, 107f.).[66] Here was an aggressor who warranted the mobilization of all resources in preparation for a potential chemical warfare attack.

Shortly before the outbreak of war, research on chemical and biological warfare accelerated at all levels. Through liaison with the ARP Subcommittee, Porton became an integral part of Britain's civil defense planning with increased access to military intelligence and hardware.[67] In 1935, an RAF "Special Duty Flight" was put on permanent stand-by to allow scientists to study the effects of airborne gas

---

[61]TNA, WO188/802, p. 51; Grayzel (2012, 181f.).

[62]TNA, WO188/802, p. 89.

[63]TNA, WO188/802, p. 51; see also Balmer (2001).

[64]TNA, CAB4/24, Committee of Imperial Defence, "Policy with Regard to the Possible Use of Gas as a Retaliatory Measure in War", 8 July 1936; also Historical Survey (2006, 209).

[65]TNA, WO188/802, p. 89.

[66]Sources have recently come to light, however, which suggest that the majority of gas masks produced for civilians and service personnel contained significant amounts of asbestos; see Schmidt (2015, 69ff.).

[67]TNA, WO188/802, p. 44.

attacks on the Porton range (ibid.). Research and development at Porton included the design of respirators for humans and animals, detector and decontamination devices, filtration units for ships, buildings (including air raid shelters) and armored vehicles, methods to prevent toxic gas from infiltrating the London Underground and government buildings, impregnated garments to protect against specific agents, and the testing of anti-gas ointments. To assist the RAF in assessing wind conditions on the ground or the Royal Navy in battleship protection, Porton conducted research on smoke, including smoke curtain installations and assessments about the relation between screening effects and meteorological conditions.[68] Offensive work involved chemical shell and aircraft gas bombs, ground mustard gas bombs to contaminate whole areas, gas-filled rocket launchers, gas-spraying devices, toxic smoke (arsenical) weapons, or substitute agents ("pseudo gases") to mislead the enemy.[69] Hand in hand with the rapid expansion of Porton's areas of responsibilities in the interwar period came the expansion of its research staff, who forged closer links with subsidiary research facilities in India and later Canada.[70] The number of scientists affiliated with Porton during these years is testimony to the way in which the government managed to integrate research and development into the planning process for future military operations (McCamley 2006, 100).

By the late 1920s, the notion of gas warfare and its associated imagery had become a powerful part of the collective memory of the European public. This applied especially to First World War memorials. Built in 1929 on Belgian soil to commemorate the suffering of thousands of victims of asphyxiation in the first chemical attack in modern history, the Steenstraate gas memorial blamed the German military for this act of inhumanity, which is why the occupying German forces duly destroyed it in 1941 (Jacobs 1996, 46–48; see also Goebel and Connelly 2017, forthcoming).

9 Ethical Relativism

At the start of hostilities, Britain, France, and Germany pledged to abide by the Geneva Protocol, yet none of the parties trusted that the agreement would be observed "a moment longer than is necessary" (Harris and Paxman 1982, 83, 107). Britain anticipated the use of chemical weapons by one or more of the belligerents.

[68]That smoke could be used as an effective weapon of war had been recognized as early as April 1915, days before Germany's first gas attack, when Winston Churchill as the First Lord of the Admiralty had commissioned a number of "wonderful smoke-making experiments"; Churchill (1923, 84f.); also Carter (2000, 53f.).

[69]TNA, WO188/802, pp. 53–66.

[70]Whereas the scientific staff had risen from 23 to 51 between 1922 and 1925, it had more than doubled to 120 by 1936/37. By 1938, Porton had a total of 152 researchers working on all aspects of defensive and offensive chemical warfare. The annual "Dominion Day" event, for example, organized by Porton between 1937 and 1942, offered visitors insight into Porton's research and development program; TNA, WO188/802, pp. 41, 89.

Millions of leaflets were distributed to all households and the BBC was on stand-by to broadcast pre-arranged gas warnings in the event of gas attacks. Following Hitler's *Blitzkrieg* campaign through the Low Countries and the surrender of France in May 1940, the threat of invasion by German forces loomed large in the minds of British officials in London.

By 1944, four years into an extraordinarily brutal and costly war, Allied military planners were growing increasingly concerned about the potential use of chemical agents by Axis forces, fearing the employment of these weapons in a desperate, last-ditch attempt to hold their positions. The large-scale decommissioning of scientific experts, who began to return to their prewar posts, added to a renewed crisis at military headquarters. Military planners warned about a lack of vigilance and the scaling down of chemical warfare preparedness which could cost Allied forces dearly in the closing stages of the war. At Porton, and elsewhere, researchers were likewise determined to finish the job at hand.[71]

At the same time, the authorities continued incessantly to prepare servicemen and civilians for the Allied invasion which might involve or trigger the use of gas warfare. In May, Porton carried out large-scale chemical warfare exercises in conjunction with beach-head operations by service and civilian authorities; at around the same time, scientists from Porton came to the conclusion that the use of mustard gas was likely to have a "big potential in the subjugation" of Japanese forces on the Pacific islands thousands of miles to the east.[72] With millions of soldiers and civilians killed and injured, families and children displaced, buildings burnt, and entire cities destroyed, there was little appetite among senior Allied officials to uphold standards of medical ethics and international morality if the end of combat operations would be delayed as a result. Mustard gas and phosgene were Churchill's chemical weapons of choice to attack deep within the German heartland and cause maximum casualties and mayhem, but also as weapons which could legitimately be deployed to defend Britain's beaches, ports and industry against an invading army. His retrospective assessment, made after the war, was that the Germans "would have used terror, and we were prepared to go all lengths" (Harris and Paxman 1982, 110; also Parker 1996, 49).

On 6 July 1944, one month after the D-Day landing of Allied forces in Normandy, Churchill returned once again to the subject by telling the House of Commons that the introduction of the German "flying bomb" raised some "grave questions" about the future conduct of the war. On the same day, dissatisfied by the negative assessment of the Joint Planning Staff (JPS) on the use of gas warfare as a retaliatory measure, he informed his Chiefs of Staff of his intention to employ chemical weapons if it were a matter of "life or death" for Britain or if it would shorten the war by a year:

> It may be several weeks or even months before I shall ask you to drench Germany with poison gas, and if we do it, let us do it one hundred per cent. In the meanwhile, I want the

[71]WL, PP/CLE/A.11, Lovatt Evans papers, Hill to Lovatt Evans, 24 April 1944; Lovatt Evans to Hill, 27 April 1944.

[72]TNA, WO188/802, p. 112.

> matter studied in cold blood by sensible people and not by that particular set of psalm-singing uniformed defeatists which one runs across now here now there. Pray address yourself to this. It is a big thing and can only be discarded for a big reason. I shall of course have to square Uncle Joe [Joseph Stalin] and the President [Franklin D. Roosevelt], but you need not bring this into your calculations at the present time. Just try to find out what it is like on its merits.[73]

Churchill considered it to be "absurd" to worry about the "morality on this topic" since all parties had used chemical weapons during the First World War. Whereas the bombing of large cities had formerly been regarded as a war crime, it was now done by the Axis and Allied forces on a day-to-day basis, he argued. Attempting to downplay any moral concerns of his senior military advisers, he noted: "It is simply a question of fashion changing as she [*sic*] does between long and short skirts for women" (ibid.). For the Prime Minister, the Geneva Protocol outlawing the use of poison gas was of no relevance if the existence of British realm were at stake.

Although Allied forces appeared to possess the capability, Churchill's senior military advisers stopped short of recommending the start of chemical warfare operations.[74] The Chiefs of Staff nonetheless expressed a high degree of confidence in respect of the state of Allied readiness to initiate chemical warfare operations. By 1944, British and American stocks located in Britain were deemed sufficient, they said, to produce a "formidable scale of gas attack on Germany during the early and most important phase after a decision has been taken to employ gas."[75] Britain alone had produced a total of 40,719 tons of mustard gas and 14,042 tons of phosgene and tear gases during the war (Carter 2000, 53). Instead of a prolonged use of some chemical agents by 20% of Bomber Command, the Chiefs of Staff recommended the concentration of all British and American long-range bombers in a "massive hammer blow," employing high explosives and phosgene and mustard gas bombs in quick succession on tactical and civilian targets. Phosgene would be dropped on 1000 tactical targets or twenty German cities, causing heavy casualties and deaths among civilians and civil defense personnel. Mustard gas, on the other hand, would be employed against 1500 tactical targets or, alternatively, against sixty specifically identified German cities covering the entire Reich that were "best calculated to bring about a collapse of German morale."[76] By causing death and destruction on a monumental scale, military commanders aimed to exercise intense pressure on the regime's leadership, but they were also acutely aware that the population was likely to lack the necessary "initiative required for active revolt" against the Nazi regime following gas attacks (ibid.).

---

[73]TNA, PREM3/89, Personal Minute Churchill to Ismay, 6 July 1944; also Harris and Paxman (1982, 127ff.).

[74]TNA, PREM3/89, "Military Considerations Affecting the Initiation of Chemical and Other Special Forms of Warfare"; Harris and Paxman (1982, 130ff.).

[75]TNA, PREM3/89, "Military Considerations Affecting the Initiation of Chemical and Other Special Forms of Warfare."

[76]TNA, PREM3/89, "Military Considerations Affecting the Initiation of Chemical and Other Special Forms of Warfare."

In France, chemical weapons could aid the war effort by helping Allied forces to "break through the German defences," but they could also slow the military advance, affect communications, unsettle civilian labor, and negatively affect the relationship with the local population. The same was the case in the east, in southern France and in the Mediterranean, where chemical warfare was seen to be counterproductive in maintaining support from civilians and partisans. Existing chemical warfare stocks in the Far East were deemed to be insufficient to allow offensive chemical warfare to be conducted simultaneously in both theatres of war, and defensive measure were inadequate to protect the military from gas under tropical conditions.

Military officials were under no illusion that Germany would immediately retaliate against the United Kingdom, with London as the principal target, if the Allies started to use gas warfare. Although the possible effects of gas on the home front were difficult to judge, they felt that the general public, after five years of war, might be resentful of being exposed to toxic agents if it could be shown that this "could have been avoided."[77] The Chiefs of Staff were also concerned about the effects on public morale of potential retaliatory measures against Allied prisoners of war who might be forced to "work in contaminated areas."[78] All things considered, and irrespective of any political, legal, or moral considerations, Britain's military planners concluded that chemical and biological weapons were not an attractive military proposition. General Hastings Lionel Ismay, one of Churchill's closest military advisers, even suggested to the Prime Minister that the use of these types of weapons was likely to be detrimental to the Allied military campaign:

> It is true that we could drench the big German cities with an immeasurably greater weight of gas than the Germans could put down on this country. Other things being equal, this would lead to the conclusion that it would be to our advantage to use the gas weapons. But other things are not equal. There is no reason to believe that the German authorities would have any greater difficulty in holding down the cowed German population, if they were subjected to gas attack, than they have had during the past months of intensive high explosive and incendiary bombings. The same cannot be said for our own people, who are in no such inarticulate condition.[79]

However impressive the plans drawn up by the Chiefs of Staff in July 1944 may appear in retrospect, we still need to be careful not to jump to any conclusions, on the basis of the above outlined memorandum, in respect of the actual state of Allied readiness to start chemical warfare operations during the closing stages of the Second World War. Given what we now know about newly developed operational research methods which allowed experts to calculate more precisely the requirements for chemical weapons stockpiles needed for a major military attack, it seems far from certain whether the Allied military would actually have been capable of delivering the kind of "massive hammer blow" to the German enemy within the

---

[77]TNA, PREM3/89, "Military Considerations Affecting the Initiation of Chemical and Other Special Forms of Warfare"; Harris and Paxman (1982, 132).

[78]TNA, PREM3/89, "Military Considerations Affecting the Initiation of Chemical and Other Special Forms of Warfare."

[79]TNA, PREM3/89, Personal Minute, Ismay to Churchill, 28 July 1944.

operational realities of war conceived by the Chiefs of Staff, had the order to employ chemical and biological weapons actually been given. As plans were drawn up in the United States to employ chemical weapons as part of an invasion of the Japanese home islands, for example, it became apparent that the quantitative requirements far exceeded existing stockpiles of chemical munitions. Yet if we assume, for a moment, that the existing chemical weapons stockpiles were likely to be insufficient for the kind of attack the Chiefs of Staff had outlined to Churchill—who at this point seems to have been, by all accounts, keen to launch chemical warfare operations—then this raises a series of questions: whether the Chiefs of Staff were aware of the fact that their chemical warfare capability might not have been quite what it seemed, and if so, why they did not communicate this fact to the Prime Minister. The following scenario is certainly possible: under considerable pressure from Churchill to confirm the viability of employing such unorthodox weapons, which up to this point had not been used in the war, senior military officials—who were keen to keep it that way—might have overstated the Allied chemical warfare capability, thus preserving the impression that the current state of readiness was such that chemical weapons could be employed on a massive scale and at any time, if necessary, whilst simultaneously arguing against the immediate use of chemical weapons in the current conflict.

Although hardly convinced by the report, Churchill decided to accept the assessment of his senior officials, at least for the time being.[80] As it happened, Britain's Chiefs of Staff, and Churchill in particular, had no need to return to the subject of chemical warfare. In April 1945, after Allied forces had crossed the Lower Rhine, the Joint Intelligence Subcommittee concluded that Germany appeared unwilling and unprepared to initiate gas warfare to defend the territory of the Reich. However, it also counseled caution: "There remains the possibility that Hitler may recklessly order its use in the final stage of disintegration."[81] He never did. At the end of the month, Hitler ended his life in his bunker beneath the Reich Chancellery. Shortly thereafter, the unconditional surrender of the German army heralded the end of one of the most murderous regimes in modern history, and with it came the uncomfortable realization that Allied intelligence agencies had almost no knowledge of one of the greatest military and scientific secrets of the Second World War: nerve gas.

References

Avery, Donald. 2013. *Pathogens for war: Biological weapons, Canadian life scientists, and North American biodefence*. Toronto: University of Toronto Press.

Balfour, Sebastian. 2002. *Deadly embrace: Morocco and the road to the Spanish Civil War*. Oxford: Oxford University Press.

[80]TNA, PREM3/89, Personal Minute, Churchill to Ismay, 29 July 1944.

[81]TNA, PREM3/89, War Cabinet, Joint Intelligence Subcommittee, "Use of Chemical Warfare by the Germans", 23 April 1945.

Balmer, Brian. 2001. *Britain and biological warfare: Expert advice and scientific policy, 1930–65*. Basingstoke: Palgrave Macmillan.
Brophy, L.P., and G.J.B. Fisher. 1959. *The chemical warfare service: Organizing for war*. Washington, DC: Department of the Army.
Buffetaut, Yves. 2008. *Ypres April 22nd 1915: The first gas attack*. Trans. C. Cook. Louviers: Ysec Editions.
Carter, G.B. 2000. *Chemical and biological defence at Porton Down, 1916–2000*. London: HMSO.
Carter, G.B., and G.S. Pearson. 1996. Past British chemical warfare capabilities. *RUSI Journal* 141: 59–68.
Charles, Daniel. 2005. *Between genius and genocide: The tragedy of Fritz Haber, Father of chemical warfare*. London: Jonathan Cape.
Chemical Corps Association. 1948. *The chemical warfare service in World War II: A report of accomplishments*. New York: Reinhold Publishing Corporation.
Chielens, Piet. 2014. *The list of names and the second battle of Ypres*. Public Lecture, University of Kent, November 5.
Churchill, Winston S. 1923. *The world crisis, 1915*. London: Thornton Butterworth Ltd.
Corrigan, Gordon. 2003. *Mud, blood and poppycock*. London: Cassell.
Cowell, F., et al. 2007. Chlorine as the first major chemical weapon. In *An element of controversy: The life of chlorine in science, medicine, technology and war*, ed. H. Chang, and C. Jackson, 220–254. London: BSHS Monographs.
Dendooven, Dominiek. 2005. Overview: 22 April 1915—Eyewitness accounts of the first gas attack. Unpublished paper given at the conference *1915. Innocence Slaughtered?* Ypres, November 17–19.
Douglas, C.G. 1936. John Scott Haldane, 1860–1936. *Obituary Notices of Fellows of the Royal Society* 2 (5): 115–139.
Ede, A. 2011. Waiting to exhale: Chaos, toxicity and the origins of the U.S. chemical warfare service. *Journal of Law, Medicine & Ethics* 39 (1): 28–33.
Evans, Rob. 2000. *Gassed. British chemical warfare experiments on humans at Porton Down*. London: House of Stratus.
Evans, Rob. 2007. Military scientists tested mustard gas on Indians. *The Guardian*, September 1.
Foulkes, C.H. 1934. *"Gas!" The story of the special brigade*. Edinburgh: W. Blackwood & Sons.
Garner, J.P. 2003. Some recollections of Porton in World War I. *Commentary. Journal of the RAMC* 149 (2): 138–141.
Gibson, C.S. 1941. Sir William Jackson Pope. 1870–1939. *Obituary Notices of Fellows of the Royal Society* 3(9): 291–324.
Goebel, Stefan P., and Mark L. Connelly. 2017. *Ypres. Great Battles*. Oxford, New York: Oxford University Press. Forthcoming.
Goodman, Martin. 2008. *Suffer and survive: Gas attacks, miners' canaries, spacesuits and the bends—The extreme life of J.S. Haldane*. London: Pocket Books.
Grayzel, Susan R. 2012. *At home and under fire: Air raids and culture in Britain from the Great War to the Blitz*. Cambridge: Cambridge University Press.
Groehler, Olaf. 1989. *Der lautlose Tod: Einsatz und Entwicklung deutscher Giftgase von 1914 bis 1945*. Reinbek: Rowohlt.
Haber, Fritz. 1924. *Fünf Vorträge aus den Jahren 1920—1923. Über die Darstellung des Ammoniaks aus Stickstoff und Wasserstoff; Die Chemie im Krieg; Das Zeitalter der Chemie; Neue Arbeitsweisen; Zur Geschichte des Gaskrieges*. Berlin: Verlag von Julius Springer.
Haber, L.F. 1986. *The poisonous cloud: Chemical warfare in the First World War*. Oxford, New York: Oxford University Press.
Haldane, John S. 1925. *Callinicus: A defence of chemical warfare*. London: Kegan Paul.
Hammond, Peter M., and Gradon B. Carter. 2002. *From biological warfare to healthcare*. Basingstoke: Palgrave Macmillan.
Harris, Robert, and Jeremy Paxman. 1982. *A higher form of killing: The secret history of chemical and biological warfare*. London: Chatto & Windus.

Harrison, Mark, 2010. *The medical war. British military medicine in the First World War*. Oxford: Oxford University Press.
Historical Survey of the Porton Down Volunteer Programme. 2006. London, MoD, unpublished manuscript.
Hobbs, Abbi, Catherine Jefferson, Nicholas Coppeard, and Chris Pitt. 2007. Ethics, public relations, and the origins of the Geneva Protocol. In *An Element of Controversy: The Life of Chlorine in Science, Medicine, Technology and War*, ed. Hasok Chang, and Catherine Jackson, 255–295. London: BSHS Monographs.
Jacobs, Mariette. 1996. *Zij, die vielen als helden… Inventaris van de oorlogsgedenktekens van de twee wereldoorlogen in West-Vlaanderen*, vol. 2. Bruges: Provincie West-Vlaanderen.
Kershaw, Ian. 1998. *Hitler, 1889–1936: Hubris*. London: Allen Lane.
Klee, Ernst 1997. *Auschwitz. Die NS-Medizin und ihre Opfer*. Frankfurt am Main: S. Fischer Verlag.
Lloyd, Nick. 2006. *Loos 1915*. Stroud, Gloucestershire: Tempus.
Marrs, Timothy C., Robert L. Maynard, and Frederick R. Sidell. 1996. *Chemical warfare agents. Toxicology and treatment*. New York: John Wiley & Sons.
McCamley, Nick J. 2006. *The secret history of chemical warfare*. Barnsley: Pen & Sword Military.
Parker, John. 1996. *The killing factory: The top secret world of germ and chemical warfare*. London: Smith Gryphon.
Peters, R.A. 1959. The faith of a master in biochemistry. *Biochemical Journal* 71: 1–9.
Richter, D. 1994. *Chemical soldiers: British gas warfare in World War I*. London: Leo Cooper.
Roughton, F.J.W. 1949. Joseph Barcroft, 1872–1947. *Obituary Notices of Fellows of the Royal Society* 6 (18): 315–345.
Schmaltz, Florian. 2005. *Kampfstoff-Forschung im Nationalsozialismus. Zur Kooperation von Kaiser-Wilhelm-Instituten, Militär und Industrie*. Göttingen: Wallstein.
Schmaltz, Florian. 2006a. Neurosciences and research on chemical weapons of mass destruction in Nazi Germany. *Journal of the History of Neurosciences* 15: 186–209.
Schmaltz, Florian. 2006b. Otto Bickenbach's human experiments with chemical warfare agents and the concentration camp Natzweiler. In *Man, Medicine and the State: The Human Body as an Object of Government-Sponsored Medical Research*, ed. Wolfgang U. Eckart, 139–156. Stuttgart: Franz Steiner Verlag.
Schmaltz, Florian. 2006c. Pharmakologische Nevengasforschung an der Militärärztlichen Akademie und an den Universitäten Marburg, Danzig und Leipzig im Zweiten Weltkrieg. In *Medizin im Zweiten Weltkrieg. Militärmedizinische Praxis und medizinische Wissenschaft im 'Totalen Krieg,'* eds. Wolfgang U. Eckart and Alexander Neumann, 171–94. Paderborn: Schöningh.
Schmidt, Ulf. 2006. Cold War at Porton Down: Informed consent in Britain's biological and chemical warfare experiments. *Cambridge Quarterly for Healthcare Ethics* 15 (4): 366–380.
Schmidt, Ulf. 2007a. *Karl Brandt. The Nazi Doctor. Medicine and Power in the Third Reich*. London: Continuum.
Schmidt, Ulf. 2007b. Medical ethics and human experimentation at Porton Down: Informed Consent in Britain's Biological and Chemical Warfare Experiments. In *History and Theory of Human Experimentation. The Declaration of Helsinki and Modern Medical Ethics*, eds. Ulf Schmidt and Andreas Frewer, 283–313. Stuttgart: Steiner.
Schmidt, Ulf, and Andreas Frewer (eds.). 2007. *History and theory of human experimentation. The Declaration of Helsinki and modern medical ethics*. Stuttgart: Franz Steiner Verlag.
Schmidt, Ulf. 2012. Justifying chemical warfare. The origins and ethics of Britain's chemical warfare programme, 1915–1939. In *Justifying War: Propaganda, Politics and the Modern Age*, eds. David Welch and Joanne Fox, 129–58. Basingstoke: Palgrave.
Schmidt, Ulf. 2013. Accidents and experiments: Nazi chemical warfare research and medical ethics during the Second World War. In *Military Medical Ethics for the 21st Century*, ed. Don Carrick, and Michael L. Gross, 225–244. Farnham: Ashgate.

Schmidt, Ulf. 2015. *Secret science: A century of poison warfare and human experiments*. Oxford, New York: Oxford University Press.
Shephard, Ben. 2000. *A war of nerves. Soldiers and psychiatrists, 1914–1994*. London: Jonathan Cape.
Spiers, Edward M. 1986. *Chemical warfare*. Basingstoke: Macmillan.
Spiers, Edward M. 2010. *A history of chemical and biological weapons*. London: Reaktion Books.
Stockholm International Peace Research Institute (SIPRI). 1971–1975. *The Problem of chemical and biological warfare*, vols. 1–6. Stockholm, New York: Almqvist & Wiksell Humanities Press.
Stolzenberg, Dietrich. 2004. *Fritz Haber—Chemist, Nobel Laureate, German, Jew*. Philadelphia, PA: Chemical Heritage Press.
Sturdy, Steve. 1987. *A co-ordinated whole: The life and work of J.S. Haldane*. Unpublished PhD thesis, University of Edinburgh.
Sturdy, Steve. 1998. War as experiment: Physiological innovation and administration in Britain, 1914–1918: The Case of Chemical Warfare. In *War, medicine and modernity*, ed. Roger Cooter, Mark Harrison, and Steve Sturdy, 65–84. Stroud: Sutton.
Sturdy, Steve. 2011. The meanings of "Life": Biology and biography in the work of J.S. Haldane (1860–1936). *Transactions of the Royal Historical Society* 21: 171–191.
Szöllösi-Janze, Margit. 1998. *Fritz Haber, 1868–1934: Eine Biographie*. Munich: C.H. Beck.
Thompson, R.H.S., and A.G. Ogston. 1983. Rudolph Albert Peters. 13 April 1889–29 January 1982. *Biographical Memoirs of Fellows of the Royal Society*, 29: 495–523.
Thorpe, J.F. 1936. Herbert Brereton Baker, 1862–1935. *Obituary Notices of Fellows of the Royal Society* 1 (4): 523–526.
Tucker, Jonathan B. 2006. *War of nerves: Chemical warfare from World War I to Al-Qaeda*. New York: Pantheon Books.
Van Bergen, Leo. 2012. Monkey-Man, Man-Monkey: Neutrality and the discussions about the "Inhumanity" of poison gas in the Netherlands and International Committee of the Red Cross. *First World War Studies* 3: 1–23.
Van Moon, Courtland, and John Ellis. 1989. Project SPHINX: The question of the use of gas in the planned invasion of Japan. *The Journal of Strategic Studies* 12: 303–323.
Van Moon, Courtland, and John Ellis. 1996. United States chemical warfare policy in World War II: A captice of coalition policy. *The Journal of Military History* 60: 495–511.
Vilensky, Joel A. 2005. *Dew of death. The Story of Lewisite, America's World War I weapon of mass destruction*. Bloomington, IN: Indiana University Press.
Wheelis, Mark, Rósza Lajos, and Malcolm Dando (eds.). 2006. *Deadly cultures. biological weapons since 1945*. Cambridge, MA, London: Harvard University Press.

Challenging the Laws of War by Technology, Blazing Nationalism and Militarism: Debating Chemical Warfare Before and After Ypres, 1899–1925

Miloš Vec

Abstract The German gas attack of April 22, 1915, took place immediately after intense efforts in international law to make war more civilized and to restrict poisonous weapons. Legal restrictions on war technologies reached a provisional peak at the Hague Conferences of 1899 and 1907. During World War I, the attitude of the German military became more radical, to the point of evading and denying international law. The silence in the face of the poison-gas attack was deafening, even among German scholars of international law. Older traditions from the history of ideas and collective mentalities played a crucial role in this, especially the idea of *raison de guerre* or military necessity, which were supposed to annul international law in case of military emergency. After the end of World War I, there was a lively international discourse on the legality of the German approach. Their debate was marked by a strong nationalist polarization of viewpoints. In subsequent agreements between states, the prohibition of poison gas was rewritten and strengthened.

1 Introduction: Chemical Weapons as the Subject of Juridification, Politicization, and Circumvention of Law

The story told in this essay has three phases and perhaps conceals a surprise. How much of a surprise it will be depends on the readers' expectations concerning historical international law around 1900 and the parties involved. Anyone who expects little will presumably not be disappointed; anyone who has high expectations of

M. Vec (✉)
Legal and Constitutional History, University of Vienna, Vienna, Austria
e-mail: milos.vec@univie.ac.at

M. Vec
Institut für die Wissenschaften vom Menschen/Institute for Human Sciences (IWM), Vienna, Austria

B. Friedrich et al. (eds.), *One Hundred Years of Chemical Warfare: Research, Deployment, Consequences*, DOI 10.1007/978-3-319-51664-6_7

international law and the parties to it may find what happened—especially one central event and its consequences—sobering. This central event was the German chlorine gas attack in the Second Battle of Flanders near Ypres on April 22, 1915, which would become the subject of a highly controversial discourse on international law.[1]

The essay that follows is about the interaction of technology and law—more precisely, about the interaction of war technology and law of war. It will discuss the changing interdependencies between them, focusing on the creation and implementation of legal norms and especially on scholarship in international law at the time. The crucial turning points in the interaction were treaties on international law and declarations, technological innovations, the outbreak of the war in the summer of 1914, the aforementioned use of poison gas in April 1915, and finally the end of the war in 1918. The actual history of events of the German poison-gas attack of April 22, 1915, is, by contrast left out—in part because it is covered by the essay by Friedrich and James (in this volume).

The various phases of the international law approach to poison gas can perhaps be characterized by three terms: juridification, politicization, and circumvention of law. In my view, the discussion of international law around World War I followed this periodization, and for extended periods, at least, it held few surprises with unpredictable points. The years immediately following the use of poison gas in World War I were perhaps the exception—they bore surprises.

It is a contribution to the history of international law whose intention can by no means be judged ex post by the legality and illegality of the historical event (the legal arguments called on by the various parties will nevertheless be presented under 4.2). The goal instead is to reconstruct and analyze from the perspective of international law the events of that time, and especially the discussion of the parties involved and from scholars of international law.

The history of the German poison-gas attack is part of the history of international law, on the one hand, and of the general history of World War I, on the other. In the framework of the latter history, however, the aspects relating to international law have played an astonishingly minor role in numerous publications. One exception is Isabel V. Hull's book of 2014.

On the other hand, there exist strong lines of continuity, especially in legal studies, of historical regulations of international law to the applicable law of today. As a result, legal historians can, to an unusually large degree, fall back on detailed, high-quality historical accounts by scholars of international law (Bothe 1973; Jaschinski 1975; Marauhn 1994). The Saint Petersburg Declaration of 1868 as well as (at least) parts of the Geneva Protocol of 1925 are considered today to be customary international law (Marauhn 1994, 52; Stockholm International Peace Research Institute 1973, 99), the Gas Declaration of The Hague of 1899 is still a valid treaty law for the signatories and their successor states (Marauhn 1994, 47). They are therefore historical standards that are of enduring significance for international law today.

[1]See the contemporaneous bibliography in Kunz (1935, 85–88).

2 Codifying War Technologies in International Law Around 1900

Four sets of international regulations are particularly relevant to the discussion of the poison gas attacks near Ypres in 1915. Their historical genesis is characterized by comprehensive first steps to the transformation of the civilizing ambitions of the time into the legal regulation of war (Dülffer 1978). This process peaked in the Hague Conferences as forums of discourse and of multilateral agreements on limiting armaments and on the juridification of international conflicts by means of international arbitration and jurisdiction (Neff 2014, 323–28). They were intended as instruments for securing peace in the age of nationalism, internationalization, and imperialism (Dülffer 1978, 73–100).

The Hague Conferences were characterized by, among other things, political activism, which was supported by the pacifist movements together with some parts of the community of international law scholars. The public response was eminent, and it reflected the high expectations of international law at the time. This attitude was an expression of the optimism toward international law at the turn of the twentieth century. It was based on institutionalizations and a positivization of international law that was historically unprecedented. The nineteenth century produced more treaties than ever before. They were often multilateral; some of them even open for accession. Certain types of treaties functionally compensated for the lack of codification of international law, and they were therefore called "law-making treaties." This juridification of international relations culminated in a "treaty-making revolution of the 19th century" (Keene 2012). International law never had it as good as it did in the final years before World War I. It had taken on new tasks and new spheres of regulation, and their structures continue to shape international society even today: administration, technology, economy, public health, and time-keeping (Vec 2006, 1–164). International law was by no means merely a technocratic instrument in this process. Often one can even make out a decided "ethicizing" of international law: moral objectives become the focus of norm setting (Lovrić-Pernak 2013).

Simplifying considerably, this often reflected a juridification of international relations. This term is a thesis in legal sociology and political science that describes extension of law by means of increasingly dense and detailed regulations, by means of institutionalizations, and ultimately by means of legal resolution of conflicts. Some jurists around 1900 expressly articulated the hope for a global law in which the progress of international law of the nineteenth century would culminate.

Within this friendly sounding panorama, however, the law of war was a particularly touchy sphere. The use of violence was, on the one hand, restricted, but, on the other hand, legitimized, in that international law itself supplied a normative order to its employment (Simon 2016, 508). Different than economic and administrative regulation, the law of war concerned questions that were understood to be more delicate and more political. All of them addressed competencies of state sovereignty, but restrictions on the law of war were subject to clearer bounds than those of the other fields of regulation. The evolution typically took the form of transforming customary law into international treaty law (e.g., the prohibition of poison in the law of war). There was also the regulatory problem that these legal standardizations around 1900, just as much as those of the twentieth and twenty-first centuries, were often challenged by new technologies. Hence it represented the problem of regulating technology by law, in which the law always seems to hobble along after the technology (in accordance with the classical "legal lag" theory). Debates in international law over the legality of the use of poison gas in World War I will underscore this problem.

2.1 Restrictions on the Means and Methods of Warfare: The Regulations in International Law of 1868, 1899, and 1907

The legal bases for restrictions on chemical weapons, the validity of which was debated in the context of the use of poison gas at Ypres and beyond, can be identified precisely. In this section, they will first simply be presented, in their original wording. The heated debates over their validity and interpretation will be addressed later.

2.1.1 The Principle of Humanity: The Saint Petersburg Declaration of 1868

The first of these legal bases was the Saint Petersburg Declaration of 1868, in full title "Declaration Renouncing the Use, in Time of War, of Explosive Projectiles under 400 Grammes Weight" of November 29 (December 11), 1868. Its preamble reads:

> Considering:
>
> That the progress of civilization should have the effect of alleviating as much as possible the calamities of war;
> That the only legitimate object which States should endeavour to accomplish during war is to weaken the military forces of the enemy;
> That for this purpose it is sufficient to disable the greatest possible number of men;
> That this object would be exceeded by the employment of arms which uselessly aggravate the sufferings of disabled men, or render their death inevitable;
> That the employment of such arms would therefore be contrary to the laws of humanity.
> The contracting parties engage mutually to renounce, in case of war among themselves, the employment by their military or naval troops of any projectile of a weight below 400 grammes [about 13 1/2 oz], which is either explosive or charged with fulminating or inflammable substances (Declaration 1868).

The declaration thus regulated and formulated not only the special, explicit prohibition of certain weapons but also general principles of the law of war. According to Thilo Marauhn (1994, 46), it prohibited specific types of weapons (projectiles), on the one hand, and, on the other hand, established the "fundamental obligation to avoid unnecessary suffering and referred to the laws of humanity." In Marauhn's assessment (1994, 46), this was not a direct prohibition of chemical (poison) weapons. Rather, it stated the first general principles of humanitarian international law from which a prohibition of certain chemical weapons can be derived indirectly. The British scholar T. J. Lawrence, who was for many years a lecturer at the Royal Naval College in Greenwich and at the Royal Naval War College in Portsmouth, saw it as "the application of the true principle, which measures the illegality of weapons, not by their destructiveness, but by the amount of unnecessary suffering they inflict" (Lawrence 1923, 529).

2.1.2 The Impotent Model: The Brussels Declaration on Land Warfare of 1874

This same principle was applied again in the "Project of an International Declaration concerning the Laws and Customs of War, Brussels, 27 August 1874", which in Article 13 included the following provision: "According to this principle are especially 'forbidden': (a) Employment of poison or poisoned weapons."

The Brussels Declaration on Land Warfare of 1874 never came into effect. It did, however, lead to analogous resolutions by the Institut de Droit International in 1880 in the form of the "Manuel des lois de la guerre sur terre" (Kassapis 1986, 10; Kunz 1927, 13), unofficially known as the "Oxford Manual," which in turn was the model for the later positive-law regulations of 1899 and 1907 (Mérignhac 1900, 197).

2.1.3 The First Poison Prohibition in International Treaty Law: The Declaration on the Use of Projectiles with Asphyxiating or Deleterious Gases and the Hague Convention on Land Warfare of 1899

On July 29, 1899, the concluding act of the First Peace Conference in The Hague followed. It included the Hague Declaration (IV, 2) concerning asphyxiating gases. The preamble states that the declaration had been "inspired by the sentiments which found expression in the Declaration of St. Petersburg of 29th November (11th December) 1868." The declaration proper is as brief as possible: "The contracting powers agreed to abstain from the use of projectiles the sole object of which is the diffusion of asphyxiating or deleterious gases." (Carnegie Endowment for International Peace 1915b, V) It goes back to Article 13 of the draft declaration of the Brussels Conference of 1874. The wording of the Brussels Declaration on Land Warfare thus became part of the appendix to the Second Hague Convention of 1899 and of the Fourth Hague Convention. Although a prohibition of poison and poisoned weapons already existed under customary international law, the Hague Declaration of 1899 was the first time international treaty law expressly referred to "poison" weapons (Marauhn 1994, 46).

The formulation of the single objective or the single goal was often a point of reference for German interpretations, which argued that projectiles whose effects included not only gas but also fragmentation were permitted. It was also controversial whether gases that were not deadly also fell under the prohibition (Jaschinski 1975, 32–34). Later, other, non-German jurists also pointed to problems of the existing version. Lawrence (1923, 531–532) found the regulation dubious in comparison to other weapons and possible uses and believed that the horror of death by asphyxiating gases was no less than that of the fate of sailors drowning. Moreover, according to Lawrence, the adjective "deleterious" was "vague."

In 1899, the Second Convention, the "Convention with Respect to the Laws and Customs of War on Land", was passed (Kunz 1927, 13). Section 2, "On Hostilities," Chap. 1, "On Means of Injuring the Enemy, Sieges, and Bombardments," Article 22 reads as follows: "The right of belligerents to adopt means of injuring the enemy is not unlimited." Definitions and specifications of what was prohibited in Article 22 followed in Article 23, in which the parties were expressly denied unlimited rights in the choice of means to harm the enemy. It reads in part:

> Article 23.
>
> Besides the prohibitions provided by special conventions, it is especially prohibited: –
> (a) To employ poison or poisoned arms;
> (b) To kill or wound treacherously individuals belonging to the hostile nation or army;
> [...]
> (e) To employ arms, projectiles, or material of a nature to cause superfluous injury;

Articles 22 and 23 established minimum standards for humanitarian warfare under existing international law. On the one hand, Article 23 provided definitions for and specifications of Article 22. On the other hand, a general principle prohibiting the use of poison in a certain sector was codified by treaty.

The final act of the Second Hague Conference of 1907 contained in Article 22 of the Fourth Convention (with identical wording) and in 23 a, b, e of the regulation the almost identical definitions of the Hague Convention on Land Warfare of 1899. The minor changes were (in *italics*)[2]:

> (a) To employ poison or poisoned weapons [instead of arms]
> (e) To employ arms, projectiles, or material calculated to cause unnecessary suffering [instead of of a nature to cause superfluous injury].

[2]Synopsis in: Carnegie Endowment for International Peace. Division of international law. 1915a. Pamphlet No. 5. The Hague Conventions of 1899 (II) and 1907 (IV) respecting the laws and customs of war on lands. Washington, D.C.: Carnegie Endowment, 17.

2.2 *No Notion of What Lay Ahead: The Intense Legal Discourse on the Hague Convention and Uncontroversial Interpretations of the Prohibition of Poison Prior to 1915*

The Hague Conventions included regulations on the law of warfare, and today it is contrasted with the Geneva Conventions, which concern the treatment and protection of people who are not or are no longer participants of the combat operations (Kassapis 1986, 1). The Hague Conferences marked a new era in the history of international law on warfare. Following the gas attacks at Ypres in 1915, the definitions established therein would become the central points of reference in the discussion of the legality or illegality of the German actions.

This led to the aforementioned problem of the relationship between legal regulations in their abstract form and the concrete events of real life to which they are supposed to be subsumed. In the period that followed, the regulations were often referred to, cited, and interpreted. But at the time of their enactment, no one could have foreseen how effective a weapon poison gas would become in just a few years and in which form it would be employed on the battlefields of the war to come (Bothe 1973, 8). For that reason, Bothe (1973, 88) argues that the intentions of the norm creators are interesting but not truly helpful when it comes to interpreting Article 23(a).

However, for the years between the passing of the norms quoted above and the beginning of World War I and even up to the German gas attack near Ypres, a lively, widespread juridical discourse on the law of war can be documented. The first and formal feature is that a considerable number of treatises appeared that addressed various subjects from the large field of regulating new warfare technologies. Their authors included jurists and specialists in international law or the military, and so did their readership.

If one considers only this thematic subset of the discussion of international law in this period, it is noticeable that the prohibition of poison is mentioned repeatedly. Nowhere, however, is this discussion focused on poisonous gas in the sense of the later events. The facts of the prohibition of poison are rather typically related to other, what could be called classical, historically experienced military strategies and given a pass (Zorn 1902, 7). In what follows, the accounts of several European scholars in international law will be laid out as examples.

The German international law specialist Hans Wehberg, who had a reputation as a pacifist, published the volume *Die Abkommen der Haager Friedenskonferenzen, der Londoner Seekriegskonferenz nebst Genfer Konvention* (The Treaties of the Hague Peace Conferences and of the London Naval War Conference and the Geneva Convention) in 1910. It only reproduces the original wording of the norm; there are no explanations of it (Wehberg 1910). The Swiss-German jurist and partisan of peace Otfried Nippold mentions poison but does not interpret or expose the problems of the term (Nippold 1911, 10). The same is true of Ernest Nys (1912, 144), professor and historian of international law in Brussels and member of the Permanent

Court of Arbitration. The German international law professor Karl Strupp wrote an account of the international law on land warfare that was published in 1914, already during World War I. In it the prohibition of poison is associated with the pollution of rivers, wells, and water pipes by infectious materials (Strupp 1914, 58).

Very similarly, Henry Bonfils, a French professor in international law teaching at the Université de Toulouse, wrote in 1908: "de contaminer les puits, les aliments, les armes, est absolument proscrit dans les guerres modernes" (poisoning wells, food, and arms is absolutely prohibited in modern wars) (Bonfils 1908, 660). A similarly limited interpretation is found in the book *Les lois de la guerre continentale*, by the French military judge Robert Jacomet, which was reprinted several times: he considers it to refer to poisoning wells and the spread of contagious diseases in the enemy's country (Jacomet 1913, 58). In the third volume of his textbook on the law of war, the French international law scholar Alexandre G. Mérignhac has an entry that again enumerates merely a classical historical arsenal of acts of poisoning, while gas as chemical weapon is not mentioned (Mérignhac 1912, 261). Mérignhac was even more monosyllabic with regard to Articles 22 and 23 in his earlier account of 1900 (Mérignhac 1900, 197). The account in the manual by J. E. Edmonds and Lassa Oppenheim, published around 1913, is also extremely terse: the prohibition is repeated word for word without any additional commentary.

To my knowledge, Albrecht Tettenborn deviates most from this widespread brevity in such accounts; in his analysis "Richtungen der einzelnen Kriegsmittelbeschränkungen" (Trends of the Individual Restrictions on Weapons), he interprets the term "Gift" (poison) (Tettenborn 1909, 22–24). In this case, his intense preoccupation with the wording of the norm that takes up the term "Gift" leads to a discussion of possible regulatory gaps. They are filled by the traditional historical practice of broad interpretation. The *ratio legis* of weapon prohibitions is extended to other fields, resulting in an expanded protection by the law of war. The discussion by Frantz Despagnet (1905, 645), professor of international law in Bordeaux and a member of the Institut de Droit International, of the possibility of poisoning air with gas also draws an analogy to the poisoning of water.

In the final years before World War I, this sort of treatment that exposes the problems of the subject was the exception, despite the codification of the prohibition of poison in treaties. The hesitant discussions that would follow in 1915 are even less evident. In the years since the prohibition of poison had been passed, modern technology had not yet posed a challenge to international law and its protective regulations. If one compares the tone and style of the comments of scholars of international law, something other than their brevity is striking as well: Scholarship on international law was not very politicized. There were no nationalistic undertones, no accusations or stereotypical discussions when it came to the establishment and application of norms of the law of war.

On the one hand, all of that is to be expected and not very surprising. On the other hand, academics working in the area of international law after 1915 will take another look at the positions previously taken by their colleagues and interpret them in light of recent events. It was not just that norms were read anew in the face of the ongoing or recently occurred gas warfare; previous positions were cited as well.

The old interpretations were reinterpreted. They took on a new weight and became in turn not only arguments, but potentially sources of international law as well (Vec 2017).

3 Militarization and Circumvention of Law: Debates on International Law During the Continental War with Gas, 1915–1918

This impression of the brevity, even inadequacy, of international-law scholars' interpretations of the prohibition of poison, combined with the not very politicizing tone of jurists' interpretations, would change after April 22, 1915, with a slight delay but a lasting effect.

As suggested in the introduction, whether this change appears surprising depends on the expectation one brings to historical international law and specifically during wartime. The lack of specific statements in the sources is also the subject of these expectations. Nevertheless, with all due caution it seems fair to conclude that there was an astonishingly weak or nonexistent response from European and American international law scholars to the gas attack of April 22, 1915.

3.1 International Law: Alive, but not Kicking

The international law of these years was often intensely preoccupied with the events of World War I. Many academics in international law adopted passionate stances on the legality or illegality of specific actions; others behaved more guardedly. This only makes all the more interesting the comparatively tepid discussion of the attack near Ypres that led me to the thesis condensed in the heading.

3.1.1 Scholarly Publications on International Law During World War I

It should be recognized first that the panorama of publications on international law during World War I is quite rich. The conflicts between states fueled the discourse on international law, opening up new themes and offering controversial issues to debate, taken up equally by scholars and nationalists (Toppe 2008, 103). Based on the type of publication, the following observations can be made: Books on international law continued to be published, especially anthologies of texts with the relevant regulations on the law of war (Carnegie Endowment for International Peace 1915a; Pohl 1915). These anthologies of texts did not, however, comment on or even mention the ongoing gas warfare (Pohl 1915). The legal journals and scholarly journals on international law that already existed during World War I published many essays and miscellanies on war and international law. The German *Zeitschrift für Völkerrecht*, for example, founded in 1906, reflected the new thematic trend in many ways (Hueck 1999). The changes in international law, the validity of specific norms, the subsumption of specific events to regulations under international law—all these issues now became the topic of intense debates among scholars of international law. It is interesting, however, that I was unable to find a single article in *Die Zeitschrift für Völkerrecht* concerning the subject of the gas attack. Monographs on international law also seem not to have mentioned gas warfare at all or at most very rarely. German scholars on international war seem to have been completely silent on the subject during World War I (Zecha 2000, 27). By contrast, poison gas was mentioned in A. A. Roberts, *The Poison War*, published in London in 1915 (Roberts 1915, 20). The statements of other international-law specialists from Allied countries will be addressed below (3.2).

3.1.2 Unclear Motives for and Few Scruples About the Use of Poison: Ex Post Justifications

The connection between the German decision in favor of the use of poison gas and international law also remained murky. According to Hull (2014, 232), the precise motive for the German employment of gas near Ypres is unclear. No documents can be found. Only subsequent justifications of the action with references to international law can be found. In these justifications the Germans argue that their use of poison gas was a reprisal for the French use. Modern scholars emphatically reject this justification as self-protection (Hull 2014, 233; Zecha 2000, 22). Instead, the Germans independently wished to employ poison gas. Nevertheless, the argument that the use of poison gas was supposedly a reprisal under international law shows that the Germans assessed its use based on the upstanding validity of the norm, since they asserted thereby that its use was a sanction for an injustice committed by the other side. That makes it clear that they assumed, at least ex post, that the use of poison gas violated international law. Other historical sources also suggest that the use on the part of the Germans had to overcome resistance that assumed it was illegal.[3] Fritz Haber (1924, 76), a chemist and the scientist responsible for the planning of German gas warfare, claimed in the 1920s that the military had conducted a legal review. Colonel General, head of the Generalstab (General Staff) and minister of war Erich von Falkenhayn had, according to Haber in a report to the investigating committee of the German Reichstag on October 1, 1923, "apparently personally reviewed the permissibility of gas weapons under international law." Haber stated: "He was convinced beyond any doubt that his orders in the area of gas warfare did not violate international law" (Bell and Schücking 1927, 13; Haber 1924, 76–77).

[3]Otto Hahn reports a conversation with Haber in Martinetz (1996, 104).

3.1.3 Was the German Employment of Poison Gas a Symptom of General Disdain for International Law?

The striking absence of evaluations based on international law or any other normative considerations raises questions. What does this say about the German attitude to international law? To what can it be traced back? Hull is very emphatic here:

> the pattern of decision making seems clear. Civilian leadership, which was chiefly in charge of applying legal considerations, faced especially strong undertow from military institutions: from the junior, and then quickly from the senior naval officer corps regarding submarines; from the war ministry, which had already bought flamethrowers without advance discussion; and from OHL [Oberste Heeresleitung], which had already bombed civilian targets from the air." (Hull 2014, 238)

In this view, international law was disdained by the very parties who made the military decisions during World War I and bore responsibility for them. Hull's verdict is even more harsh from a comparative international perspective: Germany is said to have expressed particular disdain for international law (see also Partridge 1917, 6). With regard to the legal history, it is not easy to determine from the available sources whether this was in fact the case. There are several sources, at least, in which the attitude of other states to the poison gas attacks and international law is expressed.

3.2 The Law Comes Later: The Weak Normative Discourse on Gas Warfare After the German Attack

A first approach would be to ask how the other European powers reacted to the German poison gas attack and to what extent international law played a role in that discourse. But my thesis is that considerations of international law continue to be largely absent. Legal assessments of the German use of poison gas are found only here and there: The English international law professor Coleman Philippson noted in 1915, in an addendum to his account of the law of war during World War I, that "[the Germans] diffused asphyxiating gases among their enemy; such conduct being not only unlawful under the international declaration made in 1899, but contrary to humanity and civilization." (Philippson 1915, 217). A similar verdict is also found in Hall and Higgins (Hall 1917, 569 n. 2). French law professor at the Sorbonne Antoine Pillet (1918, 218) wrote in a book published toward the end of World War I that the prohibition of gas in Article 23(a) applied only to fluid or solid poison, since the prohibition of gas had been regulated elsewhere. And the rules therein date from another era of war technology, Pillet argued, so the prohibitions should not be applied to the new German gas attacks (1918, 244–245). Several other non-German authors addressed gas warfare (e.g., Clunet 1915).

Hence the surprise that international law played hardly any role in the German decision continues to some degree with Germany's enemies as well. As will be demonstrated in what follows, there was hardly any normative discourse on gas warfare, whether among politicians or in the general public.

3.2.1 The Lack of Protest: Political Voices and Official Reports

One first point concerns the question of the extent to which political voices and official reports articulated protest over the German poison gas attack. This question has also been examined frequently in the scholarly literature thus far, and the findings are quantitatively sparse. Zecha observed in 2000 "that neither the warring nor the neutral countries, for example, the United States until 1917, protested the use of poison gas or chemical weapons" (Zecha 2000, 26).

In the period immediately following World War I, not only Haber (Martinetz 1996, 114) but also the Germans Johannes Bell and Walther Schücking (with apologetic intent) asserted with satisfaction that no warring or neutral power had protested at all (1927, 9). Jaschinski, by contrast, asserted (without specifics) that the United Kingdom had accused Germany of violating "the laws of war of civilized countries" but other Allied countries had not protested the first large use of chemical weapons. Jaschinski condensed this into the memorable formulation "the silence of the Allied forces" (1975, 115).

By contrast, Garner (1920, 284–285) mentions a charge from the British War Office dated April 21, 1915, that "Germans had violated the laws of civilized warfare during their recapture of hill 60 east of Ypres, by employing shells which emitted asphyxiating gases." Moreover, "the Belgian commission of inquiry investigated the use of asphyxiating gases at Ypres" (Garner 1920, 272). Both Garner and Kunz report that the British field marshal Sir John French (later Earl of Ypres) denounced the gas attack in a battle report: "the enemy … by the use of an entirely new war method, one contrary to engagements entered into by them at the Hague Convention" (abridged quotation in Kunz 1927, 3, 14; Palazzo 2000, 43, Zecha 2000, 27; with different wording in Garner 1920, 276). In the House of Lords, Lord Kitchener would have protested this kind of warfare on May 18, 1915 (Palazzo 2000, 43). *The Times* of London reported a number of times (Garner 1920, 275–276). Hull (2014, 235) in turn quotes four statements by politicians that express outrage at the use of poison gas weapons. These four statements were not, however, official protests but only personal remarks, which Hull cites as evidence of the authenticity of the outrage of those who made them.

The picture in France is similar. In contrast to Britain, where official protest with underpinnings in international law was supposed to have been articulated, no one would ascribe that to France. Olivier Lepick, the scholar with the best knowledge of this material (1998), responded to a request by the present author by saying that no

French protest could be found.[4] Hull (2014, 237) notes that there were no discussions worth mentioning in France of the legality of poison gas weapons. Interest focused on France's own capacities: to catch up with Germany and to deploy gas weapons.

In the end, official protest did not follow until near the end of World War I—albeit not by the countries involved but by the International Committee of the Red Cross (ICRC). On February 6, 1918, it appealed to the warring powers to renounce poison gas weapons (Jaschinski 1975, 60; Overweg 1937, 64).[5] The use of chemical warfare agents was said to violate international law. The ICRC evoked the risk of the escalation of gas warfare and proposed a treaty on the renunciation as a return to the Hague Convention (Jaschinski 1975, 60). In their reply to this note of protest, the Allied countries first referred to their own use of poison gas as a "reprisal." By doing so they implicitly admitted (as the Germans had previously) that the use of this weapon was illegal but justified it with reference to earlier violations of international law by the enemy.

The warring powers thus demonstrated their awareness of the abstract legal standard; several of them (including the United States) nevertheless employed this weapon to obtain military advantage (Jaschinski 1975, 116). It is even more interesting, of course, that the first use did not produce an official outcry that mobilized international law as a normative basis for a complaint against the enemy in question. International law was almost inaudible in the discourse between countries with regard to the years of World War I—at least in the context relevant here of poison gas. Other violations, such as that of Belgian neutrality, were publicly denounced much more strongly. The only official objection in the case of poison gas, by contrast, came not from a state and not from one of the powers involved in World War I but from the International Committee of the Red Cross. It came late, in any case much later than the first use of poison gas. Nevertheless, as a complaint it set in motion justifications from the parties who had employed this weapon. These late justifications, like the ICRC's complaint, were presented in the language of international law.

3.2.2 The Daily Press: Restraint, Disinformation, and Loud Silence

Another surprise is the almost complete lack of discussion in the daily press. Here too astonishingly few traces of and references to the poison gas attack can be found. Even when they did occur, they occurred with strange distortions. The daily newspapers of various countries reported little about their own use; sometimes

[4]Oral communication to the author on April 21, 2015 at Harnack-Haus in Berlin.

[5]Protest by the International Committee of the Red Cross, published in "Papers relating to the Foreign Relations of the United States. 1933." *1918, Supplement II: The World War*. Washington: US Gov. Print. Off., 779–781.

readers would first learn about it when the domestic press reacted to accusations from abroad or took positions justifying its own government.

Publication was subject to the conditions of wartime censorship. For that reason, it can be assumed that acts of the local government that seemed morally or legally dubious did not easily make it into the news. Those circumstances also explain why above all the possibly illegal actions in the form of the use of poison gas by one's own country were generally mirrored back via the detour of the enemy country's journalism. It remains surprising nevertheless that even the enemy's use of poison gas was not treated very prominently.

Several examples can be cited briefly for this assessment, and they are based on intense archival studies conducted by students in the summer semester of 2014–15 as part of a seminar on the history of international law in the Faculty of Law at the University of Vienna, held in Korneuburg in cooperation with the Österreichische Landesverteidigungsakademie ([Austrian] National Defense Academy) and the ABC-Abwehrschule "Lise Meitner" (Lise Meitner ABC Defense School) of the Austrian Federal Army.[6] According to these studies, the German use of gas weapons near Ypres was mentioned several times in the reporting on Ypres (Spitra 2015, 19). Often, however, the reader only learned about the use of gas from biased official reports from the Major Headquarters of the German Reich. It is even more curious that numerous reports refer first to French or English papers in order to reject their presentation of the facts or their legal views (Bischof 2015, 13–20). Moreover, such reports referred to official German announcements, according to which the enemies had been using such means for several months. In addition, there is a conspicuous silence where one would have expected reports. The first gas release by Austria-Hungary went unmentioned in the *Arbeiter-Zeitung*, the *Reichspost*, the *Neues Wiener Journal*, and the *Pester Lloyd* (Herzog 2015, 11). By contrast, there were lengthy reports on the attack on October 24, 1917, as part of the Twelfth Battle of the Isonzo; the victory was celebrated by the newspapers, but the gas that made it possible in the first place was not mentioned anywhere! The same was true of Austria-Hungary's last large gas attack as part of the Piave Offensive on June 15, 1918; no daily newspaper mentions gas.

But the poison gas attacks of the enemies also had a relatively weak journalistic response. The first English release of gas was not mentioned in any of the four Austrian-Hungarian newspapers studied. The same was true of the reporting one the first French release of gas on February 25, 1916, near Reims: the attack was not mentioned in any edition (February 21–28, 1916); nor was the attack using phosgene gas shells on February 21, 1915, near Verdun mentioned, even though this involved a new weapon (Herzog 2015, 11). In addition to the complaint of violations of international law, which by this time had become problematic, concerns about making one's own soldiers and nationals uncertain probably played the primary role in omitting such reports.

---

[6]Seminar "Giftgas im Ersten Weltkrieg: Völkerrecht, Diplomatie und chemische Kampfstoffe," hold by Colonel Dr. Wolfgang Zecha, Lieutenant Colonel Erwin Richter, and Miloš Vec.

3.3 *Possible Interpretations: Raison de Guerre as Its Own Form of Normativity?*

If we sum up the observations thus far, we can conclude that international law played an astonishingly peripheral role not only in the Germans' decision whether to employ poison gas but also in the public discourse. Justifications often followed only in response to opponents' accusations of illegality. The justifications were repeatedly based on demonstrably false information, in the cases both of the German and Austrian-Hungarian armies and of the Allied forces. In what follows I will attempt to explain this attitude, in particular on the German side. Three possible factors seem to me worth mentioning.

3.3.1 Older Traditions of Disregard for International Law

First, the disdain for international law during World War I in general can be traced back to older traditions. For example, international law became established as a scholarly discipline in the nineteenth century, having ascended important steps of institutionalization at the universities and achieved a certain autonomy in the scholarly discourse. Considerable steps toward positivization with regard to the normative order between countries can also be observed. Part of this success story is institutionalization between countries and multilateral treaties with the possibility of accession (see Sect. 2 above).

At the same time, international law had always been subjected to academic, political, and practical criticism. The keyword here is the so-called deniers of international law. This collective name brings together heterogeneous doubts about international law as a genuine normative order. On the one hand, the publishing market in Germany for international law flourished, especially in the form of textbooks and monographs. That fed on older traditions of the history of scholarship in which natural law, political science, and imperial public law had prepared the ground well already by the end of the eighteenth century (Koskenniemi 2011). On the other hand, Germany in particular produced prominent deniers of international law. The flipside of such denial was often a complementary overemphasis on the sovereignty of the nation-state. In this view, the normative order between states could be subordinated to national law and national interests. This mixed explosively with the interests of certain military and political figures, and precisely in the situation of military conflict it was able to become acute that certain regulations of behavior between states would be regarded as nonbinding. Hull comes to the following clearly contoured and sharp conclusion regarding the German attitude toward international law:

> The legal discussions of autumn 1914 inside Imperial Germany reveal no identification with international law and no sense that law might be, intrinsically, a good worth upholding or in Germany's interest to strengthen. On the contrary, it mostly appears either as an impediment to necessary action, or at most as a tool one might instrumentalize." (Hull 2014, 239)

In this view, traditions of dismissing and denying international law were particularly strong in Germany.

3.3.2 Normative Plurality and Renouncing International Law: The Nature of the Laws of War

The second factor that can be identified is a particular dismissal of international law on war that took the form of the German military circumventing the law. For example, Hull (2005) suspects there was a specifically German mentality that easily thrived in the ideological soil described above. It found its classical expression in the notorious text *Kriegsbrauch im Landkriege* (translated as *The Usages of War on Land*), which was published by the German General Staff in 1902. It states that *raison de guerre* permits any warring state to employ all means that make it possible to achieve the aim of the war. This was restricted only by "*customs, traditions, or manner of war,*" but not by international law (Großer Generalstab 1902, 2–3). This rejection of all attempts to regulate the means of war by law is justified historically: "Immersion in the history of war will protect the officer from exaggerated humanitarian views; it will teach him that war cannot do without certain hardships, that rather the only true humanity often lies in its ruthless application" (Großer Generalstab 1902, 2–3). It is hardly surprising that such publicly expressed positions caused outrage from Germany's enemies in war and were exploited for propaganda. The text was quoted critically by international law scholars (Westlake 1907, 91; Garner 1920, 280) and was translated into several languages (German General Staff 1914; Grand État-Major Allemand 1916). The Italian edition of 1915 (Grande Stato Maggiore Germanico 1915) had an eloquent title page illustration showing German soldiers using a vise to crush civilians against the backdrop of a burning city.

Andreas Toppe came to a similar conclusion, identifying a lack of implementation in the German military and a "radicalization of military doctrine" (Toppe 2008, 28, 30, 105). The systematic location of this thinking is the idea that war has its own mechanisms. They are in a position to annul law and especially international law. It is condensed in the phrase "Necessity knows no law." This very formulation makes one think of the analogy to *raison d'état*, which also permits the violation of law and morality in time of need. That is why we speak by analogy to *raison d'état* of *raison de guerre*. This term also occurs in historical sources: in an extreme case, *raison de guerre* can overturn (international) law.

Evidence of this normative stance can also be found here and there in the late nineteenth century. The results of research of recent years and decades has shown that military figures—not only in Germany—often rejected the limitations on their means of warfare that were sought or agreed on at international conferences on international law (Messerschmidt 1983, 240). The argument was that while humanity may be a feasible principle for modern international law on war, it is alien to the true nature of war (Dülffer 1978, 150). Other military figures argued that one could better pursue humanity by creating new, more effective weapons than by prohibiting them (Dülffer 1978, 150). Finally, Fritz Haber's own dictum on the rapid poisoning effect of prussic acid (which was, of course, never a weapon of war) points in this direction: "One cannot die more pleasantly" (Cassar 2014, 31; Haber 1924, 81). That was formulated by comparison to ethyl bromoacetate, which had been employed by the French in August 1914 as the first (but not lethal) gas weapon, and according to Haber it caused a truly agonizing death. The legal assessment of various gases should therefore, in Haber's view, be considered in nuanced fashion. Some effective breathing poison could be breathed without a problem, while others were excruciating to breathe and were for that very reason were less likely to be effective as poison (Haber 1924, 81). Hence the limits should themselves be limited (Huber 1913, 359). Not international law but military utility should have the last word. Prohibitions in international law were therefore eyed with distrust because the military wanted to keep open the option of better weapons in the future (Dülffer 1978, 76). Interestingly, this attitude of criticizing or annulling international law is expressly shared by some experts in international law (Cybichowski 1912, 68–69; Lentner 1880, V). Even Hans Wehberg (1910, 14–15), who is regarded as a pacifist, writes: "In the extreme case […] every principle of the law of war can be breached."

Other, non-German scholar on international rights made comparable arguments (Rivier 1896, 241–242). Thus around 1900 there was an unholy alliance in which the military and international-law scholars placed the validity of the law for certain extreme cases under the proviso of necessity. It is hardly surprising that as the war continued, the aspect of the political utility of international law was emphasized more strongly (Koellreutter 1917/1918, 500).

3.3.3 Cruel, Unmanly, and Unchivalrous: The Military's Aversion to the Use of Poison

A third and last approach to explaining the absence of discourse on the use of gas as a weapon and its permissibility under international law is the discomfort in ethos of broad swaths of the military to poison gas. This argument focuses on the perception of this weapon and contrasts it with the disposition of military actors. Gas and poison were perceived as cruel, unmanly, and unchivalrous (Encke 2015, 2006, 197–218). All poison was considered a "womanish weapon," which is in keeping with the attribution of this way of killing to women in criminal law and criminology (Weiler 1998). Gas and poison were not used in man-to-man close combat; they did not cause bleeding wounds in physical battle. Rather, it was a weapon that only worked at a distance. This had already been an argument for considering medieval crossbows to be illegitimate or even illegal. All these elements underscore the asymmetry of the debate, in which they ran counter to a chivalrous ethos that preferred beating, stabbing, and shooting weapons (Encke 2015).

The depictions of poison gas in literature and art (see the essay by Kaufmann in this volume) underscore the problems that even parties of militaristic convictions had with this particular weapon. For example, pacifists and opponents of war produced many works that have since become famous in art history denouncing the effects of poison gas, including those by Otto Dix, among others. In *Die letzten Tage der Menschheit* (translated as *The Last Days of Mankind*), Karl Kraus (1922, 337) also left no doubt about the ethical doubtfulness of poison gas. In that work he coined the pun of the "chlorious" offensive:

> For the one thing that remains inconceivable is what possible connection exists between some chemist's inspiration, in itself a disgrace to science, and heroism. How fame in battle can be attributed to a 'chlorious' offensive without choking in shame on its own poison gas." (Kraus 2015, 262)

On the other hand, revealingly, there are no mentions or descriptions of poison gas in the art and literature that glorifies or affirms war.

All this shows that poison gas was a subject above all in nonaffirmative, antiwar, and critical accounts of war. The military, by contrast, had little sympathy for this weapon, and that can be seen as another reason for its silence on the subject.

4 The Continuing Politicization of International Law: The Legal Assessment of War Crimes, 1918–1925

After the war ended in 1918, there followed a third phase in the discourse on international law concerning the use of poison gas weapons. It was an intense, retrospective debate on the legality of their use under international law. The not very surprising form taken by this third and final phase is characterized by stark nationalization and an irreconcilable polarization of the political and legal standpoints.

4.1 Crime and Argument: The Intense Discourse After the End of World War I

The intense discourse on the use of poison gas that followed the end of World War I was embedded in the general public assessments of the events of the war years. From the prehistory of the war to its outbreak and over its course up to the end: All the events were evaluated in terms of domestic and international law. The parties sought the tribunal of the public and tried to win over public opinion. Between the wars, especially, the subject of gas warfare reached a climax in the public debate.

Perspectives of history and of international law thus went hand in hand politically, and they were also associated with narratives of proper conduct. For that reason it is right to speak of "war innocence research" (Große Kracht 2004, 8). Already in the July crisis and the first months of war, the so-called "colored books" were published by official authorities or at least by sources close to the government. In these texts the various national standpoints were presented with a suggestion of historical objectivity. The colored books of the enemy were in turn accused of distortions and omissions in their descriptions of historical events (Zala 2001, 27).

What had begun with the outbreak of the war continued over the course of the war with respect to events, with the intention of demonstrating the enemy's war crimes (Dampierre 1917; Niedner 1915). Because all sides were involved and making reciprocal accusations, it has rightly been called a "war of colored books" (Kuß 2010, 334). The public fight over the question of war guilt and the politically tinged legal discourse on war crimes followed the battle with weapons (Große Kracht 2004).

Gas warfare was also assessed in a number of publications after 1918. The major international scholarly works on international law published between the wars mention gas warfare frequently. Monographs, brochures, and essays devoted exclusively to poison gas were published (Ewing 1927; Eysinga 1928; Hanslian 1934, Kunz 1935, 85–88;). Some of these specialized monographs had a more scientific and technical tone (Endres 1928; Hanslian 1927, 2009; Meyer 1926; Woker 1925); other books focused on international law (Korovine 1929; Kunz 1927, Overweg 1937).

4.2 Self-justifications: The Nationalist Polarization of International Law

A number of such works were published in Germany, investigating the conduct of the war by Germany and other countries, some expressly discussing gas warfare. *Die deutsche Kriegführung und das Völkerrecht* (Deutsches Kriegsministerium und Oberste Heeresleitung 1919, 20–26) was published already in 1919; in the 1920s followed the five-volume *Werk des Untersuchungsausschusses* (Bell and Schücking 1927). The former work claims of the use of gas on April 22, 1915:

> *When updating the historical method of smoking out in a modern form with chlorine gas at Ypres, we neither used a more harmful material nor created a new means of combat.* The defining feature of our approach was simply that we brought to bear for the first time the *mass effect* of gas as a weapon, without which a military success in the field cannot be achieved with gas weapons. (Deutsches Kriegsministerium und Oberste Heeresleitung 1919, 23, italics original)

These lines contain the core of the German defense strategy in the debate on international law over the following years. The poison gas attack at Ypres was thus placed in historical and international contexts that were intended to relativize it. The specific claim of permissibility under international law resulted from additional arguments that the German authors presented to their readers with great care. Their conclusion defines the guilt and innocence of the parties to the war in a clear black-and-white schema:

> The factual and legal procedures in gas warfare are presented by applying the critical probe; they certainly justify the conduct of Germany, while to France falls the burden of having violated a global treaty. The reproaches made against us are thereby revealed to be part of the battle of lies by which the enemy league unceasingly strives to disparage us in the public opinion of the world. (Bell and Schücking 1927, 42)

Compared with that of the war years, this discourse on international law was intense and decidedly detailed. It is characterized by a nationalist polarization of standpoints. The attributions of guilt to the respective wartime enemy were expressed in the language of law and morality. The preferred politics of international law could often be derived simply from the nationality of the authors.

The Germans loudly and energetically defended themselves against what seemed to them a form of victor's justice and corresponding assessments of international law. They appealed to a number of counterarguments purporting to justify the use of poison gas. By contrast, the assertion that the use of poison gas was illegal could not be found in a single publication from Germany. The authors presented a whole arsenal of arguments for its legality under international law: The Hague Declaration of July 28, 1899, is said not to have applied (Kunz 1927, 20, 28; Meyer 1926, 296). The Hague Convention of 1907 was also said not to be applicable in World War I (Meyer 1926, 296–297). Alternatively, it was claimed that Article 23(a) was not relevant because gas is not a "poison" (Deutsches Kriegsministerium und Oberste Heeresleitung 1919, 24; Hanslian 1927, 5; Kunz 1927, 33; Meyer 1926, 298; Overweg 1937, 48–51). It was claimed that the Hague Declaration of 1899 did not apply to the "blue and yellow cross shells of the world war" or to artillery gas shells since "the spreading of poisonous gases is not their only purpose; rather, their main purpose was to render the enemy harmless" (Kunz 1927, 26; Strupp 1922, 201). Article 23(e) was also said not to be relevant because no "unnecessary suffering" was caused (Deutsches Kriegsministerium und Oberste Heeresleitung 1919, 24; Hanslian 1927, 5; Kunz 1927, 32; Meyer 1926, 298–299).

Furthermore, it was claimed in the alternative that the Hague Declaration of July 28, 1899, had been violated first by France; Germany merely followed suit and retaliated (Deutsches Kriegsministerium und Oberste Heeresleitung 1919, 23; Haber 1924, 83; Kunz 1927, 3–4; Meyer 1926, 301). It had been a case of a "state of emergency as recognized by international law" (Deutsches Kriegsministerium und Oberste Heeresleitung 1919). In general, it was claimed that the "Hague Accords had barely touched on the essence of gas warfare," since they had failed to recognize its humane essence (Meyer 1926, 302). The prohibitions of asphyxiating gas were said to have been annulled in the world war since both sides had made use of gas, claimed the international law scholar Josef Kohler in 1918; thus gas warfare had been legal (Kohler 1918, 212–213). The German international-law professors Julius Hatschek and Arthur von Kirchenheim claimed between the wars that the use of poison gas conformed to international law (Hatschek 1923, 316; Kirchenheim 1924, 405–406).

It is probably not oversimplifying too much to say that the standpoints of German jurists were primarily legitimizing, affirmative, defensive, and militaristic. One of the few exceptions was Mendelssohn-Bartholdy. He published a critical essay in 1927 with the revealing title "Kriegsnotwendigkeiten und Repressalien: Zwei Feinde des Völkerrechts" (The Necessities of War and Reprisals: Two Enemies of International Law) (Mendelssohn-Bartholdy 1927). In this text he expressed criticism of Haber and rejected the argument that the German use of poison gas had been a reprisal. He did, however, concede that the ambiguities of the rules had left many backdoors open.

The positions of academics in international law from Allied countries, by contrast, cannot be characterized so simply. There seems to have been a greater diversity of opinion and not so much thinking in camps. For example, there are French assessments in which poison gas as such is said to violate international law (but could be justified as a reprisal) (Rolin 1920, 326–327). The Allied use had been justified if and to the extent it had been reprisal (Hall 1924, 637, footnote 2; Garner 1920, 262, 271–292). The Archbishop of Canterbury and the Bishop of London (on May 16, 1915), by contrast, appealed to their government not to employ poison gas and descend to the level of the enemy (Garner 1920, 273, footnote 1). Articles 23(a) and 23(e) had been violated by the Germans to the extent that chlorine gas had been employed; by contrast, it was argued, other types of gas should be judged less harshly under international law (Lawrence 1923, 531).

Finally, there were positions not formulated by any of the countries that had been involved in the world war. It is presumably no coincidence that these positions tended to manifest principally antimilitary and pacifist features. In the polarized climate of politics and international law between the wars, it was all the more attractive for one side or the other to co-opt these to some extent neutral authors. Authors of such texts include Franz Carl Endres and Gertrud Woker. The latter published her book on behalf of the Women's International League for Peace and Freedom. Both authors emphasized not only that the German use of poison gas violated international law, but also the ongoing threat and the alarming role of a transnationally active armaments industry (Endres 1928, 38–39). Woker (1925, 16–19) frontally attacked Haber's argumentation, polemicizing against the "magnificent militaristic logic" she saw in the arguments that gas was effective and in conformity with international law.

4.3 Politicized Scholarship: No Mediation Possible

This controversy in scholarship on international law cannot be decided *ex post*, even if the arguments defending the German use of gas seem legally inconsistent and questionable to us today. To demonstrate this, a more detailed analysis would have to examine the arguments with an eye to contemporaneous theory about legal sources and about legal argumentation. Contrary to the later German objections that the Hague Treaty had not applied, for example, one need only point out the possibility that as positivizations of previous customary law they remained valid in the form of customary law as well; the fact that the Hague Treaty was reprinted in German World War I anthologies on military law also suggests it was believed to be valid (Kramer 2003, 281). Given the highly politicized climate and the intent to legitimize past actions, it was not to be expected the scholarly discourse would get closer to the subject matter, and in fact it did not. The camps remained hostile to each other, and there was no discernible mediation by third positions or even any softening of tone during the period between the wars. This thinking as part of a camp was particular stark in German scholarship on international law.

4.4 Reforms as Affirmation of the Prohibition of Poison in International Law

Thus the scholarly community in international law was not unified and did not come together on a view of past events. It was, however, better disposed to shaping such a view in the future. After the experiences of World War I, whether or not one judged certain past acts as in conformity with or as illegal under international law, all sides appeared agreed that poison gas should not be used in the future. Legal reforms and further standardization served to affirm the prohibition of poison under international law. Four stages should be identified here in conclusion.

4.4.1 Asymmetric New Paths: The Prohibitions of the Production and Possession of Weapons in the Paris Peace Treaties of 1919

First there were the Prohibitions of the Production and Possession of Weapons in the Treaties of Paris. The Treaty of Versailles set forth legal prohibition concerning the production and import for Germany:

> Article 171
> The use of asphyxiating, poisonous or other gases and all analogous liquids, materials or devices being prohibited, their manufacture and importation are strictly forbidden in Germany.
> The same applies to materials specially intended for the manufacture, storage and use of the said products or devices.
> The manufacture and the importation into Germany of armoured cars, tanks and all similar constructions suitable for use in war are also prohibited.
>
> Article 172
> Within a period of three months from the coming into force of the present Treaty, the German Government will disclose to the Governments of the Principal Allied and Associated Powers the nature and mode of manufacture of all explosives, toxic substances or other like chemical preparations used by them in the war or prepared by them for the purpose of being so used. (Jaschinski 1975, 61)

Largely identical rules were laid out in the four other Treaties of Paris for the other countries that had lost the war: Article 135 in the Treaty of St. Germain; Article 82 in the Treaty of Neuilly; Article 119 in the Treaty of Trianon; and Article 176 in the Treaty of Sèvres (Marauhn 1994, 61, notes 110–113).

On the one hand, this legal regulation presumed an existing prohibition of use (Jaschinski 1975, 61). On the other hand, it took new paths beyond the prohibition of use and created a preventative sphere for the first time (Marauhn 1994, 63). In relation to poison gas, therefore, it represented a legal norm for the prevention of war. Admittedly, it expressed an asymmetry of power: the prohibitions applied only to those countries that had lost World War I. There was no reciprocal application to the victorious powers.

4.4.2 Pacifist Efforts: Initiatives by the League of Nations

The League of Nations was also active in the context of its arms control efforts (Schücking and Wehberg 1924, 414–416; Overweg 1937, 77–81; Jaschinski 1975, 72). These efforts were an attempt in peacetime to make the use of chemical weapons de facto impossible or at least minimize the likelihood of their use in a future war (Marauhn 1994, 72). The bodies involved were the Council, the General Assembly, the Disarmament Commission, the Nonpermanent Arms Commission, the Commission on Intellectual Cooperation, and a subcommittee. The Red Cross also made a special submission to the General Assembly about preventive measures (Schücking and Wehberg 1924, 415). The various sides pursued the limitation or prohibition of manufacture and laboratory experiments. Issuing an "appel aux savants" to all countries to publish their pertinent inventions in order to make their use in war impossible (Freytagh-Loringhoven 1926, 118) was considered (but rejected). Expert opinions and reports were commissioned and committees established. It was decided to publish a report on the horrors of a future gas war; it was intended to be distributed to as broad an audience as possible (Freytagh-Loringhoven 1926, 118). These activities of the League of Nation overlapped with those of the Washington Conference of 1922. In the end the question was raised whether the members of League of Nations should be encouraged to join the Washington Treaty. Nothing more came of this (Meyer 1926, 304). For its part, the Washington Treaty never came into effect. Nevertheless, the Fifth Assembly of the League of Nations concluded a resolution in 1924 in which gas weapons were described as a threat to civilization in future wars (Overweg 1937, 80–81).

4.4.3 An Expression of the General Opinion of the Civilized World: The Washington Treaty of 1922

At the Conference on the Limitation of Armament in Washington, which had opened on November 12, 1921, a unanimous resolution was concluded on February 6, 1922:

> The use in war of asphyxiating, poisonous or other gases, and all analogous liquids, materials or devices, having been justly condemned by the general opinion of the civilised world and a prohibition of such use having been declared in Treaties to which a majority of the civilized Powers are parties, the Signatory Powers, to the end that this prohibition shall be universally accepted as a part of international law binding alike the conscience and practice of nations, declare their assent to such prohibition, agree to be bound thereby between themselves and invite all other civilized nations to adhere thereto. (Lawrence 1923, 532)

The signatories were the United States, the British Empire, France, Italy, and Japan.

It was not just the wording that was controversial (Jaschinski 1975, 65; Overweg 1937, 74–75). It was criticized for not distinguishing between asphyxiating and harmful agents. The Washington Treaty was not ratified by the participating countries, however, and therefore never came into effect.

4.4.4 Reassuring One's Principles: The Geneva Protocol on Poison Gas of 1925

Finally, the Geneva Protocol of 1925 on poison gas provided a new positivized norm. Once again it was assumed that a prohibition already existed (Jaschinski 1975, 67; Marauhn 1994, 49). This was now affirmed in the Geneva Protocol but its scope was not significantly changed (Marauhn 1994, 49). On June 17, 1925, the Protocol was concluded with the following wording, closely following that of the Washington Convention:

> The Undersigned Plenipotentiaries, in the name of their respective Governments:
> Whereas the use in war of asphyxiating, poisonous or other gases, and of all analogous liquids, materials or devices, has been justly condemned by the general opinion of the civilized world; and
> Whereas the prohibition of such use has been declared in Treaties to which the majority of Powers of the world are Parties; and
> To the end that this prohibition shall be universally accepted as a part of International Law, binding alike the conscience and the practice of nations;
>
> Declare:
> That the High Contracting Parties, so far as they are not already Parties to Treaties prohibiting such use, accept this prohibition, agree to extend this prohibition to the use of bacteriological methods of warfare and agree to be bound between themselves to the terms of this declaration. (Marauhn 1994, 49)

The general prohibition was in that sense a kind of self-reassurance and affirmation of the content of treaties and of customary law that had been valid prior to World War I but had often been breached in praxis between states during the war. Today the Geneva Protocol of 1925 is still valid and has been judged "probably the most significant special standard thus far prohibiting the use of chemical weapons" (Marauhn 1994, 47). In 1997, the "Convention on the Prohibition of the Development, Production, Stockpiling and Use of Chemical Weapons and on their Destruction" enhanced the prohibition of chemical and gas weapons and enforced the elimination of such weapons. Until today (February 2017), 192 states have signed the treaty.

5 Summary: Expectations Regarding International Law

The history of the struggle over international legal rules on the prohibition of the use of chemical weapons is thus far from over. On the contrary, countless violations of legal norms are etched into the history of the twentieth century. Time and again, new inventions have posed challenges to the prohibition as well. Nonetheless, the Geneva Protocol on Poison Gas seems a potential terminus for the narratives about the events of Ypres in 1915.

The prohibition of perfidious means and unnecessary suffering was violated many times in April 1915 and up until the end of the war in 1918, in the face of express agreements and despite the ongoing validity of customary international law. Even if one were to regard the detailed special prohibitions under international law as somehow inapplicable, there would remain the spirit of the norms as a further point of reference. After all, alongside the specific prohibitions there were also older general legal principles proscribing perfidious means and unnecessary suffering under international law deriving from the principle of humanity. That general principles of international law were disregarded to such an extent in the First World War relates to the particular moral values expressed in these principles. By this expression, legal principles can endow values with more vigor and prompt surprising decisions, but are also particularly fragile. It is no coincidence that certain legal disciplines in which the norm setting is not yet complete tend towards legal principles. The international law of the nineteenth century was one such area. Alongside the juridification of international relations, it relied heavily on general legal principles as well as on natural law (Vec 2012, 2017). The conflict surrounding poison gas in the law of war shows how fragile certain rules can become in practice. Under the conditions of the world war, under pressure from blazing nationalism and militarism, international law is drawn into the undertow of politicization, an undertow that weakened it so severely that one is inclined to speak of its almost total absence from certain areas of regulation in these years. Certain actors, including both military figures and international lawyers, practiced the circumvention or even denial of the law of war. Whether one is disappointed by this depends—as we have said—on the expectations of a contemporary reader contemplating the historical events of earlier epochs. The law of war in World War I was, especially when compared to other matters, a highly political area of regulation. In any event, the consideration of this time should always take into account the political circumstances and the larger context when it accentuates the *Realpolitik* and the relativism of this international law (Orakhelashvili 2011, 454–455).

References

Bell, Johannes, and Walther Schücking. 1927. *Das Werk des Untersuchungsausschusses der Deutschen Verfassungsgebenden Nationalversammlung und des Deutschen Reichstages 1919–1928, Völkerrecht im Weltkrieg*. Vol. 4. Berlin: Deutsche Verlagsgesellschaft für Politik und Geschichte 1927.

Bischof, Daniel. 2015. *Reaktionen und öffentliche Debatten zum Gaskrieg in vergleichender Perspektive (Deutsches Reich–Österreich-Ungarn)*. Unpublished seminarpaper, University of Vienna.

Bonfils, Henry. 1908. *Manuel de droit international public (Droit des gens). Destiné aux étudiants des Facultés de Droit et aux aspirants aux fonctions diplomatiques et consulaires*. 5th edition by Paul Fauchille. Paris: Arthur Rousseau.

Bothe, Michael. 1973. *Das völkerrechtliche Verbot des Einsatzes chemischer und bakteriologischer Waffen. Kritische Würdigung und Dokumentation der Rechtsgrundlagen*. Köln: Carl Heymanns.

Carnegie Endowment for International Peace. Division of International Law. 1915a. *Pamphlet No. 5. The Hague Conventions of 1899 (II) and 1907 (IV) respecting the laws and customs of war on lands. Washington, D. C.: Carnegie Endowment.*

Carnegie Endowment for International Peace, Division of International Law. 1915b. *Pamphlet No. 8. The Hague declaration (IV,2) of 1899 concerning asphyxiating gases*. Washington, D. C.: Carnegie Endowment. https://ihl-databases.icrc.org/applic/ihl/ihl.nsf/Article.xsp?action=openDocument&documentId=2531E92D282B5436C12563CD00516149. Accessed 14 Dec 2016.

Cassar, George H. 2014. *Trial by gas. The British Army at the Second Battle of Ypres*. Washington: Potomac books.

Clunet, Edouard. 1915. La Guerre Allemande par la Combustion, l'Asphyxie et l'Empoisonnemet de l'Adversaire, Extrait revue et augmenté du journal l'Information du 3 et du 8 mai 1915.

Convention (IV) respecting the Laws and Customs of War on Land and its annex: Regulations concerning the Laws and Customs of War on Land. 18 Oct 1907. https://ihl-databases.icrc.org/applic/ihl/ihl.nsf/ART/195-200033?OpenDocument. Accessed 14 Dec 2016.

Convention on the Prohibition of the Development, Production, Stockpiling and Use of Chemical Weapons and on their Destruction. 29 April 1997. https://www.opcw.org/chemical-weapons-convention/articles/. Accessed 31 Jan 2017.

Convention with Respect to the Laws and Customs of War on Land (HAGUE, II). 29 July 1899. http://avalon.law.yale.edu/19th_century/hague02.asp. Accessed: 14 Dec 2016.

Cybichowski, Sigmund. 1912. Notwendigkeit und Völkerrecht. In *Studien zum internationalen Recht*, ed. Sigmund Cybichowski, 21–71. Berlin: Vahlen.

Dampierre, Jacques Marquis de. 1917. *German imperialism and International Law, based upon German authorities and the archives of the French government, by Jacques Marquis de Dampierre, archiviste-paleographe, with illustrations, charts, and facsimiles*. London: Constable & Company Ltd.

Declaration Renouncing the Use, in Time of War, of Explosive Projectiles Under 400 Grammes Weight. Saint Petersburg, 29 November/11 December 1868. https://ihl-databases.icrc.org/applic/ihl/ihl.nsf/Article.xsp?action=openDocument&documentId=568842C2B90F4A29C12563CD0051547C. Accessed 14 Dec 2016.

Despagnet, Frantz. 1905. *Cours de Droit international public*. 3rd edition. Paris: L. Larose & L. Tenin.

Dülffer, Jost. 1978. *Regeln gegen den Krieg? Die Haager Friedenskonferenzen von 1899 und 1907 in der internationalen Politik*. Frankfurt am Main: Ullstein Verlag GmbH.

Edmonds, J.E., and L[assa] Oppenheim. ca.1913. *Land Warfare. An exposition of the laws and usages of war on land for the guidance of officers of His Majesty's Army*. London: His Majesty's Stationary Office.

Encke, Julia. 2006. *Augenblicke der Gefahr*. Wilhelm Fink: Der Krieg und die Sinne. München.

Encke, Julia. 2015. Die gemeine Waffe. Interview mit Herfried Münkler. *Frankfurter Allgemeine Sonntagszeitung*, April 12: 41.

Endres, Franz Carl. 1928. *Giftgaskrieg die grosse Gefahr*. Zürich: Rascher & Cie.

Ewing, Russell H. 1927. The legality of chemical warfare. *American Law Review* 61: 58–76.

Eysinga, Jonkheer Willem Jan Mari van. 1928. La guerre chimique et le movement pour sa repression, Academie de Droit International. Rec. des Cours 1927, Bd.16, Paris 1928, 329–381.

Freytagh-Loringhoven, Axel. 1926. *Die Satzung des Völkerbundes. Mit Einleitung und Erläuterungen*. Berlin: Stilke.

Garner, James Wilford. 1920. *International Law and the World War*, vol. I. London: Longmans, Green und Co.

German General Staff. 1914. Kriegsgebrauch—The Customs of War. In *Britain as Germany's vassal*. Together with Kriegsbrauch–The Customs of War. Published by the German General Staff, Berlin, 1902 and Extracts from Regulations adopted by the Hague Conference 1907 and subscribed to by Germany, ed. by Bernhardi, Friedrich v., trans. by Barker, J. Ellis. London: Wm. Wawson & Sons Ltd.

Grand État-Major Allemand. 1916. *Les lois de la guerre continentale (Kriegsbrauch im Landkriege)*, ed. and trans. by Carpentier, Paul. Paris: Librairie Payot & Cie.
Grande Stato Maggiore Germanico. 1915. *Le Leggi della guerra (Kriegsbrauch im Landkriege)*, trans. by Bianchi, Icilio. Milano: Rava & C.-Editori.
Große Kracht, Klaus. 2004. Kriegsschuldfrage und zeithistorische Forschung in Deutschland. Historiographische Nachwirkungen des Ersten Weltkriegs, *Zeitgeschichte-online*. http://www.zeitgeschichte-online.de/thema/kriegsschuldfrage-und-zeithistorische-forschung-deutschland. Accessed 15 Dec 2016.
Großer Generalstab (ed.). 1902. *Kriegsbrauch im Landkriege*. Kriegsgeschichtliche Einzelschriften: 31. Berlin: Ernst Siegfried Mittler und Sohn.
Haber, Fritz. 1924. Zur Geschichte des Gaskrieges. Vortrag, gehalten vor dem parlamentarischen Untersuchungsausschuß des Deutschen Reichstages am 1. Oktober 1923. In *Fünf Vorträge aus den Jahren 1920–1923*, ed. Fritz Haber, 76–92. Berlin: J. Springer.
Hall, William Edward. 1917. *A Treatise on International Law*. 7th edition by A[lexander] Pearce Higgins. Oxford: Clarendon Press.
Hall, William Edward. 1924. *A Treatise on International Law*. 8th edition by A[lexander] Pearce Higgins. Oxford: Clarendon Press.
Hanslian, Rudolf. 1927. *Der chemische Krieg*, 2nd ed. Berlin: Mittler und Sohn.
Hanslian, Rudolf. 1934. *Gaskrieg! Der deutsche Gasangriff bei Ypern am 22. April 1915, eine kriegsgeschichtliche Studie aus dem Jahr 1934*, ed. Tümmler, Holger, 2009 Berlin: Verlag Gasschutz und Luftschutz.
Hatschek, Julius. 1923. *Völkerrecht als System rechtlich bedeutsamer Staatsakte*. Leipzig: A. Deichert.
Herzog, Amelie. 2015. *Die Darstellung des Giftgaskrieges anhand zeitgenössischer österreichisch-ungarischer Tageszeitungen*. Unpublished seminarpaper, University of Vienna.
Huber, Max. 1913. Die kriegsrechtlichen Verträge und die Kriegsraison. *Zeitschrift für Völkerrecht* 7: 351–374.
Hueck, Ingo J. 1999. Die Gründung völkerrechtlicher Zeitschriften in Deutschland im internationalen Vergleich. In *Juristische Zeitschriften. Die neuen Medien des 18.–20. Jahrhunderts*, ed. Michael Stolleis, 379–420. Frankfurt am Main: Klostermann.
Hull, Isabel V. 2005. *Absolute Destruction*. Military Culture and the Practices of War in Imperial Germany. Ithaca: Cornell University Press.
Hull, Isabel V. 2014. *A Scrap of Paper. Breaking and making International Law during the Great War*. Ithaca: Cornell University Press.
Jacomet, Robert. 1913. *Les lois de la guerre continentale. Préface de M. Louis Renault*, 3rd ed. Paris: L. Fournier A. Pedone.
Jaschinski, Herbert. 1975. *Neuartige chemische Kampfstoffe im Blickfeld des Völkerrechts. Der Einsatz nicht tödlich wirkender sowie Pflanzen schädigender chemischer Kampstoffe in bewaffneten Konflikten und das Völkerrecht. Ein Beitrag zur Auslegung und Ermittlung kriegsrechtlicher Normen*. Berlin: Duncker & Humblot 1975.
Kassapis, Georg. 1986. *C-Waffen: der völkerrechtliche Hintergrund der Genfer Verhandlungen über ihre Eliminierung*. München: Florentz.
Keene, Edward. 2012. The treaty-making revolution of the nineteenth century. *The International History Review* 34: 475–500.
Koellreutter, Otto. 1917/1918. Kriegsziel und Völkerrecht. Betrachtungen aus der Front. Abhandlungen. *Zeitschrift für Völkerrecht* 10: 493–503.
Kohler, Josef. 1918. *Grundlagen des Völkerrechts. Vergangenheit, Gegenwart, Zukunft*. Stuttgart: Verlag von Ferdinand Enke.
Korovine, E.A. 1929. La guerre chimique et le droit international. *Revue générale de Droit International Public* 36: 646–668.
Koskenniemi, Martti. 2011. Between Coordination and Constitution: International Law as German Discipline. In *Redescriptions*. Yearbook of Political Thought, Conceptual History and Feminist Theory: 15, ed. K. Palonen, 45–70. Münster: Lit Verlag.

Kramer, Alan. 2003. Kriegsrecht und Kriegsverbrechen. In *Enzyklopädie des Ersten Weltkriegs*, ed. Gerhard Hirschfeld, Gerd Krumeich, and Irina Renz, 281–292. Paderborn: Schöningh.

Kraus, Karl. 1922. *Die letzten Tage der Menschheit*. Vienna: Verlag 'Die Fackel'.

Kraus, Karl. 2015. *The last days of mankind*. trans: Fred Bridgham, and Edward Timms. New Haven: Yale University Press.

Kriegsministerium, Deutsches, and Oberste Heeresleitung (eds.). 1919. *Die deutsche Kriegführung und das Völkerrecht. Beiträge zur Schuldfrage*. Berlin: Ernst Siegfried Mittler und Sohn.

Kunz, Josef L[aurenz]. 1927. *Gaskrieg und Völkerrecht* (Erweiterter Sonderabdruck aus "Zeitschrift für öffentliches Recht," Band VI, Heft 1), Wien: Springer.

Kunz, Josef L[aurenz]. 1935. *Kriegsrecht und Neutralitätsrecht*. Berlin: Springer.

Kuß, Susanne. 2010. *Deutsches Militär auf kolonialen Kriegsschauplätzen. Eskalation von Gewalt zu Beginn des 20. Jahrhunderts*. Berlin: Ch. Links Verlag.

Lawrence, T[homas] J[oseph]. 1923. *The principles of international law*, 7th edition by Percy H. Winfield. Boston: D.C. Heath & Co.

Lentner, Ferdinand. 1880. *Das Recht im Kriege. Kompendium des Völkerrechtes im Kriegsfalle. Dargestellt auf Grund der Brüsseler Deklaration vom 27. August 1874 über die Kriegssatzungen und Kriegsgebräuche*. Wien: L. W. Seidel und Sohn.

Lepick, O. 1998. *La Grande Guerre Chimique 1914–1918*, 2nd ed. Paris: Presses Univ. de France.

Lovrić-Pernak. Kristina. 2013. *Morale internationale und humanité im Völkerrecht des späten 19. Jahrhunderts: Bedeutung und Funktion in Staatenpraxis und Wissenschaft*. Baden-Baden: Nomos.

Marauhn, Thilo. 1994. *Der deutsche Chemiewaffen-Verzicht. Rechtsentwicklungen seit 1945*. Heidelberg: Springer.

Martinetz, Dieter. 1996. *Der Gaskrieg 1914–1918. Entwicklung, Herstellung und Einsatz chemischer Kampfstoffe. Das Zusammenwirken von militärischer Führung, Wissenschaft und Industrie*. Bonn: Bernard & Graefe.

Mendelssohn-Bartholdy, A[lbrecht]. 1927. Kriegsnotwendigkeiten und Repressalien: Zwei Feinde des Völkerrechts. *Europäische Gespräche* 6/1927, 319–328.

Mérignhac, Alexandre G. 1900. *La conférence internationale de la paix. Étude historique, exégétique et critique des travaux et des résolutions de la conférence de la Haye de 1899 par A. Mérignhac professeur de droit enternational a la faculté de droit de l'Université de Toulouse adjoint a l'intendance du cadre auxiliaire avec une préface de M. Léon Bourgeois premier délége de la France a la conference de la paix*. Paris: Librairie Nouvelle de Droit et de Jurisprudence Arthur Rousseau.

Mérignhac, Alexandre G. 1912. *Traité de Droit Public International*, Vol. 3: *Le Droit de la Guerre*. Paris: Librairie Générale de Droit et de Jurisprudence.

Messerschmidt, Manfred. 1983. Völkerrecht und „Kriegsnotwendigkeit" in der deutschen militärischen Tradition seit den Einigungskriegen. *German Studies Review* 6(2): 237–269.

Meyer, Julius. 1926. *Der Gaskampf und die chemischen Kampfstoffe*, 2nd ed. Leipzig: S. Hirzel.

Neff, Stephen C. 2014. *Justice among nations. A history of International Law*. Cambridge: Harvard University Press.

Niedner, Johannes. 1915. *Der Krieg und das Völkerrecht, Vortrag gehalten in der Staatswissenschaftlichen Gesellschaft zu Jena*. Jena: Verlag von Gustav Fischer.

Nippold, Otfried. 1911. *Die zweite Haager Friedenskonferenz*. 2. Teil. *Das Kriegsrecht, unter Mitberücksichtigung der Londoner Seerechtskonferenz*. Leipzig: Duncker & Humblot.

Nys, Ernest. 1912. *Le droit international. Les principes, les theories, les faits*. Vol. 3. 2nd edition. Paris: Librairie des sciences politiques et sociales Marcel Rivière.

Orakhelashvili, Alexander. 2011. The 19th-century life of international law. In *Research Handbook on the Theory and History of International Law*, ed. Alexander Orakhelashvili, 441–455. Cheltenham: Edward Elgar Publishers.

Overweg, Adolf-Boelling. 1937. *Die chemische Waffe und das Völkerrecht. Eine rechtshistorische und rechtskritische Studie*. Berlin: E.S. Mittler & Sohn.

Palazzo, Albert. 2000. *Seeking victory on the western front. The British Army and Chemical Warfare in World War I*. Lincoln: University of Nebraska Press.

Papers relating to the Foreign Relations of the United States. 1933. 1918, *Supplement II. The World War*. Washington: US Gov.Print.Off.
Partridge, Frank C. 1917. *The Future of International Law. An Address delivered at Middlebury College, Charter Day, November 1, 1917. On the Occasion of the 117th Anniversary of the Granting of the Charter of the College.*
Philippson, Coleman. 1915. *International Law and the Great War. Introduction by John MacDonald*. London: T. Fisher Unwin.
Pillet, Antoine. 1918. *Les conventions de La Haye du 29 juillet 1899 et du 18 octobre 1907. Étude juridique et critique*. Paris: A. Pedone.
Pohl, Heinrich. 1915. *Deutsches Landkriegsrecht. Quellensammlung mit Sachregister*. Berlin: Carl Heymann.
Project of an International Declaration concerning the Laws and Customs of War, Brussels, 27 August 1874, Means of Injuring the Enemy. https://ihl-databases.icrc.org/ihl/INTRO/135. Accessed 14 Dec 2016.
Rivier, Alphonse. 1896. *Principes du Droit des Gens*, vol. II. Paris: Arthur Rousseau.
Roberts, A[lfred] A. 1915. *The Poison War*. London: William Heinemann.
Rolin, Albéric. 1920. *Le Droit moderne de la Guerre. Les Principes—Les Conventions—Les Usages et Les Abus*. Bruxelles: Albert Dewit.
Schücking, Walther, and Hans Wehberg. 1924. *Die Satzung des Völkerbundes, kommentiert*, 2nd ed. Berlin: Franz Vahlen.
Simon, Hendrik. 2016. Das Recht des Krieges. *Rechtsgeschichte - Legal History* 24: 508–510.
Spitra, Sebastian. 2015. *Rechtfertigungsnarrative & Gaskrieg. Eine theoretische Betrachtung des Begriffs ‚Rechtfertigungsnarrativ' & eine praktische Untersuchung seiner Anwendung beim Gaskrieg bei Ypern 1915*. Unpublished seminarpaper, University of Vienna.
Stockholm International Peace Research Institute. 1973. *The problem of chemical and biological warfare. A study of the historical, technical, military, legal and political aspects of C.B.W., and possible measures*. Vol. III. C.B.W. and the Law of War. Stockholm: Almqvist & Wiksell.
Strupp, Karl. 1914. *Das internationale Landkriegsrecht*. Frankfurt am Main: J. Baer & Co.
Strupp, Karl. 1922. *Grundzüge des positiven Völkerrechts*, 2nd ed. Bonn: Ludwig Röhrscheid.
Tettenborn, Albrecht. 1909. *Prinzipien und Richtungen der Kriegsmittelverbote des Landkrieges*. Dissertation, University of Würzburg.
Toppe, Andreas. 2008. *Militär und Kriegsvölkerrecht. Rechtsnorm, Fachdiskurs und Kriegspraxis in Deutschland 1899–1940*. München: R. Oldenbourg Verlag.
Vec, Miloš. 2006. *Recht und Normierung in der Industriellen Revolution. Neue Strukturen der Normsetzung in Völkerrecht, staatlicher Gesetzgebung und gesellschaftlicher Selbstnormierung*. Frankfurt am Main: V. Klostermann.
Vec, Miloš. 2012. Principles in 19th century International Law doctrine. In *Constructing International law—The birth of a discipline*, ed. Luigi Nuzzo, and Miloš Vec, 209–227. Frankfurt am Main: Klostermann.
Vec, Miloš. 2017. Sources in the 19th Century European Tradition. The Myth of Positivism. In *Oxford handbook on the sources of International Law*, ed. Samantha Besson and Jean d'Aspremont, Seiten. Oxford: Oxford University Press. [in preparation for 2017].
von Kirchenheim, Arthur. 1924. Geschosse. In *Wörterbuch des Völkerrechts und der Diplomatie, Erster Band*, ed. Karl Strupp, 403–406. Berlin und Leipzig: W. de Gruyter.
Wehberg, Hans. 1910. *Die Abkommen der Haager Friedenskonferenzen, der Londoner Seekriegskonferenz nebst Genfer Konvention. Mit einem Vorwort von Prof. Dr. Zorn*. Berlin: Guttentag.
Weiler, Inge. 1998. *Giftmordwissen und Giftmörderinnen. Eine diskursgeschichtliche Studie*. Tübingen: Max Niemeyer Verlag.
Westlake, John 1907. *International Law*. Part II: *War*. Cambridge: University Press.
Woker, Gertrud. 1925. *Der kommende Giftgaskrieg*. Leipzig: E. Oldenburg.

Zala, Sacha. 2001. *Geschichte unter der Schere politischer Zensur. Amtliche Aktensammlungen im internationalen Vergleich*. München: R. Oldenbourg.
Zecha, Wolfgang. 2000. *"Unter die Masken!" Giftgas auf den Kriegsschauplätzen Österreich-Ungarns im Ersten Weltkrieg*. Wien: ÖBV&hpt.
Zorn, Albert. 1902. *Kriegsmittel und Kriegführung im Landkriege nach den Bestimmungen der Haager Conferenz*. Königsberg in Preußen: Ostpreuss. Druckerei u. Verlagsanst.

Military-Industrial Interactions in the Development of Chemical Warfare, 1914–1918: Comparing National Cases Within the Technological System of the Great War

Jeffrey Allan Johnson

Abstract This chapter examines the development of chemical warfare on the Western Front in the context of the large-scale technological systems developed by each of the major powers—Germany, France, Britain, and later the United States—in order to coordinate their industrial, academic, and military resources. As chemical warfare intensified from the tentative, small-scale experiments of 1914–1915 to the massive bombardments of 1918, it also changed qualitatively. Each side's innovations forced similar responses from their opponents, in an escalating arms race in which military exigencies increasingly overrode ethical concerns while tending to institutionalize chemical warfare. This process exemplified the war's increasingly "total" nature as a technological meta-system integrating the fighting fronts and home fronts on each side and across the lines. On the verge of permanently institutionalizing chemical warfare and militarizing its supporting industries, the process abruptly ended as the German system collapsed. But by then the war had transformed the image of chemical science and technology from a progressive force to one associated with the horrors of war.

1 Introduction

Although there is a great deal of literature on the role of scientists, especially Fritz Haber, in the development of chemical warfare during the First World War, relatively little has appeared on the role of the chemical industries and their collaboration with the military authorities. Chemical warfare was admittedly only a part, in some ways a relatively insignificant part, of the wartime activities of academic and industrial chemists. Yet in the popular mind it rightly looms much larger. For Germany's introduction of poisonous gas to the battlefield clearly violated the spirit

J.A. Johnson (✉)
Villanova University, Villanova, USA
e-mail: jeffrey.johnson@villanova.edu

B. Friedrich et al. (eds.), *One Hundred Years of Chemical Warfare: Research, Deployment, Consequences*, DOI 10.1007/978-3-319-51664-6_8

of the 1899 and 1907 Hague Conventions, thereby introducing an era of literally unconventional warfare. Wilfred Owen's well-known poem "Dulce et Decorum Est" created an unforgettable image of the horrifying effect of gas on an unprotected soldier, which was also used to great effect in the commemorative ceremonial concert concluding the symposium at which the present chapter was presented. Images such as these highlight the role of chemical warfare in making the Great War increasing "total," in the process blackening the image of chemistry, especially the German variety. This chapter examines on this transformation after examining the development of interactions between the chemical industry and the military (as well as the interactions of both of these with academically-trained experts) in Germany, France, Britain, and later the United States, as a special case of the broader technological meta-system created by these opposing national systems on the Western Front from 1914 to 1918.

2 The Western Front as a Technological Meta-System

The Western Front can be viewed as a large technological system, or rather meta-system composed of several interacting national systems, within which military, industrial, and academic subsystems interacted in various ways. The idea of studying the chemical industry as a large technological system in a wartime setting goes back to Thomas Parke Hughes. Hughes conceptualized the development of high-pressure hydrogenation processes during and after the First World War as a case of "technological momentum," whereby a large-scale technological system tends to grow by maximizing existing productive capacities (when necessary, by adapting them to new uses) and by applying the experience of its scientists and engineers with previously successful approaches to the development of new products and productive capacities (Hughes 1969, 111–112). This pattern of growth is thus creative but also conservative, making fundamental changes in direction only when influenced by external forces. In a later full-length study of the electrical industry 1880–1930, Hughes developed valuable conceptual insights and a comparative regional approach, but with little attention to the First World War, whose impact on the electrical industry was far less than on the chemical industry. Hughes identified three sorts of technological systems: a purely technical, a technical-institutional, and a more "loosely structured system" whose components interconnect and interact, but which is "neither centrally-controlled nor directed toward a clearly defined goal" (Hughes 1983, 6). The present chapter examines this latter type as a meta-system and examines the development of chemical warfare as part of the larger meta-system of the Western Front—in which three and ultimately four large-scale systems operated and interacted in response to each other's initiatives: that of the Germans on the east side of No Man's Land, and those of the Allies to the west—the French, British, and finally the Americans—which cooperated increasingly well yet never grew into a seamlessly operating single system.

The growth and interactions of these systems thus shaped chemical warfare during the First World War.

From a military-economic perspective, the war was an extension of prewar competition among technological systems originating during the Second Industrial Revolution, in which since the 1860s large firms had utilized systematic innovation by teams of academically trained chemists, physicists, and engineers to carve out oligopolistic positions in the world market. With the advent of war, the opposing systems carried on—with apologies to Clausewitz—a kind of "economic competition by other means." That is, the war refashioned the process of systematic technological innovation, shifting it to military settings, whereby the oligopolistic opponents now stood on the other side of No Man's Land, and it was on the front rather in the marketplace that product testing took place. Depending upon results in the "battlefield marketplace," each side might expand its production, modify its product, or imitate or improve upon competing products of the opposition. Success could thus depend upon the ability of each system to function effectively as an innovative system, potentially on a very large scale, in a manner not very different from the process of peacetime competition, albeit without regard to questions of intellectual property at least for the duration of the war.

From a systems perspective, the war on the Western Front was actually two successive wars, which one could call the "Great War" and the "Total War" (Chickering and Förster 2000). Each of these involved a type of mobilization; the first, beginning on both sides in August 1914, was limited and based mainly on prewar structures and capacities, in proportion to the degree of technological momentum in the peacetime systems. New problems and constraints posed in particular by the advent of static trench warfare at the end of 1914 created growing pressures that led to a "second mobilization" in each nation, representing a much more "total" and more innovative utilization of their resources and marking the advent of a wartime system with its own technological momentum. The timing of these second mobilizations depended in part upon a variety of unanticipated developments such as the continuing German occupation of a significant part of the industrial region in northwest France, as well as the inadequacy of prewar tactics and equipment for achieving breakthroughs. Thus by the spring of 1915, pressures toward a second mobilization already existed on the Allied side, marked in the British case by the creation of the Ministry of Munitions in May 1915, which took over from the War Office the coordination of national production, began systematically to mobilize scientific and technical expertise, and notably departed from the British tradition of free enterprise by supervising the construction of a series of National Factories for munitions production, initially intended only to cover wartime needs (Simmonds 2012, 67–96). The French, reeling from the loss of industrial capacity to the German invasion, had already begun the task of remobilizing their economy for war in the fall of 1914, but a major transition also came in May 1915 with the appointment of Albert Thomas as Under-Secretary of State for Artillery and Munitions (Hardach 1992). It can hardly be coincidental that these major innovations as well as several others occurred shortly after the first German attacks with chlorine gas at Ypres on April 22, which represented a move toward total war in two senses, not only toward unconventional warfare but also a greater mobilization of the chemical industry, which was now beginning to discover the possibility of "dual-use" chemicals. Because this initially entailed only the adaptation of existing capacities and products to wartime uses, it did not yet represent a full second mobilization. That finally came with the Hindenburg Program of September 1916, in response to the British-French offensive on the Somme in summer 1916, which had been made possible by the Allied innovations since spring 1915 (Herwig 1997, 259–266).

The advent and development of chemical warfare could be said to constitute a special case of this broader systemic interaction and the "totalizing" process it produced, which began to have a major impact on the conduct of the war in 1916 and was still gaining momentum on the Allied side—especially with the addition of the American system and its thorough-going mobilization beginning in 1917 (Steen 2014, 75–112)—when the German system collapsed in November 1918, in part because its limited resources could not sustain a total mobilization (Herwig 1997, 440–450).

3 Chemical Weapons as an Illustrative Case

Chemistry in 1914 was already a highly industrialized science, marked by well-developed, institutionalized academic-industrial relations. Primarily because of the development of high explosives based on organic compounds during the late nineteenth century, there also existed elements of a prewar military-industrial symbiosis, albeit on a relatively small scale. All of the major countries had relatively small testing facilities in their arsenals, and all had contracts with civilian companies to produce munitions and other items that could not be produced in the arsenals themselves. Far more significant would be the prewar academic-industrial relationships that had emerged outside the military system. These relationships could to some extent be carried over, or at least serve as a model for the developing military-industrial and academic system during the war. Thus a potentially decisive military advantage for the Germans, albeit unrecognized before the war, was the highly innovative academic-industrial symbiosis developed by their coal-tar dye industry (Johnson 2000, 15–23). A half-dozen large, research-intensive firms, organized in two oligopolistic alliances, had obtained a quasi-monopoly amounting to almost ninety percent of world synthetic dye production. Nearly all of the dye factories or sales outlets in Britain, France, Russia, and the United States were actually German, using German chemists and mostly German-made chemicals for key processes. With the outbreak of war, the Germans would find themselves with a "chemical weapon"—thousands of research-trained, technically-experienced industrial chemists—which the Allied system would find it very difficult to match until the Americans began systematically mobilizing chemists for war work in 1918 (MacLeod 1998; Steen 2014, 96–97). Following the logic of technological momentum, most wartime innovations redirected known technologies in novel ways. This however gave a decided advantage to systems including large, well-established firms with longstanding traditions of technical expertise and good connections to academic institutions—precisely the characteristics of the German dye industry, which would thus find itself especially suited for the chemical war.

Moreover, by August 1914 the concept of "dual use," common today in discussions of chemical warfare, was already inherent in the nature of the chemical industry, especially in regard to synthetic organic chemicals. It was easy to modify chemical production processes so that with slight variations in raw materials, reagents, intermediates, and operating conditions, one could produce a wide variety of different final products for a wide variety of purposes, some of which could be military. At the outset of the war there were already three categories of products that included examples of what one might characterize, using a type of later American military jargon, as the "three D's" of *unplanned* dual use: disinfectants (chlorine), dyes (phosgene), and drugs (arsenicals) (cf. Haber 1986, 15–16, 21, 159). Chlorine had long been used for disinfecting municipal water supplies, among other things; phosgene (a deadly compound of carbon monoxide and chlorine) was an intermediate in the coal-tar dye industry, used for producing several different dyes; and most recently the dye corporation Farbwerke vorm. Meister Lucius and Brüning—Höchst (henceforth, Höchst) had begun to market organic arsenical compounds (developed in collaboration with the 1908 Nobel laureate for medicine Paul Ehrlich) as the first effective drugs for treating syphilis. Dual use in these cases was unplanned, because none of these products had originally been intended for military purposes. But the experience and expertise gained from systematic innovation in these fields—especially dyes and drugs, for which the largest firms together had about a thousand chemists in 1914, synthesizing and testing thousands of potential products—could easily be redirected. The Farbenfabriken vorm. Bayer-Leverkusen (henceforth: Bayer) and Höchst in particular had been working with synthetic drugs for decades, and in the process they had developed medical testing facilities and collaborative relationships with physicians to test the physiological effects of their compounds. In the pharmaceutical industry as such, there were also several larger firms such as Merck-Darmstadt that had developed similar combinations of chemical and medical expertise (Baumann 2011, 36–194). Thus the basic structure of the system was already in place in 1914, especially in the German context. The Allies' chemical industries, with less diverse product assortments (especially for organic chemicals) and less intensive processes of innovation, were less suitable for adaptation to chemical warfare; the Allies would thus require more fundamental changes in their prewar industrial systems. Nevertheless even on the Allied side there were possibilities for dual use; for example, both the British and French explosives industries produced picric acid as a high explosive; combined with chlorine, this would produce chloropicrin, which could be used as a chemical agent. Moreover, Allied producers of chlorine, for example for bleach, could also (in principle) fairly easily produce phosgene gas from chlorine and carbon monoxide, which required no organic-chemical expertise. In practice, however, inexperience and incompetence led to delays and inefficiencies, especially on the British side (Haber 1986, 83–86, 162–163).

4 Industrial Mobilization for Chemical Warfare: The Experimental Phase, 1914–15

At the outset of what was widely expected to be a short war, the German dye firms by and large did not expect to supply the military with much besides dyes for uniforms. They did produce some nitrates and nitrated products for dye manufacture, selling their surplus to the explosives industry (nitrotoluene and dinitrotoluene could be used to produce the high explosive TNT or trinitrotoluene), but they lacked the safeguards in their plants that were required by insurance regulations for producing the actual explosives. For this reason the leading German dye companies in August 1914 rejected appeals by the Prussian War Ministry to produce explosives, though they did however agree to produce nitrates as raw materials for explosives. As with the toluene products, these were not explosives as such but products for the explosives industry, however, and thus not fundamentally different from what the dye companies had already been doing before the war (Johnson 2006, 4–8).

Instead, chemical weapons became the bridge away from peacetime production patterns to the "weaponizing" of the dye chemical industry. This came about because of excess capacity in the dyeworks brought about first by the German embargo on the export of dyes imposed at the outbreak of the war, followed later by the tightening of the British blockade on the Central Powers, which cut the dye companies off from most of their global markets and forced them to consider other ways to use their idle facilities. Despite the induction of a large proportion of their staffs into the military, they wanted to use their remaining staff and facilities to produce something of value. Discovering the logic of dual use, as early as October 1914 both Carl Duisberg at Bayer (working with the physical chemist Walther Nernst as part of a secret military commission that followed up on unsuccessful secret prewar military experiments with ideas such as aerial phosgene bombs) and Albrecht Schmidt at Höchst (who had also tried to sell a chemical fog generator to the Imperial Navy before the war) began experimenting initially with non-lethal irritants that would not violate the Hague conventions, but when packed into artillery shrapnel shells could serve to drive enemy troops out of protected shelters such as cellars in buildings where the shrapnel alone could not reach them—at this point trenches were not yet the issue (Johnson 2003, 92–99; Baumann 2011, 195–271). It was relatively easy for the Germans to test these agents, as the process of synthesizing and testing such mainly organic compounds required no significant modification of their existing system. The Allies had greater difficulties, despite some prewar experimentation with chemical weapons. The French had actually entered the war with limited quantitates of non-lethal irritants packed in rifle grenades (Lepick 1998, 54–56). Although the early initiatives on each side had no significant impact on the early months of the war, the advent of trench warfare at the end of 1914 and the ensuing military stalemate fundamentally transformed the situation. The new German high commander Erich von Falkenhayn now demanded more lethal chemical agents, initiating the process of escalation that would lead to the emergence of the new system on both sides (Szöllösi-Janze 1998, 324; Johnson 2003, 94; Baumann 2011, 312–313, 738–739).

5 Scaling up, Innovation and Integration, 1915–17

The German introduction of chlorine cloud attacks at the Second Battle of Ypres in April 1915 was both the catalyst for the development of the new military-industrial-academic system on all sides, as well as—given the war's ultimate emphasis on delivery by artillery—a false start, albeit an inevitable one. After all, Fritz Haber (who had from the beginning of the war put his expertise and that of his Kaiser Wilhelm Institute (KWI) for Physical Chemistry and Electrochemistry at the service of the military) had originally proposed the chlorine cloud because of the shortage of shell casings and propellants—chlorine clouds would not require artillery shells—as well as the relative abundance of domestically-produced chlorine and not least the German efforts to remain at least technically within the Hague conventions (Haber 1924, 76–77, 87; Haber 1986, 27, 41–42). A decisive shift did not come until 1916, when they began using a toxic agent in artillery shells in response to a similar initiative by the French at Verdun. It is worth noting however that the Germans used diphosgene, somewhat less toxic than phosgene as such, apparently chosen because the chemical companies producing it found it easier and less dangerous to produce and load into shells (Haber 1986, 86). Thus although Haber's KWI became militarized in 1916 and substantially expanded its staff, the German chemical-warfare system in this period still largely depended upon the expertise, capabilities, and initiatives of its private industrial component, which had redirected its dye and pharmaceutical laboratories to systematically synthesize and test hundreds of potentially lethal compounds. And it was private industry that in 1916 established a loose "community of interests" (Interessen-Gemeinschaft, IG) encompassing all eight principal dye manufacturers in order to minimize internal competition while fostering the exchange of technical expertise and experience for war work and an expected "war after the war" with the rapidly growing and diversifying Allied systems (Abelshauser 2004, 171–173).

The British, whose domestic production of chlorine was about a tenth that of the Germans, initially chose to respond in kind to the German initiative, albeit after a characteristic delay of several months needed to produce just three-quarters of the amount their generals ordered for the unsuccessful chlorine cloud attacks at the Battle of Loos in September 1915. The British military authorities had been so unfamiliar with their own nation's chemical industry that it had taken them two months to find a suitable supplier (Haber 1986, 150, 162; Palazzo 2000, 62–63). Ultimately the British developed a considerable productive capacity for chemical agents, they established an effective testing range at Porton in 1916, and they produced an outstanding gas mask in the small box respirator of 1917; but they remained chiefly dependent on the French for phosgene, and their system too long remained decentralized and poorly integrated, with weak communication between its academic, business, national-factory, and military components (Haber 1986, 144–147, 162–170). These weaknesses resulted in such errors as developing cyanide compounds in 1916 to counter expected use by the Germans, who had already rejected them as ineffective (Girard 2008, 105–209). The British may here have been following the French, who relied heavily on cyanide products, but in general communication with the French was weak until the British sent a formal liaison officer to Paris in August 1916 (Lepick 1998, 118). Perhaps this was not surprising, as the British had consigned gas offensives to a peripheral, harassing role, mainly using clouds (and by 1917 drums of phosgene fired at close range from Livens projectors), whereas the French took a very different approach.

The French response had begun from an even weaker position than the British. Having lost a significant part of their chemical industry when the Germans occupied the northwestern part of the country, they were forced to commit massive resources toward reconstructing lost plants and establishing new ones, just to meet the requirements of the war economy in general. Moreover, for many critical substances (such as chlorine and bromine) they had been dependent on German imports. Responding to the German chemical warfare initiative would thus require fundamental changes to the French system, which they initiated immediately following the Ypres attacks. By July 1915 they had created a central Service du matériel chimique de guerre for the overall coordination of the key academic, industrial, and military functions, including a research and development section for gas offense (under the chemist Charles Moureu) and defense, a technical and industrial section to expand production and create new factories as needed, and a logistical section, all in Paris. The city's many laboratories allowed the French to effectively utilize their limited stock of technical expertise in chemical warfare work. The chemical service became part of the Under-Secretariat for Artillery under Albert Thomas, and from December 1916 of the new Ministry of Armaments (Lepick 1998, 109–110; also Lepick chapter, this volume). The connection to artillery reflected a tactical choice against cloud gas attacks, a logical choice because French domestic production of chlorine was insignificant. But because most chemical alternatives that the chemists initially tested (culminating in phosgene) also contained chlorine, and because they could obtain only limited amounts from the British, it was still necessary to construct a series of new chlorine plants beginning in August 1915, reaching a total of ten by 1917. Moreover, although the French could use gas shells in their rapid-firing 75-mm field gun, probably the best weapon of its kind in the war, in 1915 they still lacked heavy artillery with its higher-capacity shells. Thus to use gas effectively they first had to accumulate large quantities of 75-mm phosgene shells, which they did not begin firing until a critical situation arose with the German offensive against Verdun beginning in February 1916. In doing so they took the Germans by surprise, however, achieving the first effective Allied initiative of the chemical war (Lepick 1998, 113–119; Lepick, in this volume).

Because the ensuing German production of chemical shells was still only a tiny percentage of overall shell production in 1916, its impact on the war was still insignificant, and as yet the German chemical industry could meet demand by modifying existing plant (Johnson 2006, 12; Haber 1986, 157–159). Nevertheless, broader developments changed the balance between conventional explosives and chemical shell on the German side; in response to the Allied Somme campaign, the Hindenburg munitions program of September 1916 called for a massive increase in production of propellants and shells, but limited resources for high explosives production meant that the Germans would necessarily need to increase their dependence on chemical shells (Herwig 1997, 259–266; Johnson 2006, 13–14; Haber 1986, 260–261). Moreover, the increasing effectiveness of Allied gas defense would force the Germans to introduce new offensive weapons in 1917–1918, which would bring about the culmination of the chemical war—and help precipitate the German collapse (Haber 1986, 226–229, 275).

6 Culmination of the Chemical War, 1917–1918

The chemical war began to reach its culmination in mid-1917, when innovations on all sides, along with the addition of the United States to the Allied side, further magnified the "totalizing" tendencies that had begun to take effect in the previous year. These also produced significant institutional changes, further integration of the chemical war into the broader war effort, and the introduction of several new types of chemical agents with novel, increasingly insidious properties. These were products of the increasingly sophisticated research facilities and increasingly close academic-industrial-military collaboration that had begun to develop in the previous period. It now appeared that chemical warfare would be fully institutionalized on all sides.

In Germany Haber's KWI expanded into many of the other institutes in Dahlem, as well as several other institutions around Berlin, while mobilizing scientists from all over Germany to become a multifunctional center for all aspects of chemical warfare under Haber's department A10 in the War Ministry (see chapters of Bretisav Friedrich and Jeremiah James, and of Margit Szöllösi-Janze in this volume). By mid-1917 Haber saw arsenicals and sulfur compounds as the key to a new German chemical offensive, but he cautioned the High Command that Germany must win the war within a year. Once the Allies could produce the same weapons, Germany's situation would become "hopeless" (cited in Szöllösi-Janze 1998, 332). Elaborating on French approaches, the Germans had developed artillery tactics using a variety of chemical shell types, which they called *Buntschiessen*—varicolored shooting—after the spectrum of chemical shells designated by colored crosses. Blue Cross arsenicals, introduced in July 1917, were intended to penetrate the Allied mask filters and cause so much irritation that soldiers would be unable to keep on their masks, thus making them vulnerable to toxic Green Cross (diphosgene) that the masks could otherwise block. The artillerists welcomed this theoretically effective weapon, and the IG produced 8,000 tons in 1917–1918, but the KWI's scientists had not solved the practical problem of achieving fine enough particle size, so that Blue Cross caused relatively few Allied casualties. The Allies responded with their own arsenicals in 1918, but their system produced scarcely a hundred tons by the Armistice (Haber 1986, 261–269).

In contrast to the arsenicals, the Germans' near-simultaneous introduction of Yellow Cross shells had an immediate, dramatic effect. Yellow Cross contained the sulfur compound and strong vesicant (blister agent) bis(2-chloroethyl) sulfide, misleadingly known as mustard gas. In fact, it was not a gas but an aerosol, nor did it have any chemical relation to mustard. This oily liquid's delayed action and persistent nature (adhering to surfaces and remaining potent for days) temporarily burned and blinded thousands, producing the dramatic image captured in John Singer Sargent's classic painting *Gassed*. The substance was difficult and dangerous to produce and fill into shells; thus the French could not counter with their own version, *ypérite*, until June 1918, and it took the British (and Americans) until September to begin mass production. As Haber had warned in 1917, Germany's failure to win the war with Yellow Cross before the Allies produced it themselves now ensured German defeat, because it had produced an even more drastic destabilization of the German system. Although the IG had solved the production problems and the military had built new depots at Adlershof and Breloh for filling the shells, with their limited resources the Germans could not solve the decontamination problem posed by mustard gas (Haber 1986, 189–190, 265). The "only effective counter-measures" such as rubberized protective suits or even simply replacing contaminated uniforms were "practically infeasible" according to Haber (Haber 1924, 38). The Germans had no effective defense once the Allies achieved large-scale production of mustard gas shells. Moreover, even though in 1918 more than a quarter of German shells contained chemical agents, a significantly higher proportion than on the Allied side, this reflected German limitations in producing high explosives as much as their ability to produce chemical weapons. By late 1918, total Allied shell production (all types) was twice that of the Germans, and the American mobilization was about to eliminate even the German advantage in chemical warfare (Johnson 2006, 15–16).

British concerns about the German innovations of mid-1917 finally brought about the long-delayed centralization of their system with the founding within the Ministry of Munitions of the Chemical Warfare Department in October. Its able leader, General Henry F. Thuillier, finally began to coordinate the academic research, industrial production, and military testing efforts, yet even his influence could not avoid serious delays in mustard gas development, as previously noted. But by late 1918, the British had overcome these problems and had finally integrated chemical weapons into their artillery, finding (like the Germans and French) that mustard gas made an excellent counter-battery weapon against enemy artillery during the final offenses of the war. Nevertheless the British produced fewer than half as many gas shells as their French allies, and both together reached only two-thirds of the German total (Haber 1986, 147–149, 260–261; Palazzo 2000, 173–186).

There was now greater coherence within the Allied system as a whole, particularly after inter-Allied coordination of chemical supplies began in spring 1918, and its overall scale dramatically increased as the Americans began to mobilize. Whereas a "quasi-mobilization" in 1914 by private companies like DuPont to produce explosives for the Allies had developed that side of the system, an American military-industrial system for chemical warfare remained to be created in 1917. Here the government (the U.S. Bureau of Mines) and the military necessarily took the lead in gathering information from the British and French, coordinating academic research, and recruiting several thousand chemists for war work. Although direct experience was lacking, ambition was not; Americans planned for war on a large scale. The same was true on the production side, where after unsatisfactory dealings with private suppliers, the Army Ordnance Department constructed a large military chemical complex, Edgewood Arsenal in Maryland. Edgewood came under the newly organized Chemical Warfare Service in June 1918, and plants for the principal war gases and for filling shells came on line in summer 1918. Capacities continued to increase, supplemented by additional production contracts elsewhere. At the Armistice the American system could produce 900 tons of mustard gas monthly, but the planned capacity by January 1919 was 4,000 tons—vs. a German output of less than 8,000 tons for the entire war (Steen 2014, 98–110; Haber 1986, 149, 261).

By the time of the Armistice in November 1918, the German system was collapsing in mutiny and revolution, while the Allied system was still gaining momentum. Had the German military fought on into the spring, the Allies were prepared to launch massive attacks using chemical weapons, including aerial bombardments with tons of mustard gas. The American military believed that they had achieved the ultimate weapon in Lewisite, an arsenical with the vesicant properties of mustard gas, but this did not reach France before the Armistice (Steen 2014, 109–110). The Germans were thus spared the horrors that "total" chemical warfare might have produced in 1919 (Table 1).

7 Concluding Reflections

The chemical war presents an almost ideal case of a technological meta-system, whose dynamics are perhaps best illustrated in Table 1, which highlights the timing and diversity of offensive innovations on both sides. Here one can see the way in which each system responded (or not) to its opponent's innovations, usually after a considerable delay required to adjust to the production of a new chemical agent or to develop a new form of delivery. One can also see that the German system kept the initiative (with some exceptions) until mid-1918, whereas the Allied system was just reaching its full potential as the war ended. Henceforth chemists confronted a very different world than they had known in 1914, a change reflected in Fig. 1.

The war transformed modern chemistry as an international industrial system and as an industrialized scientific discipline. Both became increasingly multi-centered, following the wartime recognition that dual-use chemicals were vital to national security, as indicated by the physical chemist and British general Harold Hartley in

Table 1 Innovation in the meta-system of the Western Front: introduction of chemical agents by Germans and Allies, 1914–1918 (based in part on the figure in Martinetz 1996, 98). **The most significant innovations are indicated in bold type**. I have not listed each of several different types of irritants introduced on both sides in the first two years of the war, nor have I listed all of the less significant war chemicals developed or used in the last two years (for a comprehensive list see Martinetz 1996, 69–71)

Germans introduced:	**Year**	**Allies introduced:**
Oct.: irritants (shrapnel shells)	**1914**	Aug.: irritants (rifle grenades, Fr.) Nov.: irritants (hand grenades, Fr.)
Jan.: irritants (shells, Russ. front) **April: chlorine (cloud)** **May: chlorine-phosgene (cloud)**	**1915**	
June: irritants (trench mortars)		June: irritants (trench mortars) **Sept.: chlorine (cloud, Br.)** Fall: various irritants/toxic agents (shells, hand grenades)
March: diphosgene [phosgene-like, less toxic] (green cross—shells) Aug.: chlorine-chloropicrin (cloud)	**1916**	**Feb.: phosgene [highly toxic](shells, Fr.)** July: hydrogen cyanide [toxic](shells) Aug.: chloropicrin [strong irritant] (shells) Oct.: cyanogen chloride [cyanide-like, less toxic] (shells, hand grenades)
Spring: Phenylcarbylamine chloride (green cross—shells) **July: diphenylarsine chloride (blue cross arsenical—shells)** **July: Bis(2-chloroethyl) sulfide (yellow cross/mustard gas—shells)** **Oct.: phosgene (projected drums at Caporetto, It.)**	**1917**	**Apr.: phosgene (projected drums, Br.)**
Spring-summer: blue and green cross variants (shells)	**1918**	**June: mustard gas (shells, Fr.)** **Sept.: mustard gas (shells, Br./Am.)** Fall: dimethyl sulfate (toxic agent—shells) Brombenzyl cyanide (irritant—shells) In preparation (not introduced before Armistice): Lewisite, Adamsite (arsenicals—shells)

his report on the first Allied inspection tour of German chemical factories in January-February 1919: "In the future … every chemical factory must be regarded as a potential arsenal, and other nations cannot … submit to the domination of certain sections of chemical industry which Germany exercised before the war" (Great Britain. Ministry of Munitions 1919, 12). Chemists in the Allied nations thus sought to carry over their wartime gains into peacetime, if possible with government support, which also meant promoting academic research and education as well as academic-industrial collaboration, both consciously emulating and seeking to weaken their German rivals (cf. Steen 2014). Thus aside from the chemical disarmament and reparations provisions of the Treaty of Versailles, leading chemists among the victorious Allies created a postwar global organization that excluded the

Fig. 1 Power of a symbol: "The Spirit of German Science" (Raemaekers 1918, 195). The German scientist shown in this cartoon is clearly based on the organic chemist Adolf von Baeyer (Nobel Prize for Chemistry, 1905), and most of the objects of war shown in his laboratory are products of chemistry, including poison gas, tear shells, a flame thrower, incendiaries, bombs, and poison (cf. Johnson 2011, 99–100)

Germans throughout the 1920s, thus greatly delaying the process of international scientific reconciliation and probably hindering diplomatic efforts to end chemical warfare research and development (MacLeod 2003; Szöllösi-Janze 1998, 590–597; cf. other chapters in this volume).

The paradoxical result was that the greater peacetime support for and general interest in chemistry came at the cost of global industrial overcapacities and oversupplies of trained chemists, which subsequent economic crises only exacerbated. Moreover, the war had tarnished the prewar image of chemistry. Thus "The Spirit of German Science," a cartoon by the Dutch artist Louis Raemaekers (Fig. 1), appeared with the following commentary by the Princeton psychologist, J. Mark Baldwin:

> The moral revulsion of the world against the Germans is justified by their use of science ... They have abolished the distinction between the knight and the brute, between the man and the snake, between pure science and foul practice ... To future generations this will damn the German race" (Raemaekers 1918, 194).

The primary target here is clearly German chemistry, but the demonization cut both ways, as the Allies had of course replied with many of the same agents. If anything, the war had effectively demonized all chemists, a curse that arguably still haunts the public image of the discipline (cf. Johnson 2011, 99–100).

References

Abelshauser, Werner, et al. 2004. *German industry and global enterprise. BASF: The history of a company*. Cambridge, U.K., New York: Cambridge University Press.

Baumann, Timo. 2011. *Giftgas und Salpeter: Chemische Industrie, Naturwissenschaft und Militär von 1906 bis zum ersten Munitionsprogramm 1914/15*. Düsseldorf: Universitäts- und Landesbibliothek der Heinrich-Heine-Universität Düsseldorf.

Chickering, Roger, and Stig Förster. 2000. *Great War, Total War: Combat and mobilization on the Western Front, 1914–1918*. Washington, D.C: German Historical Institute.

Girard, Marion. 2008. *A strange and formidable weapon: British responses to World War I poison gas*. Lincoln: University of Nebraska Press.

Great Britain. Ministry of Munitions. Department of Explosives Supply. 1919. *Report of the British mission appointed to visit enemy chemical factories in the occupied zone engaged in the production of munitions of war*. London: Ministry of Munitions. Department of Explosives Supply. Copy in The National Archives (Kew, UK), MUN 4/7056.

Haber, Fritz. 1924. *Fünf Vorträge aus den Jahren 1920–1923*. Berlin: Springer.

Haber, Ludwig F. 1986. *The poisonous cloud: Chemical warfare in the First World War*. Oxford, U.K.: New York: Clarendon Press.

Hardach, Gerd. 1992. Industrial mobilization in 1914–18. In *The French Home Front, 1914–1918*, ed. Patrick Fridenson, 57–88. Providence [R.I.]: Berg.

Herwig, Holger H. 1997. *The First World War: Germany and Austria-Hungary, 1914–1918*. London, New York: Arnold.

Hughes, Thomas Parke. 1969. Technological momentum in history: Hydrogenation in Germany, 1898–1933. *Past & Present* 44 (August): 106–132.

Hughes, Thomas Parke. 1983. *Networks of power: Electrification in Western Society, 1880–1930*. Baltimore: Johns Hopkins University Press.

Johnson, Jeffrey Allan. 2000. The academic-industrial symbiosis in German chemical research, 1905–1939. In *The German Chemical Industry in the twentieth century*, ed. John E. Lesch, 15–56. Dordrecht, NL: Kluwer Academic Publishers.

Johnson, Jeffrey Allan. 2003. La mobilisation de la recherche industrielle allemande au service de la guerre chimique. In *Le sabre et l'éprouvette. L'invention d'une science de guerre, 1914–1939*, ed. David Aubin and Patrice Bret, *14–18 Aujourd'hui-Today-Heute* 6: 89–103.

Johnson, Jeffrey Allan. 2006. Technological mobilization and munitions production: Comparative perspectives on Germany and Austria. In *Frontline and factory: Comparative perspectives on the chemical industry at war, 1914–1924*, ed. Roy MacLeod and Jeffrey Allan Johnson, 1–20. Dordrecht, NL: Springer.

Johnson, Jeffrey Allan. 2011. Preface: Reflections on war and the changing perception of chemistry. *Ambix*, Special Issue: *Chemistry in the Aftermath of World Wars*, 58/2 (July): 99–100.

Lepick, Olivier. 1998. *La Grande Guerre chimique: 1914–1918*, 2d ed. Paris: Presses universitaires de France.

MacLeod, Roy M. 1998. Chemistry for King and Kaiser: Revisiting chemical enterprise and the European War. In *Determinants in the evolution of the European Chemical industry, 1900–1939: New technologies, political frameworks, markets and companies*, ed. Anthony S. Travis, et al., 25–49. Dordrecht, NL: Kluwer.

MacLeod, Roy M. 2003. L'entente chimique: l'échec de l'avenir à la fin de la guerre. In *Le sabre et l'éprouvette: l'invention d'une science de guerre, 1914/1939*, ed. David Aubin and Patrice Bret, *14–18 Aujourd'hui-today-heute* 6: 134–151.

Martinetz, Dieter. 1996. *Vom Giftpfeil zum Chemiewaffenverbot: Zur Geschichte der chemischen Kampfmittel*. Thun. Frankfurt am Main: Verlag Harri Deutsch.

Palazzo, Albert. 2000. *Seeking victory on the Western Front: The British Army and chemical warfare in World War I*. Lincoln, Neb.: University of Nebraska Press.

Raemaekers, Louis. 1918. *America in the War*. New York: The Century Co.

Simmonds, Alan G.V. 2012. *Britain and World War One*. Abingdon, U.K., New York: Routledge.

Steen, Kathryn. 2014. *The American synthetic organic chemicals industry: War and Politics, 1910–1930*. Chapel Hill: The University of North Carolina Press.

Szöllösi-Janze, Margit. 1998. *Fritz Haber, 1868–1934: Eine Biographie*. Munich: C.H. Beck.

Part II
Contexts and Consequences of Chemical Weapons

The Gas War, 1915–1918: If not a War Winner, Hardly a Failure

Edward M. Spiers

Abstract Contemporary claims that gas warfare proved "a failure" during the First World War would have baffled wartime adversaries, who invested heavily in the research, development, and production of gas warfare. If poison gas, like other conventional weapons, never broke the stalemate of the trenches, it evolved into a weapon of harassment that compounded the effects of conventional weapons and degraded the effectiveness of enemy forces compelled to wear gas masks for protracted periods of time. The introduction of mustard gas in July 1917 greatly increased the number of gas casualties, and set the scene for a steady increase in the use of chemical weapons during the later stages of the war. Like the tank and aircraft, gas was not strategically decisive, but continuing investment in this form of warfare underscored its potential utility.

1 Introduction

The onset of chemical warfare in the First World War produced not only major scientific, industrial, and military challenges to the principal belligerents but also a legacy that has been fiercely debated. After the first major use of chlorine gas by German forces, when they dispersed chlorine from 5,730 cylinders along a 6-km front at Ypres on April 22, 1915 (McWilliams and Steel 1985, Chaps. 5 and 6), the gas war expanded prodigiously as the main belligerents introduced new and more potent gases and sought to deliver them more efficiently. Although the French and Germans had used irritant agents before April 22 (Trumpener 1975, 461–465), they later employed lethal agents such as chlorine, phosgene, and, above all, mustard gas as the primary instruments of gas warfare. Like the British, they enhanced their methods of gas protection, and dispersed gas by various means, including cylinders,

E.M. Spiers (✉)
University of Leeds, Leeds, UK
e-mail: e.m.spiers@leeds.ac.uk

B. Friedrich et al. (eds.), *One Hundred Years of Chemical Warfare: Research, Deployment, Consequences*, DOI 10.1007/978-3-319-51664-6_9

mortars, projectors, and gas shells. The gas war, argued Major Victor Lefebure, became "one of continual attempts on both sides to achieve surprise and to counter it by some accurate forecast in protective methods. It is a struggle for the initiative" (Lefebure 1921, 109–110).

2 Debate

Assessing the significance of this struggle has produced a wide array of judgments. The allies, eying the response in neutral United States (Peterson 1939, 63; Read 1941, 195–199), denounced the first use of chlorine gas as "an atrocious method of warfare" which would "fill all races with a new horror of the German name" (*The Times*, April 29, 1915). Nevertheless, when they retaliated in kind, a reaction described "as just and necessary" (Brown 1968, 15), they did so without analyzing its effectiveness. Only at the end of the war were several British and American gas officers able to comment on the impact of chemical warfare. Major Samuel J. M. Auld, a former Chemical Advisor, British Third Army, argued that "the use of poisonous and irritating gases is as fundamental as the introduction of gunpowder, and probably even more so," and he extolled the effects of mustard gas during the last year of the war (Auld 1922, 58, 66). Amos A. Fries, an American engineer, who became Chief of the Chemical Warfare Service, described gas as "one of the most powerful means of offense with which the American troops had to contend" (Fries and West 1921, 386). Major-General Charles H. Foulkes, the former British Director of Gas Services (1917–18), insisted that gas had "changed the whole *character* of warfare." Gas, he observed, had seriously affected German morale during the last months of 1918, contributing "to the Allied victory": it was of "increasing" importance towards the end of the war, and "might have played a decisive part in 1919" (Foulkes 1936, 334, 336, 345). Brigadier-General Harold Hartley, a future Fellow of the Royal Society, claimed, too, that "gas is a very valuable weapon, as it supplements other weapons, offers great opportunities for surprise, and is a most effective means of achieving many tactical objects" (Hartley 1919–20, 504).

Revisionism followed the Second World War, a conflict in which poison gas was stockpiled massively but not used between the principal adversaries in Europe, in contrast to the Pacific Theater where the Imperial Japanese Army used chemical weapons from 1937 onwards in occupied China. In his final volume of the official history of the Great War, produced in 1947, Brigadier J. E. Edmonds contradicted much of his earlier writing (Edmonds et al. 1937, vol. 2, 163–164, 383, 390, 412) by observing that "[g]as achieved but local success; it made war uncomfortable, to no purpose. It was not used 1939–45" (Edmonds et al. 1947, vol. 5, 606 n.2). Julian Perry Robinson subsequently argued that the Germans had bungled the strategic potential of poison gas at Ypres, and that the introduction of respirators ensured that gas achieved only a "limited tactical success." The military establishments, he argued, became "decidedly lukewarm" towards this unpredictable weapon (Robinson 1971, 51–52, 59).

Several scholars have disputed the recollections of Foulkes, claiming that he exaggerated the effects of gas at Loos, overrated the value of cylinder-based gas-cloud attacks, and relied excessively upon the unreliable testimonies of prisoners of war (Haber 1986, 57, 279; Richter 1994, 3, 91–92; Griffith 1994, 116–119). L. F. Haber even claimed that gas was a failure, despite some tactical successes as in precipitating the Italian defeat at Caporetto, October 24, 1917. He contended that gas failed at Second Ypres and Loos, and that it proved unpredictable as respirators blunted its effects: by complicating warfare, gas, unlike the tank, aircraft and the light machine gun, failed to change "the face of war in 1918" (Haber 1986, 264, 270, 278).

3 Gas: Not a War Winner

Poison gas was certainly not a war winner. Its use at the Second Battle of Ypres (April 22–May 25, 1915) followed the unexpected onset of trench deadlock and the failure of conventional weapons to break the stalemate. Professor Fritz Haber, who was then only an unofficial advisor to the German Ministry of War, had pressed the case for experimenting with chlorine gas. General Erich von Falkenhayn never appreciated the potential effects of releasing 149,000 kg of chlorine from 5,730 cylinders (McWilliams and Steel 1985, 41): he saw it as primarily a diversionary move "to cloak the transportation of the [German] troops to Galicia," and admitted that the "surprise effect was very great. Unfortunately we were not in a position to exploit it to the full. The necessary reserves were not ready. The success achieved, however, was considerable" (Falkenhayn 1919, 84–87).

This gas release was far from universally popular. While German infantrymen resented the labor of installing the cylinders in front-line trenches and the days spent waiting for favorable winds, Crown Prince Rupprecht of Bavaria worried that retaliation in kind by the Allies would benefit from the prevailing westerly winds (von Frauenholz 1929, vol. 1, 304–305), and Rudolph Binding wrote in his diary on April 24, 1915, "I am not pleased with the idea of poisoning men. Of course, the entire world will rage about it first and then imitate us" (Binding 1929, 64). This proved true, and Hartley claimed that the Germans had

> made almost every possible mistake in their earliest gas attacks. They chose a gas against which protection could be obtained with comparative ease, they used it in small quantities on narrow fronts in discharges of long duration and low concentration, thus losing the effect in depth, and finally they failed to exploit the partial advantage they gained. Within three weeks we were protected (Hartley 1919–20, 493).

However correct in hindsight, this judgment overlooks "the fact" that the British "were aware" of German preparations for the gas attack for "several days previously" but assumed that "the enemy's attempt would certainly fail", and so "the terrible effect of the gas came to us as a great surprise" (TNA, WO 32/5483, "Account"). The gas release also caused mass panic among its victims, enabling the Germans to capture over 1,800 prisoners, "more than 51 guns, of which four were heavy, and about 70 machine guns" (Duguid 1938, vol. 2, 320) as well as ground they would hold for another two and a half years. Subsequent attacks were less productive; on April 24 the gas encountered a resolute defense led by Canadians, using improvised protection, to ensure "only a moderate dent in the line" (McWilliams and Steel 1985, 86, 155), and further gas attacks on May 1, 6, 10, and 24 failed to dislodge the Allied grip on the Ypres salient (Spiers 1986, 16–17).

The first designed British response was the Black Veil Respirator, a pad of cotton waste soaked in sodium thiosulphate, sodium carbonate, and glycerol held in place by a long piece of veiling. Issued in May 1915, it afforded only limited protection against chlorine, leaving Yorkshire soldiers to complain about the "rotten gas" that "nearly choked and blinded us" ("Letters" May 28 1915, 6). Much more effective was the Hypo Helmet, a bag with eyepieces and made of flannel, soaked in the impregnating solution, which was put over the head and tucked into the collar. The issue of 2.5 million copies during June 1915, prompted Driver E. Broadley to affirm: "our respirators kept us all right" during a subsequent gas attack ("Letters" June 25 1915, 6). The British later issued the P helmet, impregnated with phenate solution, in anticipation of the enemy using phosgene as they did on December 19, 1915. By adding hexamethylenetetramine, a Russian idea, the PH helmet gave enhanced protection against phosgene, but all helmets were unpleasant to wear. The British eventually devised a large box respirator (LBR) followed by the small box respirator (SBR), in which a flexible rubber tube connected the mask to a filter containing charcoal and sodium permanganate-lime granules. This afforded extra protection against prussic acid, and the SBR became the standard British respirator, issued to all troops by January 1917.[1]

Yet Sir John French, then commander-in-chief of the British Expeditionary Force (BEF), never regarded defensive protection as an adequate response to chemical attacks. As he informed Lord Kitchener, the Secretary of State for War: "We are taking every precaution we can think of but the most effective would be to turn their own weapon against them & stick at nothing" (TNA, Kitchener 1915). Many regimental officers and men agreed. The Germans were "dirty devils," wrote Lieutenant the Hon. William Fraser, "[b]ut we must play their own dirty game as far as the gas goes."[2] Another Gordon Highlander, Lance-Corporal George Ramage, contended: "All war is foul. Why object to gas & not to bullets" (NLS, Ramage 1915). Meanwhile, once Major-General Henry Rawlinson, commander of IV Corps, learned about the depth of the German underground bunkers, he maintained: "What we want is a favourable wind and plenty of good strong chlorine & bromine gas which will sink right down into the deep trenches" (NAM, Rawlinson 1915).

[1]TNA, WO 32/5483, "Diary" and "Account"; WO 142/99, Fergusson 1915.

[2]Fraser, Lt. the Hon. William. 1915. Letter to his Father, May 3. In Fraser 1990, 52.

The British cabinet required several meetings and "considerable discussion" before it approved retaliation in kind (TNA, CAB37/127 and 37/128, 1915). This followed on the first day of the battle of Loos, an offensive demanded by France. Sir Douglas Haig, commander of I Corps, who had reservations about the location and timing of the proposed battle, saw a successful trial discharge of gas at Helfaut on August 22 (IWM, Ashley August 22 1915). Henceforth he declared:

> On the one hand, with gas, decisive results were to be expected. On the other hand, without gas, the fronts of the attacks must be restricted, with the result of concentrated hostile fire on the attacking troops, considerable loss, and small progress! In my opinion the attack ought not to be launched except with the aid of gas! (NLS, Haig 1915)

After successive postponements at the behest of General Joseph Joffre, the BEF launched its first gas assault on the morning of September 25, 1915, when Foulkes, who commanded the newly formed Special Companies of trained chemists, planned to release 150 tons of chlorine from 5,500 cylinders. He recorded in his diary: "Wind was almost calm—SSW—very unfavourable for a gas attack but the battle could not be postponed" (LHCMA, Foulkes 1915). Following the release of chlorine gas and smoke "alternately" (IWM, Ashley September 24 1915), the gas cloud facilitated the advance of the 15th and 47th Divisions on the right of the British attack and assisted the brief capture of the Hohenzollern Redoubt. Elsewhere it moved far too slowly, veering and hanging around the British trenches, with the effects on British forces compounded by leakages, faulty connections, and inadequate training. Even Foulkes, who always insisted that gas caused surprise at Loos,[3] admitted that this attack had been one of "peculiar difficulty" and "extemporization and creation" (LHCMA, Foulkes 1933). Far from breaking through at Loos, the British suffered 59,000 casualties over three weeks, including 2,639 casualties and seven deaths from their own gas in the first three days (Palazzo 2000, 75–77).

4 The Challenge of Chemical Warfare

An inability to cause a breakthrough did not distinguish chemical weapons from any other weapon in 1915. Nevertheless, once used, gas could not be ignored and all the principal belligerents invested in its development. On the Western Front they undertook offensive and defensive research & development, tested new gases and various delivery systems, produced the gases, munitions, and ancillary equipment for an expanding gas war, and formed special gas forces. Under Haber's leadership, the Kaiser Wilhelm Institute for Physical Chemistry and Electrochemistry became committed to the military in February 1916 and expanded steadily into nine departments in various locations around the suburbs of Berlin. By the end of the war it employed 150 academically trained staff, 1,300 non-commissioned officers,

[3]LHCMA, Foulkes 1915 and Foulkes n.d., 5: ch. 20; Foulkes 1936, ch. 5; Foulkes 1962, 179–80.

soldiers, other workers, and additional support personnel. Meanwhile the centralization of French science enabled scientists and laboratories to forge close links between offensive and defensive research in Paris, with another department responsible for application and pilot-plant operation, and a third for purchases and dispatch. Conversely, the French reliance upon state-controlled and smaller independent gas enterprises, often located far from the front, never matched the economies of scale secured by the larger German industrial combines (Stoltzenberg 1993, 133–142; Coleman 2005, 25–26; Hartcup 1988, 105–106).

The British developed entirely separate groups working on offensive and defensive research but, in 1916, the War Office purchased 2,886 acres at Porton Down, where it established an experimental ground. The site expanded steadily over the remainder of the war until it occupied 6,196 acres and enabled Porton to examine 147 toxic substances and conduct field trials of new munitions. Porton also examined individual and collective protection when it acquired the Anti-Gas Department from the Royal Army Medical College in 1917. British industrial support, despite expanding to involve 70 factories by the end of the war, remained less productive than its German counterpart (Carter 1992, 7–25; Haber 1986, 172).

These organizational innovations, coupled with the formation of dedicated chemical corps, testified to the continuing interest in chemical warfare. Although disagreements recurred between scientists and senior military officers, improvisation flourished (Haber 1986, 174–175, 208, 273–274; Hartcup 1988, 106).[4] Faulty cylinder connections, which caused gas leakage at Loos, and were described by R. C. Gale as "a ghastly failure" (IWM, Gale 1915, 70), were corrected with the use of rubber connections. Thereafter the British persisted with cylinders despite their weather dependence, the infantry's dislike of installing them at night in front-line trenches, fear of accidents, and the counter barrage from German gunners whenever the gas was released.[5] As employing cylinders exploited the prevailing westerly winds, and offset the shortage of shells in Britain, the Special Brigade (as Foulkes's expanded force became known) launched 150 of the 220 gas-cloud attacks between April 1915 and November 1918, while the Germans launched 50 and the French 20 (Prentiss 1937, 52; Foulkes 1936, 184–186).

Quite apart from employing gas cylinders, the British introduced the 4-inch Stokes mortar at Loos. On account of its caliber and rate of fire (about twenty rounds a minute), this was an ideal gas weapon, as each round delivered 3 to 4 kg of agent at ranges up to 1,000 m (Prentiss 1937, 362–364). Foulkes praised the Stokes mortar as a versatile weapon that could be brought into action quickly and deliver concentrated amounts of gas over the target area. It could also project smoke barrages during an assault and bombard advanced enemy positions with thermit, bursting the bombs in the air and showering anyone below with globules of molten iron at "white heat" (LHCMA, Foulkes 1917).

---

[4]On disputes involving Foulkes, see IWM, Hodgkin August 2 1918 and Richter 1994, 183.

[5]TNA, WO 158/270, Barrow 1916; Winter 1979, 126.

Another British invention was the projector, designed by Captain William H. Livens, which was first used at the Somme before the discharge in a mass formation of 2,340 projectors at the opening of the battle of Arras (April 4, 1917). Easily and cheaply produced, the projector fired canisters holding about 15 kg of agent and was not weather dependent. Installed in the ground in batteries just behind the front lines at 45-degree angles, 4,000 projectors could be fired simultaneously by an electrical device. As the Livens projector set up sudden, very high concentrations of agent that neither mortars nor gas shells could emulate, large-scale usage ensued: 4,200 drums (and 3,100 Stokes mortar bombs) were discharged on the eve of the battle of Cambrai (November 1917) and 2,960 drums were fired into St. Quentin on March 19, 1918.[6] Less accurate than mortars or shells, projectors were limited in range (about 1,550 m), but the German 111th Division testified to their impact:

> By this new procedure, the enemy has combined the advantages of gas cloud with those of gas shell, obtaining the density of the former with the surprise effect of the latter. Our losses have hitherto been heavy, because the enemy, in most cases, successfully took us unawares, and gas masks were put on too late (TNA, WO 158/294, July 8 1917).

Nevertheless, gas shells became the primary mode of gas delivery. Although Germany had pioneered the employment of lachrymatory *Ni-Schrapnell* and *T-Stoff* shells in the winter of 1914–15, and then *K-Stoff* shells in the Argonne sector on July 16, 1915, the French proved more innovative in fuse design. By using only a small bursting charge to open their phosgene shells, the French increased both the payload of the shells and the concentration of gas on target (February 1916). As the Germans responded with Green Cross shells, filled with the lethal agent, diphosgene (March 9, 1916), both adversaries exploited the advantages of artillery: accuracy, flexibility, and less dependency on the weather.[7] Gunners ultimately delivered some 85% of the toxic gases during the First World War, and caused about 85% of the gas casualties (Prentiss 1937, 657, 660). Gas shells grew from a negligible proportion of artillery ammunition in 1915 to about 50% of the German, 35% of the French, 25% of the British, and 20% of the American ammunition expenditure by the Armistice. Had the war continued into 1919, all the belligerents planned to employ even more chemical shells.[8]

Gas shells complemented other forms of shelling, as the gases released were multi-purpose area weapons. In preparing for the battle of the Somme, the BEF borrowed 60,000 phosgene shells from France for use in "surprise" salvoes "on front-line trenches, when parapets are manned, and also for counter-battery work" (TNA, WO 158/234). Colonel Georg Bruchmüller refined German artillery tactics on the Eastern Front to neutralize hostile batteries and fire short but highly intensive bombardments at "gas squares." Thereafter his batteries attacked infantry and artillery strong points with as much as 50 or 80% gas shells (TNA, WO I33/1072, Hartley n.d., 8–9). Gas shells were also used to intercept enemies moving at night or

[6]REM, R15, Crowden Report, 1917; *History of the Corps* 1952, 5: 522.

[7]TNA, WO 188/213, Hartley n.d.; Lepick 1998, 182, 184–185.

[8]Prentiss 1937, 657–658, 660, 683–684; TNA, WO 188/213, Hartley n.d.

in advance of night-time raids: as Captain A.E. Hodgkin observed, it was easier to take a German prisoner "when he has got his gas mask on" (IWM, Hodgkin May 23 1918). Latterly, when on the defensive, Germans exploited the persistent effects of mustard gas by saturating areas with Yellow Cross liquid and compelling attacking parties to avoid the contaminated terrain (TNA, WO 158/128, Foulkes 1918).

Similar developments occurred on the Eastern Front. Having experimented unsuccessfully in the cold weather at Bolimov with 18,000 *T-stoff* shells (January 31, 1915), the Germans mounted large-scale poison-gas attacks with cylinders, beginning at Rawka (May 31, 1915). Exploiting the Russian lack of preparedness and protection, they delivered at least ten cylinder attacks over the next eighteen months before introducing more accurate chemical shelling (Main 2015, 116–20, 136). Although estimates of Russian gas casualties are disputed (ibid., 130–131; Prentiss 1937, 653; Krause and Mallory 1992, 16–17), the Russian chemist Professor Vladimir N. Ipatieff, who chaired the Commission for the Preparation of Explosives and later the Chemical Committee of the Chief Artillery Administration, claimed that in the first attack "seven to eight thousand men were poisoned in one night, the majority of whom died" (Eudin et al. 1946, 221). Heavy gas casualties persisted, reflecting recurrent lapses in anti-gas discipline, transport, and distribution problems, and delays in producing effective gas masks through a Tsarist bureaucracy, riven with corruption and incompetence (ibid., 218–226, 230–231).

Over a year elapsed before Russia, hampered by its industrial shortcomings, could retaliate in kind, and, in the second attack, all the casualties were Russian (Main 2015, 126; Krause and Mallory 1992, 26–27). Despite both sides mounting gas operations with cylinders, they used shells predominantly in 1917, where Bruchmüller earned his nickname *"Durchbruchmüller"* ("breakthrough Müller") for artillery assaults, using "gas squares" at the Stochod, East Galicia, and Riga in April, July, and September respectively. In establishing a bridgehead across the Stochod, German gas shells incapacitated the 27th Artillery Brigade and accounted for the capture of two Tsarist divisions. German artillery also employed intense gas barrages in the capture of Seret in East Galicia and in the forcing of the Dvina River in the Riga operation.[9]

What sustained the development of the gas war was the search for more effective chemical warfare agents. The criteria of battlefield effectiveness meant that only several dozen of the thousands of toxic substances examined were employed militarily. This process was still underway in the last few weeks of the war, when the Germans began looking for an agent with more persistence than mustard gas (Robinson 1971, 38–51; Coleman 2005, 26). These substances had to be produced in significant quantities from available materials, thereby confirming the huge advantage that the large German dyestuffs industry had over its rivals. Chemical agents had to be stable in storage, non-corrosive in munitions, relatively safe to handle and transport, and deliverable from a practical military device in sufficient concentration to produce the desired effect on target. Understanding all these

[9]TNA, WO 42/195, Hicks 1917, 204–205; TNA, WO 33/1072, Hartley n. d., 9.

properties took time, fuelling friction between scientists and the military, and between the military and those in charge of wartime supply.

Of the harassing agents used, lachrymators (tear-gases) were employed throughout the war. In 1917, the Germans introduced a new class of solid arsenical sternutators (causing sneezing). Dubbed Blue Cross agents, diphenylchloroarsine (DA) and diphenylcyanoarsine (DC) proved highly irritating but good mask discipline could blunt their effects. The British hoped that adamsite (DM), developed during 1918, would prove a mask-breaking agent in 1919, if used extensively (Robinson 1971, 38–51; Coleman 2005, 26). The Germans led the way in the use of chlorine and phosgene (lethal lung agents), followed by the Allies, with the British employing 50/50% combinations in their White Star gas clouds. The French hoped that the highly toxic blood gases, hydrogen cyanide (prussic acid) and cyanogen chloride, would become effective war gases, but these highly volatile substances had to be mixed with stabilizers, diluents and smoke-markers, and so lethal dosages proved difficult to deliver. The Germans introduced mustard gas (bis (2-chloroethyl) sulfide) at Third Ypres on July 12, 1917 and, by only using it when they had a vast stock available, delivered 2,500 tons of mustard gas in the first ten days from over one million Yellow Cross shells.[10]

5 Effectiveness of the Gas War

The so-called "king of the war gases" transformed the gas war. Highly effective in low concentrations, mustard gas had a slight odor and delayed action, which, coupled with its persistence and capacity to burn and blister through clothing, defied the defensive precautions of the day. Wearing a respirator could thwart its fatal effects, and spreading decontaminants—chloride of lime on guns and bleaching powder on the ground—helped, but mustard gas produced a massive number of casualties. In the first three weeks of Yellow-Cross shelling, the British incurred more casualties (14,276), and almost as many deaths (nearly 500), as they had suffered from all previous gas engagements. From July 12, 1917 to November 23, 1918, British casualty clearing stations admitted 160,970 gas casualties, 1,859 of whom died. 77% of these were victims of mustard gas.[11]

All of the casualties had to be removed for treatment to the rear; the more lightly blistered with swollen eyelids, like Captain Richard Foot, might recover in a week (IWM, Foot n.d., 98); others like T. H. Holmes, gassed on August 22, 1918, found their fighting career over (IWM, Holmes n.d.). As Tim Cook asserts, the best weapons were those that "remove fighting men and leave fear and unrest among the survivors" (Cook 1999, 215): in other words, mustard gas had a psychological effect upon everyone in a gassed area. Its slight odor was difficult to detect amongst

[10]Robinson 1971, 46; TNA, WO 188/213, Hartley n.d.

[11]Macpherson et al. 1923, 2: 294 and 304–308; TNA, WO 142/99, 1917.

the odors of the battlefield and its delayed action caught soldiers unawares, especially those newly deployed at the front. Mustard gas penetrated former places of safety—shell craters and trenches—, demoralizing the tired and exhausted. In rear areas, too, doctors, nurses, and orderlies had to learn how to treat their patients without suffering from cross-contamination.[12]

Having seen 1,400 men gassed in the Villers Bretonneux area, Rawlinson wrote to Winston Churchill, then Minister of Munitions, on April 22, 1918:

> Can you give me any idea when we may expect to have available shells filled with mustard gas? I ask because we have had very severe casualties lately from this form of projectile [...] The men naturally feel that the enemy has a distinct advantage over us in possessing mustard gas and the contention of our chemists that our own lethal shells are still more effective, a contention with which I do not agree, is no satisfactory answer. We feel that we are at a disadvantage in this respect and morale suffers as a consequence (NAM, Rawlinson 1918).

Chemical weapons had already become established as a weapon of harassment. In preparations for the battle of the Somme, British army commanders dispersed gas and smoke amidst the preliminary artillery bombardments at "selected places along the whole British front," compelling the enemy "to wear his gas helmets," induce "fatigue," and cause casualties (TNA, WO 256/10, Haig 1916). The ensuing 110 cylinder discharges, mainly dispersing White Star gas clouds at night, harassed the enemy and caused operational degradation. Foulkes later insisted that a cylinder gas cloud was "far more searching in its effects than the cloud produced by projectiles," since it swept over a much more extensive area, penetrating "every nook and cranny," and tested the enemy's defenses more extensively than any other means of discharge (LHCMA, Foulkes, 1917).

Foulkes sought to maximize these benefits by introducing the retired cylinder or beam attack in 1918. Less hazardous to the infantry, who were withdrawn from the front lines when the attacks occurred, the operations involved thousands of cylinders loaded onto flatcars, brought up by rail (or in one case by lorries and horse-drawn wagons) to rear-area positions, and then releasing the gas simultaneously by electrical detonators. Despite losing cylinders in transport accidents, and suffering delays due to lack of wind, the Special Brigade launched ten beam attacks, releasing gas from 27,000 cylinders and achieving greater concentrations of gas than in previous cylinder operations.[13]

Most gas commanders, though, preferred the flexibility of gas shells, projector drums, and mortar bombs. On all sides gunners experimented with the different types of shell and variations in the volume and rate of fire to achieve surprise, inflict casualties, and neutralize enemy batteries or at least reduce the rates of artillery fire. When faced with British gas shells at the battle of Arras, a German commander emphasized the "complete protection" of the German respirators, but acknowledged

---

[12]Cook 1999, 149–154, 216–17; Heller 1984, 65, 80; Winter 1979, 122–123.

[13]TNA, SUPP 10/292, Foulkes 1918; IWM, Hodgkin July 4–23 1918; LHCMA, Bunker 1965; Foulkes 1936, 291. For a critique of these operations, see Richter 1994, 200–230.

that the "fighting resistance of the men suffered considerably from wearing the mask for many hours." Even worse, horses suffered severely from gas and so the ammunition supply faltered and "the timely withdrawal of batteries could not be affected" (TNA, WO 158/294, von Below 1917).

Various forms of harassment occurred. Once able to fire projectors and 4-inch Stokes mortars in combination, the British bombarded enemy front-line strong points, combining gas with smoke and thermit. Using eleven different fillings in the Stokes mortar bombs and projector drums, the Special Brigade attacked at "all hours of the day and night, and in all wind velocities," even dead calm (TNA, SUPP10/292, Foulkes 1918, 3). They occasionally repeated attacks from the same front after a few hours' interval or disguised attacks by feints with smoke. All this ingenuity reflected the effectiveness of the German respirator once it received a 3-layer drum in June 1916 and further fillings in April 1918. As the Allies were unlikely to penetrate this respirator unless it was damaged or defective,[14] they tried to catch the enemy unawares, distract him, or degrade his fighting efficiency. Even a diversionary bombardment, as took place south-east of Lens in July 1917, involved the delivery of 3,564 drums of gas and 909 mortar bombs across a 3.7 km front over five nights (REM, Crowden, August 3 1917).

During March and April 1918, including Operation Michael (March 21–April 5) the Germans discharged a massive volume of gas. As early as March 9, their "150,000 to 200,000 rounds of Yellow Cross shell [...] caused heavy casualties" and, on the morning of March 21, "some millions of rounds of gas shell" targeted forward posts, trenches, strong points, batteries to a depth of 4.8 km, and villages to a depth of 12.9 km. Although the Blue Cross and Green Cross shell combination failed, as the British SBR blocked the penetration of Blue Cross agents, respirators had "to be worn for many hours, thus adding greatly to the strain and fatigue, and hampering movement and communication" (Hartley 1919–20, 499). Intensive gas shell bombardments continued into April 1918, with estimates of 30,000 to 40,000 rounds poured into Armentières on the night of April 7–8, reportedly leaving the gutters running with mustard gas (ibid.; Edmonds et al. 1918, vol. 2, 163).

The SBR minimized fatalities but protection came at a price in fighting efficiency. As Captain Arthur A. Hanbury Sparrow (Royal Berkshires) observed:

> We gaze at one another like goggle-eyed, imbecile frogs. The mask makes you only half a man. You can't think. The air you breathe has been filtered of all save a few chemical substances. A man doesn't live on what passes through the filter—he merely exists. He gets the mentality of a wide-awake vegetable (Hanbury-Sparrow 1932, 309).

German soldiers subsequently bore the brunt of such harassment when the Allies moved onto the offensive in August 1918. The Canadians employed gas shells, up to 20%, in preliminary assaults upon German batteries, command posts, assembly trenches, observation points, and lines of communication (Cook 1999, 189–90). The American Expeditionary Force (AEF) emulated French artillery tactics, both rapid intense gas bombardments at short range to catch the enemy by surprise, and

[14]TNA, SUPP 10/292, Foulkes 1918, 2; *History of the Corps* 1952, 5: 523.

longer, slower barrages to induce fatigue and lower the enemy's physical resistance and morale. They discharged gas in support of infantry attacks, involving 25% of the ordnance delivered. They also employed Stokes mortars in groups, firing phosgene (as well as smoke and thermit) to attack machine-gun nests, weaken resistance, and inhibit counter-attacks (Heller 1984, 86, 88). As soon as the French acquired mustard-gas shells in March, and the British in September, they incorporated them into their fire-plans: over four days from September 26, 1918 onward, the British Fourth Army fired 750,000 shells at the Hindenburg Line, including some 30,000 mustard-gas shells (Lloyd 2014, 181).

Neither at the time nor subsequently has it been possible to evaluate the exact impact of these chemical attacks. Foulkes lacked evidence about the effects of over half the attacks mounted by the Special Brigade, and the remaining evidence from British observation, the testimony of prisoners and deserters, and German letters, diaries and official documents found on the battlefield related largely to the "losses of small units" (TNA, SUPP 10/292, Foulkes 1918, 3). Both Haber and Richter rightly questioned whether much of this evidence withstood scrutiny, both the value of wartime testimony by prisoners of war and deserters, and post-war evidence from incomplete medical records, including the suspicions of malingering among a disproportionate number of the American gas casualties.[15] Yet Richter, unlike Haber, accepts that gas was perceived as a valuable means of harassment (Richter 1994, 224). The steadily increasing use of poison gas by all belligerents on the Western Front, coupled with plans to use it on an even greater scale in 1919, had the war continued, underscored this perception (Palazzo 1999, 39–50; Prentiss 1937, 684).

Finally, any assessment of poison gas during the Great War has to accept that gas was only used because conventional weapons had failed to break the deadlock of the trenches. Thereafter the belligerents relied primarily upon conventional ordnance, namely 2 million tons of high explosives and 50,000 million rounds of small arms ammunition (Prentiss 1937, 656, 662). Nor did the other novel weapons—the tank and airplane—"change the face of war" as Haber alleged (Haber 1986, 270). The Germans employed only nine tanks in their spring offensive and the British used tanks in masses on only two days during the entire war. Neither Cambrai on November 20, 1917 nor Amiens on August 8, 1918 proved decisive because the British lost the vast majority of their machines (of the 414 sent into battle on August 8, only 145 were available one day later and, by August 12, a mere six machines were able to continue). As John Terraine argued: "The German empire was not going to be overthrown by six tanks, any more than by Trenchard's ten bomber squadrons at Nancy" (Terraine 1978, 116). Although the British and French employed aircraft in unprecedented numbers in 1918, dropping 543 tons of bombs on German targets from June 6 until November 11, 1918, these aircraft had only a supportive role. Limited by meteorological conditions during

[15]Haber 1986, 246–248; Richter 1994, 92; US Department of War, Surgeon General's Office, 1926, XIV: 65.

the autumn, these aerial operations were not decisive in 1918 (Edmonds et al. 1947, vol. 5, 577; Terraine 1982, 274–275, 304–306).

In short, chemical weapons were only one of several novel weapons introduced during the Great War. None of these weapons proved war winners in and of themselves, and none of them broke the deadlock of the trenches. All the major belligerents experimented with new gases and means of delivery; they ensured thereby that chemical warfare evolved in scope and method and grew steadily in tonnage, albeit within a largely supportive role. For the British army, the Special Brigade discharged 87,968 cylinders and fired 196,940 projector drums as well as 177,408 Stokes mortar bombs, delivering some 5,700 tons of gas (TNA, SUPP 10/292, Foulkes 1918, 3). By 1919, the British mounted aerial gas attacks against the Bolsheviks, hardly evidence of any failure shrouding the sense of inquiry and experimentation with poison gas (TNA, WO 106/1148, Ironside 1919).

Anti-gas defenses may have been a priority throughout the war, but if the respirators saved lives, they did so at the price of operational degradation and proffered scant protection against the burning and blistering properties of mustard gas. The desperate desire of the Allies to retaliate in kind as soon as they acquired mustard gas in 1918 demonstrated their concerns about the psychological effects and the perceived operational utility of poison gas. As Hanbury-Sparrow observed:

> It wasn't so much the harm it did to the body, which was always much over-estimated in the popular imagination, as the harm it did to the mind [...] this harmless-looking almost invisible stuff would lie for days on end lurking in low places waiting for the unwary. It was the Devil's breath (Hanbury-Sparrow 1932, 309–310).

This was hardly a weapon that failed.

References

Auld, Major Samuel. 1922. Chemical warfare. *Royal Engineers Journal* 35: 57–71.

Binding, Rudolph. 1929. *A fatalist at war*. Trans. Ian F.D. Morrow. London: G. Allen & Unwin.

Brown, Frederick J. 1968. *Chemical warfare. A study in restraints*. Princeton, NJ: Princeton University Press.

Carter, Gradon B. 1992. *Porton Down: 75 Years of chemical and biological research*. London: HMSO.

Coleman, Kim. 2005. *A history of chemical warfare*. Basingstoke: Palgrave Macmillan.

Cook. Tim. 1999. *No place to run: The Canadian Corps and gas warfare in the First World War*. Vancouver and Toronto: University of British Columbia.

Duguid, Colonel A. Fortescue. 1938. *Official history of the Canadian Forces in the Great War 1914–1919*. 2 vols. Ottawa: J. O. Patenaude.

Edmonds, Brigadier-General Sir, E. James, et al. 1937. *Military operations: France and Belgium, 1918*, vol. 2. London: HMSO.

Edmonds, Brigadier-General Sir, E. James, et al. 1947. *Military operations: France and Belgium, 1918*, vol. 5. London: HMSO.

Eudin, Xenin, Helen D. Fisher, and Harold H. Fisher (eds.). 1946. *The life of a chemist. Memoirs of Vladimir N. Ipatieff*. California: Stanford University Press.

Falkenhayn, General Erich von. 1919. *General Headquarters 1914–1916 and its critical decisions*. London: Hutchinson.
Foulkes, Major-General Charles H. 1936. *'GAS!' The story of the special brigade*. Edinburgh and London: Blackwood.
Foulkes, Major-General Charles H. 1962. Chemical Warfare now, and in 1915. *Royal Engineers Journal* 76: 177–84.
Fraser, David (ed.). 1990. *In good company: The First World War letters and diaries of the Hon. William Fraser, Gordon Highlanders*. London: Michael Russell.
Fries, Amos A., and Clarence J. West. 1921. *Chemical warfare*. New York: McGraw-Hill Book Co.
Griffith, Paddy. 1994. *Battle tactics of the Western Front: The British Army's Art of Attack, 1916–18*. New Haven: Yale University Press.
Haber, Ludwig F. 1986. *Poisonous Cloud: Chemical warfare in the First World War*. Oxford: Clarendon Press.
Hanbury-Sparrow, Lt.-Col. Arthur A. 1932. *The land-locked lake: Impressions of active service in the European War*. London: Arthur Baker.
Hartcup, Guy. 1988. *The war of scientific invention: Scientific developments, 1914–18*. London: Brassey's Defence Publishers.
Hartley, Brig.-Gen. Harold. 1919–20. A general comparison of British and German methods of gas warfare. *Royal Artillery Journal* 46: 492–509.
Heller, Major Charles E. 1984. *Chemical warfare in World War I: The American experience, 1917–18*. Leavenworth Papers No. 10. Fort Leavenworth.
The history of the corps of royal engineers. 1952. vol. 5. Chatham: The Institution of Royal Engineers.
Imperial War Museum (IWM). 85/22/1. Lt. Charles Allen Ashley. 1915. Diary, August 22 and September 24.
IWM. 86/57/1. Brigadier-General Foot. n.d. Once a Gunner.
IWM. 66/163/1. R. C. Gale. 1915. Diary, September 28.
IWM. 06/30/1. T. H. Holmes, n.d. Memoirs.
IWM, P. 399. Captain Adrian E. Hodgkin. 1918. Diary, May 23, July 4–23, and August 2.
Krause, Joachim, and Charles K. Mallory. 1992. *Chemical weapons in Soviet Military doctrine: Military and historical experience, 1915–1991*. Boulder: Westview Press.
Lefebure, Victor. 1921. *The riddle of the Rhine: Chemical strategy in peace and war*. London: Collins.
Lepick, Olivier. 1998. *La grande guerre chimique: 1914–1918*. Paris: Presses Universitaires de France.
Letters From The Front. 1915a. *Bridlington Free Press*, May 28.
Letters From The Front. 1915b. *Bridlington Free Press*, June 25.
Liddell Hart Centre for Military Archives (LHCMA). King's College London. 2/16. Foulkes Mss. 1915. Diary, 25 September.
LHCMA. 6/13. Foulkes. 1917. Notes on the offensive use of gas and methods of gas discharge. Lecture delivered to commanders and staff officers of the Second Army at Helfaut, April 28.
LHCMA. 6/61. Foulkes. n.d. Draft of his chapter for the *History of the Royal Engineers* 5: Chap. 20.
LHCMA. 6/78. Lt.-Col. S.W. Bunker. 1965. 50th Anniversary Re-union Dinner, 25 September.
LHCMA. G/119. Foulkes. 1933. At a Gas Services Reunion Dinner, February 4.
Macpherson, Walter Grant, et al. 1923. *History of the Great War: Medical services*. Diseases of War. London: HMSO.
McWilliams, James, and R. James Steel. 1985. *Gas! The Battle for Ypres, 1915*. St. Catharine's, ON: Vanwell.
Lloyd, Nick. 2014. *Hundred Days: The campaign that ended World War I*. New York: Basic Books.
Main, Steven J. 2015. Gas on the Eastern front during the First World War (1915–1917). *Journal of Slavic Military Studies* 28: 99–132.

National Army Museum (NAM). Rawlinson Mss. Add Mss. 1952-01-33-18. Rawlinson. 1915. Letter to Maj.-Gen. W.P. Braithwaite, June 25.
NAM. Rawlinson Mss. Add. No. 1952-01-33-21, Rawlinson. 1918. Letter to W.S. Churchill, April 22.
National Library of Scotland (NLS). Acc. 3155. Haig Mss. 1915. Diary, September 16.
NLS. Ms 945. Ramage. 1915. Diary, May 15.
New Technical Weapons. 1915. *The Times*, April 29.
Palazzo, Albert. 1999. Plan 1919—The other one. *Journal of the Society for Army Historical Research* 77: 39–50.
Palazzo, Albert. 2000. *Seeking victory on The Western front: The British Army and Chemical Warfare in World War I*. Lincoln and London: University of Nebraska Press.
Peterson, Horace C. 1939. *Propaganda for war: The campaign against American Neutrality, 1914–1917*. Norman, OK: University of Oklahoma Press.
Prentiss, Augustin M. 1937. *Chemicals in war: A Treatise on chemical warfare*. New York, NY: McGraw-Hill.
Read, James, M. 1941. *Atrocity Propaganda 1914–1919*. New Haven, CT: Yale University Press.
Richter, Donald. 1994. *Chemical soldiers: British gas warfare in World War I*. Barnsley: Leo Cooper.
Robinson, Julian Perry. 1971. The rise of CB weapons. In *The problem of chemical and biological warfare*, ed. Stockholm International Peace Research Institute, vol. 1. Stockholm: Almqvist & Wiksell.
Royal Engineers Museum (REM), R15. Special Brigade re. Gas in World War 1 1915–18.
Crowden. 1917. Report on Operations of the Special Companies, R.E. attached to Canadian Corps, July 15th to August 15th.
REM. R15. Crowden. 1917. Operation Order No. 4/C, August 3.
Spiers, Edward M. 1986. *Chemical Warfare*. London: MacMillan Press.
Stoltzenberg, Dietrich. 1993. *Fritz Haber: Chemist, Nobel Laureate, German, Jew*. Philadelphia: Chemical Heritage Press.
Terraine, John. 1978. *To win a war*. London: Sidgwick & Jackson.
Terraine, John. 1982. *White Heat: The New Warfare 1914–18*. London: Book Club Associates.
The National Archives (TNA). Cabinet Mss. CAB 37/127. 1915. Henry H. Asquith to King George V, 27 April and CAB 37/128. 1915. Asquith to King George V, May 5.
TNA. PRO 30/57. Kitchener Mss. French. 1915. Sir John French to Lord Kitchener, April 27.
TNA. SUPP 10/292. Brigadier-General C. H. Foulkes. 1918. Report on the Activity of the Special Brigade During the War, December 19.
TNA. War Office. WO 32/5483. n.d. An Account of German Cloud Gas Attacks on British Front in France.
TNA. WO 32/5483. n.d. Diary of Development of British Respirator.
TNA. WO 33/1072. Brigadier-General H. Hartley. n.d. Report on German Chemical Warfare Organisation and Policy, 1914–1918.
TNA. WO 42/195. Captain W. L. Hicks. 1917. Gas Services Reports from Russian Front.
TNA. WO 106/1148. Maj.-Gen. Edmund Ironside. 1919. Letter to Gen. Rawlinson, September 16.
TNA. WO I33/1072. Brigadier-General Harold Hartley. n.d. Report on German Chemical Warfare Organization and Policy, 1914–1918.
TNA. WO 142/99. Lt.-Gen. Charles Fergusson. 1915. Gas Attack on VI Corps, December 22.
TNA. WO 142/99. 1917. Report on the Gas Shell Bombardment at Nieuport on Night of 21/22nd July.
TNA. WO 158/128. Brigadier-General C. H. Foulkes. 1918. Summary of Hostile Gas Activity, October 18.
TNA. WO 158/234. Lt.-Gen. Lawrence Kiggell. 1916. Memorandum, June 19 and Maj.-Gen. Archibald A. Montgomery. 1916. Report of the Army Commanders remarks at the conference held at Fourth Army headquarters, 22 June 1916, June 24.
TNA. WO 158/270. Gen. George de S. Barrow. 1916. Report, First Army No. 852(G), August 18.

TNA. WO 158/294. 1917. Effect of British Gas Projectors, containing extract from an order of the 111th Division, July 8.
TNA. WO 158/294. Gen. Fritz von Below. 1917. Experiences Derived From the Arras Battle, April 11 and Hartley. n.d. British and German Methods of Gas Warfare.
TNA. WO 188/213. Brigadier-General H. Hartley. n.d. Chemical Warfare.
TNA. WO 256/10. Haig Diary. 1916. Army Commanders Conference at HQ Third Army, May 27.
Trumpener, Ulrich. 1975. The road to Ypres: The beginnings of gas warfare in World War I. *Journal of Modern History* 47: 460–480.
United States Department of War, Surgeon General's Office, 1926, *Medical department of the U.S. Army in World War. Medical aspects of gas warfare.* XIV. Washington, D.C.: War Department.
von Frauenholz, Eugen (ed.). 1929. *Kronprinz Rupprecht von Bayern, Mein Kriegstagebuch*, vol. 1. Munich: Deutscher National Verlag.
Winter, Denis. 1979. *Death's Men: Soldiers of the Great War.* London: Penguin.

"Gas, Gas, Gaas!" The Poison Gas War in the Literature and Visual Arts of Interwar Europe

Doris Kaufmann

Abstract The gas attacks during the First World War stood for a new kind of warfare and shaped the soldiers' experience of living through an apocalypse never before imagined. This article examines the literary and artistic topics and forms used to express this ordeal by German, British and French writers, poets and painters, the majority of whom had fought in the war. There are striking similarities in their representation of the gas war: the impersonality of this enemy, the feeling of helplessness in gas attacks, the shock of seeing one's comrades "guttering, choking, drowning" and not least the exposure to an infernal landscape. Nearly all of the authors and painters condemned the waste and pointlessness of the ongoing or past war, but their vision of the future often differed according to their national background. The second part of this article addresses the public battle over the interpretation and collective remembrance in the war's aftermath. Particularly at the end of the 1920s, a wave of publications mainly in England and Germany displayed a renewed public interest in the preceding war. The written recollections and paintings of the gas warfare played a significant role here.

In his 1929 war novel *Death of a Hero*, the English writer Richard Aldington depicts at one point how his protagonist—a soldier stationed on the Western Front but a modern painter in civilian life—attempts in vain to sketch a military engagement he once experienced, a combat operation that included heavy artillery shelling, a long-lasting barrage of gunfire, and a gas grenade attack. Although he sees "the ruined village" and "the broken desecrated ground" in front of him and hears "the 'claaang' of the heavies dropping reverberantly into M—," "his hand and brain" fail him (Aldington 2013, 315–316). He destroys both of his pre-war drawings, along with an old self-portrait. Aldington describes here one of the answers to the problem that was inevitably posed for the European avant-garde artists immediately after the outbreak of the war. In 1914 many of them had volunteered enthusiastically to go to war, a war from which they hoped to

D. Kaufmann (✉)
University of Bremen, Bremen, Germany
e-mail: dkaufman@uni-bremen.de

B. Friedrich et al. (eds.), *One Hundred Years of Chemical Warfare: Research, Deployment, Consequences*, DOI 10.1007/978-3-319-51664-6_10

experience a purification and also a destruction of societies they regarded as outdated, decrepit, and suffocating, and from which they expected the birth of a "New World of Art" and, even more, of a "New Age." Soon after the turn of the century, the avant-gardists had already commenced with their artistic "fragmentation of reality" (in the words of Gottfried Benn) and were creating a new kind of art both in form and content. But how should the unprecedented and initially incomprehensible experiences in the industrialized war of materials be dealt with and expressed artistically: the mass killings and mass deaths, the new demands imposed on perception and behavior induced by the long-range artillery rounds, machine gun deployments, and drumfire, by gas, grenades, aerial bombs, and the first tank attacks? (For a survey, see Jürgens-Kirchhoff (1993), Cork (1994), Kunst- und Ausstellungshalle der Bundesrepublik Deutschland in Bonn (2013).)

Part I

As early as September 1914, the war volunteer Franz Marc wrote with a sense of amazement:

> It is incredible that there were times in which one represented war by painting campfires, burning villages, falling horses or riders on patrol and the like. This idea strikes me as downright comical, even if I think about Delacroix [...], we must do this completely differently, completely differently! (Marc 1989, 102–103)

The French writer Léon Bloy also asked himself, after visiting an exhibition of war paintings, how one could express the reality of this war "without making oneself ridiculous and without becoming a liar" (Robichon 1994, 296). A clear answer was provided in 1917 by the art historian Richard Hamann. Modern battle, Hamann wrote, had become impossible to portray. A representational portrait of a large battlefield would be unable to depict human beings at all. It could only display a vast field, ruins, vapors, clouds, and sky. Above all—according to Hamann —"the mass, the quantum of suffering that such a war has brought on the world cannot even be intimated by condensing it onto the narrow space of a picture" (in Jürgens-Kirchhoff 1993, 18).

But this was precisely what visual artists from a number of countries, almost without exception combatants and veterans, were attempting as they created important works—works they were already making during the war as well as in the 1920s and 1930s. These paintings are important or relevant because they had a major impact on contemporaries, an impact the viewer of today still senses. They often facilitate a deeper insight into the World War beyond factual knowledge. Therefore this group of sources is of particular relevance for my inquiry, namely the paintings of visual artists who dealt with the war in a critical way. Their reception by contemporaries shows that these works expressed dimensions and interpretations of the war that reverberated in postwar societies and have also found their way into our contemporary cultural memory. Pictures that were apologetic about the war are not examined here. This is due to the thematic aspect that is the focus of attention here. As a pictorial subject, the gas war does not lend itself—saying this is almost a banal statement—to affirming war, and it is poorly suited as a motif for trivializing war, though in anti-German war cartoons after 1915 it certainly lent itself to an "irrefutable illustration of 'Hun' barbarity"[1]

Anyone examining the immense inventory of pictures from the First World War will notice that the European avant-garde artists, as well as the painters of other art movements critical of academicism, did not make the gas war one of their favored subjects. The "lack of pictorialness" in modern war lamented by contemporaries,[2] especially because the individual soldier's achievement had become invisible—for which the London *Times* found the apposite phrase "the butchery of the unknown by the unseen"[3]—became particularly apparent in depictions of the gas war. What might initially seem to be the obvious and conventional approach—painting the ostensibly visible, that is, the emptiness of the ravaged battlefield, with swathes of gas—left the viewer in the dark, since the swathes might signify anything from poison gas to artificial fog or the smoke of artillery and grenade fire. This is obvious when looking at the two paintings by Ferdinand-Joseph Gueldry and Georges Leroux (Figs. 1 and 2).

Fig. 1 Ferdinand-Joseph Gueldry, Le ravin de la mort à Verdun, 1916

[1]This refers to the Dutch cartoonist Louis Raemekers's work (Das 2012, 398–399).

[2]According to the sculptor Erich Stephani, in Leonhardt (2014, 598).

[3]*The Times*, November 1914, in Bogacz (1986, 661).

Fig. 2 Georges Leroux, L'enfer, 1921, Imperial War Museum London

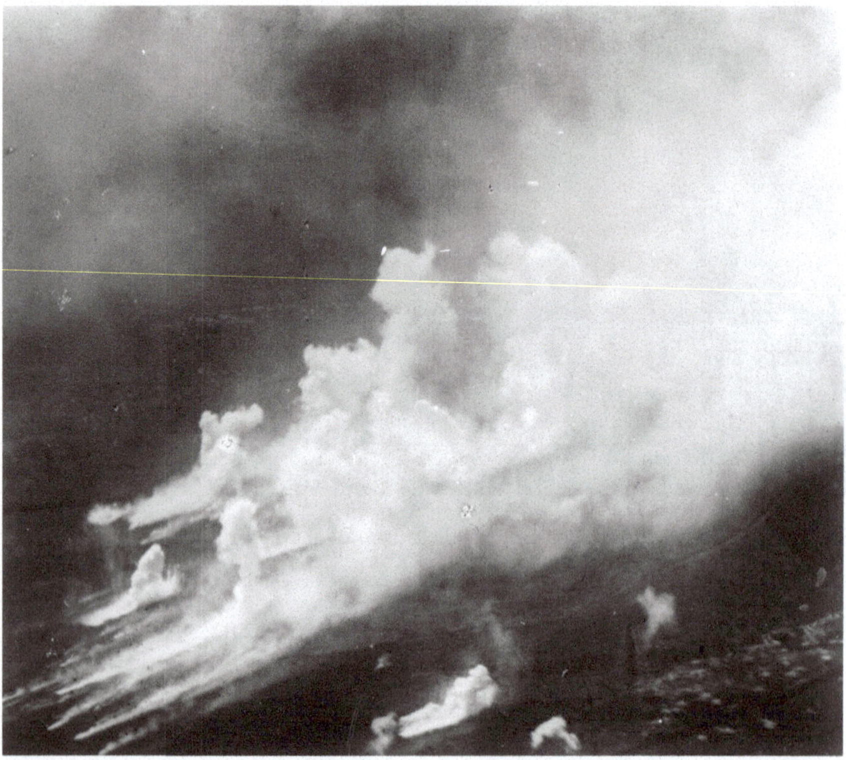

Fig. 3 Gas attack photographed from the air, Imperial War Museum London

Leroux's painting recalls one of the battles at Verdun in 1916 where phosgene was used for the first time. The view of the scene by these two artists here is competing with photography, the frequently used new medium that could presumably capture the moment more accurately and with a claim to "authenticity" (see Hüppauf 1992, 2004a) (Fig. 3).

Most importantly, the terror of gas attacks is rather absent in the two paintings. Yet works of art were created in which painters took up the challenge that a British art critic had posed to them, namely "to recover the 'truth' of modern war" (Bennett). Since this "truth" was hidden behind the visible reality, it seems logical that elements of cubism and futurism (rather than the more naturalistic conventions of pre-war painting) dominate the two works of art by Otto Dix and Robert Williams introduced below (Fig. 4).

Fig. 4 Otto Dix, Lichtsignale, 1917

In 1917 the war volunteer and machine-gunner Otto Dix painted a gouache entitled *Lichtsignale* ("Light Signals"), (Fig. 4). Green, red, and white flares warned of gas attacks, acting as a kind of "Gas S.O.S" (Spear and Summersgill 1991, 310). They signaled the beginning of the terror whose end Dix had painted here. In the literature written by war veterans, too, flares are a constantly repeated theme. Thus, Edlef Koeppen writes in his novel *Heeresbericht* ("Army Communiqué") from 1930:

> 'Lieutenant, Sir!' He cannot say more than that. Red flares dance in front before his eyes. The green against the morning sky can only be seen dimly, which makes the red more menacing. Green-red everywhere. From Loos to the dump, like a veil, green-red is dancing, whirling. At the same time, mind-boggling gunfire. 'Gas!' All three of them shout this at the same time" (Koeppen 1992, 191).

The Vorticist William Roberts, an artist from the English prewar avant-garde, served until the end of 1917 as a machine gunner before he painted "The First German Gas Attack at Ypres" in the spring of 1918 as official military artist for the Canadian War Memorials Fund (see Gough 2010, 278–290; Malvern 2004, 122–124) (Fig. 5). The picture was shown in 1919 at the London Royal Academy of Arts in the exhibit "The Nation's War Paintings," where it generated controversy. In a panic-like flight from the gas clouds, Franco-Algerian soldiers in blue-red

Fig. 5 William Roberts, The First German gas attack at Ypres, 1918, National Gallery of Canada, Ottawa

uniforms come up against a rearward position occupied by Allied Canadian troops. A chaotic tangle of suffering men, not exactly a heroic narrative for a victorious nation—the Canadians ultimately did hold their emplacement—is what Roberts had painted.

Like William Roberts, other English artists also wanted to convey the horrendous front-line experiences they had gained in a war in which technology had long since carried any kind of heroic romanticism to the point of absurdity. In the last year of the war, the painter Paul Nash wrote:

> I am no longer an artist interested and curious. I am a messenger who will bring back word from the men who are fighting to those who want the war to go on for ever. Feeble, inarticulate, will be my message, but it will have a bitter truth, and may it burn their lousy souls" (in Jürgens-Kirchhoff 1993, 382, fn 110).

How artists could express this "bitter truth" of the gas war is something the English public was able to view as early as the spring of 1918 in several nationwide exhibitions (Malvern 2004, 37–55; Hynes 1990, 198–202) (Figs. 6 and 7).

None of the relentless pictures of artists like William Rogers, Eric Kennington, Gilbert Rogers or Paul Nash became a "corner stone" in "the public's imagination

Fig. 6 Eric Kennington, Gassed and Wounded, 1918

Fig. 7 Gilbert Rogers, Gassed: In Arduis Fidelis, 1919

in the decades after the Great War."[4] This key position in the public imagination was occupied by another work, the monumental painting "Gassed" by John Singer Sargent (Fig. 8).

The British War Memorials Committee had commissioned Sargent, a very well-known American portraitist and salon painter of the late nineteenth century, to contribute a mural for the Hall of Remembrance the Committee had been planning. Sargent visited the Western Front from July through September 1918, where he witnessed the impact of a German mustard gas attack near Arras. The huge oil painting, measuring six by three meters, was selected by the Royal Academy as the picture of the year for 1919 and was admired by Winston Churchill "for its brilliant genius and painful significance."[5] Yet there were also critical voices. After visiting the exhibition in the Royal Academy, Virginia Woolf saw Sargent's painting as testimony to the belief in soldierly suffering as something that has to be counted as the price that must be paid for the "greater good of the Empire" (Harvey 2010, 149).[6] The pathway of the apparently more lightly wounded blond men with clean head bandages walking upright —two of them still carrying a gun—to the dressing station (indicated on the far right of the picture by tent poles) is depicted as a sunlit sacrificial path.

[4]Gough (2010, 197–200), on Sargent, p. 197.

[5]*The Times*, 5 May 1919, in Harvey (2010, 148).

[6]On the shifting reception of Sargent's painting—exhibited in the Imperial War Museum—between 1920 and 1939, see also Malvern (2000).

Fig. 8 John Singer Sargent, Gassed, 1919

Five years after Sargent's picture, Otto Dix published two etchings of gas victims that are far removed from causing such an impression (Figs. 9 and 10).

Fig. 9 Otto Dix, Die Schlafenden von Fort Vaux (Gas-Tote), in: Mappe "Der Krieg," 1924

Fig. 10 Otto Dix, Gas Tote (Templeux-La-Fosse, August 1916), in: Mappe "Der Krieg," 1924

There is no need for any statement from Dix about how he wanted to show the war "without all the propaganda."[7] The two etchings come from his 50-sheet portfolio *The War*. It was issued with an afterword by Henri Barbusse, the author of the first autobiographically inspired anti-war novel *Le Feu* in 1916. Its publication was accompanied by an exhibition tour in Germany taking place in 1924—the very year the pacifist movement had declared as the Antiwar Year (Schubert 2002, 39–46). Dix's pictures of the First World War need to be placed in the political context of the antiwar movement that gained influence in the mid-1920s (Riesenberger 1990, 250–275; Holl 1988, 138–204). In Germany it competed with a national-conservative, militaristic camp and its interpretation of the World War as an opportunity for the rebirth of an antidemocratic nation, ready for war and structured along authoritarian lines (to summarize this political vision in very abbreviated fashion). Ten years after the World War's end, the debate about how to interpret its meaning picked up noticeably, especially in Germany and England (Hynes 1990, 423–459; Hüppauf 2004b; Eksteins 1989, 275–299). At the center of this debate were also literary works, often with an autobiographical background, in which the authors tried to come to terms with their frontline experience during the war.

Part II

The second part of this article examines the different ways in which the gas war was recalled in the anti-war and pro-war literature, the contexts in which the gas war appeared, and which general interpretation of the war was promoted by the gas war narratives.

As in the visual artworks, suffering and death by gas constitute an important topic in the war literature. In his poem "Dulce et Decorum Est," Wilfred Owen, the famous English war poet who died in battle shortly before the end of the war, put the agonizing physicality of death by gas into haunting words:

Gas! GAS! Quick boys!—An ecstasy of fumbling,
Fitting the clumsy helmets just in time;
But someone still was yelling out and stumbling,
And flound'ring like a man in fire or in lime…
Dim, through the misty panes and thick green light,
As under a green sea, I saw him drowning.
In all my dreams, before my helpless sight,
He plunges at me, guttering, choking, drowning.
If in some smothering dreams you too could pace

[7]Thus Otto Dix in an interview from 1957, in Dix (2014).

Behind the wagon that we flung him in,
And watch the white eyes writhing in his face,
His hanging face, like a devil's sick of sin;
If you could hear, at every jolt, the blood
Come gargling from the froth-corrupted lungs, […][8]

In her moving autobiography, Mary Britnieva, who was employed as a voluntary nurse on the Eastern front, likewise describes the shock of having to stand by helplessly and watch the agony of gas poisoning:

> They lay on their backs mostly, their upturned faces terribly swollen and livid—some almost blue—choking and coughing, their bloodshot eyes protruding, unable to utter a word, yet fully conscious, only their eyes and their occasional spasmodic feeble movements proclaiming the supreme agony that they were enduring. Some were even coughing up pieces of their lungs that the cruel gas had disintegrated in their living bodies. […] The realization of our helplessness was almost unbearable; a wound can be dressed and the flow of blood from a hemorrhage can be staunched, but this fiendish weapon had got science and surgery beaten."[9]

In the last volume of his serial novel *Les Thibault*, Roger Martin du Gard, the French writer who was awarded the Nobel Prize in Literature in 1937, confronts his readers with a lengthy depiction, replete with all the medical details, of how it takes months for his protagonist, a physician who himself becomes the victim of a mustard gas attack during the inspection of a hospital for gas poisoning cases, to die (Martin du Gard 1972). If Owen, Britnieva, and Martin du Gard are plainly articulating their subjective concern and compassion in the face of the gas poisoning victims, these kinds of emotions are more noticeable by their absence from other writers' accounts. In *Storm of Steel*, Ernst Jünger is rather detached as he describes the following scene:

> [I]n Monchy we saw a lot of men affected by gas, pressing their hands against their sides and groaning and retching while their eyes watered. It was a sad business, because a few of them went on to die over the next several days, in terrible agony. […] Henceforth, I resolved never to go anywhere without my gas mask. (Jünger 2004, 81; 1993, 92)

The steel-hard Stormtroop stoicism praised a bit later by him—"If a man falls, he's left to lie. No one can help. No one knows if he'll return alive" (Jünger 2004, 92; 1993, 104)—is illustrated by another author who also belonged to the camp of soldierly nationalism. Werner Beumelburg writes in his best-selling novel of 1930, *Die Gruppe Bosemüller* ("The Bosemüller Group"):

> The fire trench is crashed to a pulp. They stumble across a couple of figures who are sitting there. Why aren't they going any further? […] 'Comrade,' one of them whispers, holding fast to the lieutenant's leg. 'What's wrong? What are you doing here?' 'Gas…' In the spraying flares of a hundred flames one sees their ghostly-yellow faces, their elongated necks, their circularly lacerated eyes. They regurgitate. They gasp for breath. Somebody, out of sheer

[8]Wilfred Owen, "Dulce et Decorum Est," in Barlow (2014, 40). Owen's poem is the focus of Das (2012).

[9]Mary Britnieva, *One Woman's Story*, in Hallett (2010, 75).

> helplessness, has wrapped bundles of bandages around their throats […] 'Comrade' is whispered, as the lieutenant's leg is clutched […] He tears his leg away with force. 'Comrade…' 'Forward!' screams the lieutenant. 'Form groups of four men […,] wait over there in the fire trench in front of the Vaux-Cross […]—Move on in' (Beumelburg 1930, 271–272).[10]

In her book *Augenblicke der Gefahr: Der Krieg und die Sinne* ("Moments of Danger: The War and the Senses"), the literary scholar Julia Encke has tellingly explored the measures undertaken to armor the individual against the imperceptible gas weapon (Encke 2006, 197–218). Indeed, a recurrent topic in the war literature are the emphatic descriptions of soldiers' attempts to locate gas before, during, and after the attack by way of hearing, smell, and vision. Whistling hisses, the peculiar way that gas grenades pop up, the specific formation and color of the gas clouds, the smell of bitter almonds, sweet onions, of apples, mustard, and garlic, and in particular the way comrades are watched—are there any consequences as soon as they take off their gas masks?—are frequent themes.[11] In order to make this barely discernible danger of gas describable, a menacing bodily shape is often ascribed to the gas. In Erich-Maria Remarque's *Im Westen nichts Neues* ("All Quiet on the Western Front"), the puffs of gas become a soft jellyfish animal that lays itself down in the craters and stays there to loll (Remarque 1980, 54). As depicted by Werner Beumelburg, gas is a creature that has dead eyes and frozen hands, that longingly extends its frozen hands, is not forgetful of a single crack in the ground, trickles and flows and spreads (Beumelburg 1930, 286–287).

The anonymity of the modern battlefield becomes all-encompassing. Not only is the enemy invisible; people on the own side become indistinguishable behind the gas mask. During a gas attack, the Russian officer Fedor Stepun remembers "the terrible unrecognizability of all the people all around, the loneliness of an accursed, tragic masquerade: white rubber skulls, quadratic glass eyes, long green snouts" (Stepun 1963, 318–19) (Fig. 11). There is a similar description by Richard Aldington in his war novel *Death of a Hero*. His protagonist, Winterbourne, undergoes a gas attack's aftermath. He

> stood at the end of the trench to help out the groping, half-blinded men. As they filed by, grotesques with india-rubber faces, great, dead-looking goggles, and a long tube from their mouths to the box respirators, Winterbourne thought they looked like lost souls expiating some horrible sin in a new Inferno. (Aldington 2013, 279)

How much the experience of such an unprecedented kind of inferno was shared by combatants from all countries is also demonstrated by the following passage from *Storm of Steel* by the German writer Ernst Jünger:

> With weeping eyes, I stumbled back to the Vaux woods, plunging from one crater into the next, as I was unable to see anything through the misted visor of my gas mask. With the

---

[10]On Beumelburg, see also Krumeich (2011).

[11]For an example, see Dorgelès (1988, 261): "Bouffioux lay huddled in a corner and no longer even wanted to remove his gas mask; the smallest little cloud of powder pressing down on us frightened him. For a whole hour we heard him stammering: 'That smells like apples … That smells like mustard … That smells like garlic …' and each time he anxiously slipped on his pig snout. Now he was no longer even taking it off, and the way he hunkered down in his hole with his wagging head made him resemble a carnival monster."

Fig. 11 Henri de Groux, Masques à gaz, 1916

extent and inhospitableness of its spaces, it was a night of eerie solitude. Each time I blundered into sentries or troops who had lost their way, I had the icy sensation of conversing not with people, but with demons. We were all roving around in an enormous dump somewhere off the edge of the charted world. (Jünger 2004, 114; 1993, 129) (Fig. 12)

Modern weapons technology has transformed the landscape into a gigantic scrapyard. So it is not surprising that the central figure in Edlef Koeppen's *Heeresbericht*, Lieutenant Reisinger, heads off not unwillingly, together with a noncommissioned officer, on a reconnaissance mission beyond the immediate combat zone. They marvel at the sunlit green grass, the many poppy flowers, and the young birch forest nearby. Then all of a sudden the warning cry: "The leaves aren't green, but lilac" (Koeppen 1991, 356). The woods have been gassed, and even the white stems are sprayed with a greasy lilac-red fluid. Nature itself has become a weapon. The crossing turns into a problem. "For heaven's sake, don't bump into just any tree. Don't touch any leaf. Hands in your pockets. Make yourself as tight and small as possible" (ibid., 358). It is not about to turn out well. The noncommissioned officer fails to notice—with the open field already in front of him—an overhanging birch branch, which tears off his gas mask. He dies right in front of Reisinger, who is as helpless as he is shocked. Not coincidentally, Koeppen has placed this scene almost at the end of his book. It is jointly responsible in a

Fig. 12 Otto Dix, Die Sturmtruppe geht unter Gas vor, Kriegsmappe 1924

fundamental way for Reisinger's ultimate "breaking point"—as the English war veteran Robert Graves called it in his book *Goodbye to All That* (Graves 2000, 164). Koeppen's Reisinger brings his personal war to an end and ends up in a psychiatric hospital. Aldington's hero too shares a similar experience after a day-long battle with a massive use of gas. Toward the end of the war he senses "a cut in his life and personality," and "a sense of fear he had never experienced," which allows him to continue fighting only with a huge expenditure of coercion (Aldington 2013, 293–294). Finally, he commits a hidden suicide. An open admission like that of Fedor Stepun—"but hovering over all this, the insane fear of a difficult and disgusting death (by gas)" (Stepun 1963, 319)—is seldom found in the war literature. In any event, such a confession would only be expected in the pacifist war literature, though even here the nearly ineluctable dictate that doubtless prevailed among contemporaries was bravery against the enemy under all circumstances. However, the aforementioned Robert Graves, author of the best-known 1929 English war book with its telling title, links in this memoir his condition of suffering from bad nerves with his experience of poison gas attacks on the Western Front. "Since 1916, the fear of gas obsessed me: any unusual smell, even a sudden strong scent of flowers in a garden, was enough to send me trembling" (Graves 2000, 220; see also 217–218).

The autobiographically guided front literature largely screens out what caused and who was responsible for the gas war. These questions are, however, at the center of some plays, novels, and science fiction literature that appear around the same time. These works of art focus on the war-inducing nexus linking the chemical gas-producing armaments industry, the military, and government policy to each other and contemplate different ways to break up this military-industrial complex (Fig. 13).

Fig. 13 Gerd Arntz, Fürs Vaterland, 1936

In this context, the expressionist playwright Georg Kaiser uses two of his plays that were Europe-wide hits, *Gas* and *Gas 2* from 1918 and 1920, to warn against the way technological knowledge and technological processes create rules and an order of their own (Kaiser 1978). At the end of *Gas 2*, a poison gas explosion destroys all of civilization. In 1926 Johannes R. Becher, who later becomes the first Minister of Culture in the German Democratic Republic, published his gas warfare novel *CHCl = CH)*$_3$ *As (Levisite) oder Der einzig gerechte Krieg* ("Levisite or The Only Just War"), which was immediately banned by the censor. Becher creates an apocalyptic global class war that is conducted by capitalist governments with aerial poison gas bombings against the population and workers' armies. In his novel he alludes to existing literary narratives about the recent gas war. Building scientific treatises about chemical agents and the injuries caused by poison gas into his horror story, Becher goes on to contrast this doomsday scenario with a bright Communist future that he illuminates for the reader. In one part of that future world, the proletariat, protected by the Soviet air fleet, will own and control the industries that make chemical dyes and weapons (Becher 1985).[12]

With his assumption that the war of the future would be an aerial war using gas and bacteria, in which cities would be gassed and the population thereby drawn into what is now a total war, Becher falls into line with the new horror scenario invoked by literature in the 1920s and 1930s. In France, Germany, England, Italy, and the USA, pacifist writers, but also authors from military circles and authors of science fiction in postwar Europe, conducted a *Zukunftskrieg* (future-war fiction) in which the inhabitants of Paris, London, and Berlin as well as entire tracts of land were sometimes destroyed from the air by poison gas. If these scenarios as employed by pacifist writers were meant to warn against a new war, in the hands of the military

[12]For a detailed discussion, see Berman (1985); Vollmer (2003).

authors they served to promote a new arms buildup or rearmament for each country's air force (Schütz 2005).

In the war literature of all the countries that participated in the First World War, there are hardly any differences in how the new perceptional and behavioral impositions caused by the gas war are represented. Gas warfare intensified the impression of the enemy's anonymity as well as magnifying the unrecognizability —in Stepun's words—"of all people all around" (Stepun 1963, 318). Furthermore, in many of the European literary and pictorial works on the war the destruction of nature and of the landscape by modern weapon systems is addressed and condemned. The artists are aware that this devastation had been heightened even more by the gassing of the environment, which becomes a weapon in its own right. Helplessness in the face of agonizing dying and of death by gas is also a common theme in literature and visual arts of interwar Europe, frequently coupled with horror visions of a future total gas and aerial warfare.

Yet there are differences: That soldiers often reacted to the "new Inferno" of the gas war with a mental breakdown—suffering a gas shock as a variation of shell shock—is addressed almost only by English writers, whose military career prohibited the verdict of cowardice often associated with breakdown.[13] Robert Graves and Richard Aldington, gas victims themselves, used their descriptions of psychological injuries at the end of the 1920s to affirm their overall interpretation of the war as completely meaningless. This places them in the ranks of those European writers and visual artists who had already come to the same conclusion during the war and in the 1920s like Wilfred Owen, Henri Barbusse, Roland Dorgelès, Erich Maria Remarque, Edlef Koeppen, Roger Martin du Gard, and Fedor Stepun, to mention only the authors quoted above.[14] In the German spectrum of writers, however, a completely different interpretation was propagated by the advocates of soldierly nationalism who glorified the war experience like Ernst Jünger and Werner Beumelburg. The former Stormtroop officers saw an armored and heroic "New Man" emerging from behind the gas clouds of the war. Their combat continued after 1918—this time against the Weimar Republic.

References

Aldington, Richard. 2013 [1929]. *Death of a hero*. London: Penguin.

Barlow, Adrian, ed. 2014. *Six poets of the Great War*. Cambridge: Cambridge University Press.

Becher, Johannes R. 1985 [1926]. *$(CHCl = CH)_3$ As (Levisite) oder Der einzig gerechte Krieg*. Berlin, Weimar: Aufbau.

Bennett, Lloyd. Richard Jack: Canada's Battle Painter. http://visualarts.inside.tru.ca/files/2014/06/Richard-Jack-Canadas-Battle-Painter1.pdf. Accessed 6 Oct 2016.

[13]This point has been emphasized in Cook (1999, 233–238).

[14]In this context, the historian Jay Winter emphasizes a "communality of European cultural life in the aftermath of the war" (Winter 1995, 227).

Berman, Russell A. 1985. The politics of form and the discourse of war in Johannes R. Becher's Levisite. *Modern Language Studies* 15 (4): 110–116.
Beumelburg, Werner. 1930. *Die Gruppe Bosemüller*. Gerhard Stalling: Oldenburg.
Bogacz, Ted. 1986. 'A Tyranny of Words': Language, poetry, and antimodernism in England in the First World War. *Journal of Modern History* 58 (3): 643–668.
Cook, Tim. 1999. *No place to run: The Canadian corps and gas warfare in the First World War*. Vancouver: University of British Columbia Press.
Cork, Richard. 1994. Das Elend des Krieges: Die Kunst der Avantgarde und der Erste Weltkrieg. In *Die letzten Tage der Menschheit: Bilder des Ersten Weltkrieges*, ed. Rainer Rother, 301–396. Berlin: Ars Nicolai.
Das, Santanu. 2012. 'An Ecstasy of Fumbling': Gas warfare, 1914–18 and the uses of affect. In *The Edinburgh companion to twentieth-century British and American war literature*, ed. Adam Piette and Mark Rawlinson, 396–405. Edinburgh: Edinburgh University Press.
Dix, Otto. 2014. 'Schade um die Kerle': Was Otto Dix in den Schützengräben des Ersten Weltkriegs erlebte—und wie er dieses Trauma als Künstler verarbeitete. Hier kommt er selbst zu Wort. *Die Zeit*, April 16. http://www.zeit.de/2014/17/otto-dix-feldpost/.
Dorgelès, Roland. 1988 [1919]. *Die hölzernen Kreuze* [*Les croix de bois*]. Leipzig, Weimar: Gustav Kiepenheuer.
Eksteins, Modris. 1989. *Rites of spring: The Great War and the birth of the modern age*. Boston, MA: Houghton Mifflin.
Encke, Julia. 2006. *Augenblicke der Gefahr: Der Krieg und die Sinne*. Munich: Wilhelm Fink.
Gough, Paul. 2010. *A terrible beauty: British artists in the First World War*. Bristol: Sansom and Company.
Graves, Robert. 2000 [1929]. *Goodbye to all that*. London: Penguin.
Hallett, Christine E. 2010. Portrayals of suffering: Perceptions of trauma in the writing of First World War nurses and volunteers. *Canadian Bulletin of Medical History* 27 (1): 65–84.
Harvey, Benjamin. 2010. Virginia Woolf, art galleries and museums. In *The Edinburgh companion to Virginia Woolf and the arts*, ed. Maggie Humm, 140–159. Edinburgh: Edinburgh University Press.
Holl, Karl. 1988. *Pazifismus in Deutschland*. Frankfurt am Main: Suhrkamp.
Hüppauf, Bernd. 1992. Kriegsfotografie an der Schwelle zum Neuen Sehen. In *Geschichte und Psychologie*, ed. Bedrich Loewenstein, 205–233. Pfaffenweiler: Centaurus.
Hüppauf, Bernd. 2004a. Ikonographie. In *Enzyklopädie Erster Weltkrieg*, ed. Gerhard Hirschfeld, Gerd Krumeich, and Irina Renz, 569–570. Paderborn: Ferdinand Schöningh.
Hüppauf, Bernd. 2004b. Kriegsliteratur. In *Enzyklopädie Erster Weltkrieg*, ed. Gerhard Hirschfeld, Gerd Krumeich, and Irina Renz, 177–191. Paderborn: Ferdinand Schöningh.
Hynes, Samuel. 1990. *A war imagined: The First World War and English culture*. London: Pimlico.
Jünger, Ernst. 1993 [1920]. *In Stahlgewittern*. Stuttgart: Klett-Cotta.
Jünger, Ernst. 2004 [1920]. *Storm of steel*. Trans. Michael Hofmann. New York, London: Penguin.
Jürgens-Kirchhoff, Annegret. 1993. *Schreckensbilder: Krieg und Kunst im 20. Jahrhundert*. Dietrich Reimer: Berlin.
Kaiser, Georg. 1978. *Gas: Schauspiel*. Frankfurt am Main: Ullstein.
Koeppen, Edlef. 1992 [1930]. *Heeresbericht*. Reinbek: Rowohlt.
Krumeich, Gerd. Zwischen soldatischem Nationalismus und NS-Ideologie: Werner Beumelburg und die Erzählung des Ersten Weltkriegs. In *Burgfrieden und Union sacrée: Literarische Deutungen und politische Ordnungsvorstellungen in Deutschland und Frankreich 1914–1933*, eds. Wolfram Pyta and Carsten Kretschmann, 295–312. Munich: Oldenbourg.
Kunst- und Ausstellungshalle der Bundesrepublik Deutschland in Bonn, ed. 2013. *1914: Die Avantgarden im Kampf*. Cologne: Snoeck.
Leonhardt, Jörn. 2014. *Die Büchse der Pandora: Geschichte des Ersten Weltkriegs*. Munich: C.H. Beck.
Malvern, Sue. 2000. War, memory and museums: Art and artifact in the Imperial War Museum. *History Workshop Journal* 49: 177–203.

Malvern, Sue. 2004. *Modern art, Britain and the Great War: Witnessing, testimony and remembrance*. New Haven, CT, London: Yale University Press.
Marc, Franz. 1989. *Briefe, Schriften und Aufzeichnungen*, ed. Günter Meißner. Leipzig, Weimar: Kiepenheuer.
Martin du Gard, Roger. 1972. *Les Thibault*, vol. 5: *L'été 1914. Épilogue*. Paris: Gallimard.
Remarque, Erich-Maria. 1980 [1928]. *Im Westen nichts Neues*. Frankfurt am Main: Ullstein.
Riesenberger, Dieter. 1990. Der Kampf gegen den Gaskrieg. In *Lehren aus der Geschichte? Historische Friedensforschung*, ed. Rainer Steinweg, 250–275. Frankfurt am Main: Suhrkamp.
Robichon, François. 1994. Ästhetik der Sublimierung: Die Französische Kriegsmalerei. In *Die letzten Tage der Menschheit: Bilder des Ersten Weltkrieges*, ed. Rainer Rother. Berlin: Ars Nicolai.
Schubert, Dietrich. 2002. Otto Dix: Die Radierungen ‘Der Krieg’ von 1924. *Kunsthistorische Arbeitsblätter* 9: 39–46.
Schütz, Erhard. 2005. Wahn-Europa: Mediale Gas-Luftkrieg-Szenarien der Zwischenkriegszeit. In *Krieg in den Medien*, ed. Heinz-Peter Preußer, 127–147. Amsterdam: Rodopi.
Spear, Hilda D., and Sonya A. Summersgill. 1991. Poison gas and the poetry of war. *Essays in Criticism* 4: 308–323.
Stepun, Fedor. 1963 [1929]. *Als ich russischer Offizier war*. Munich: Kösel.
Vollmer, Jörg. 2003. Gift/Gas oder das Phantasma der reinigenden Gewalt. Johannes R. Becher: $(CHCl = CH)_3$ As (Levisite) oder der einzig gerechte Krieg (1926). In *Von Richthofen bis Remarque: Deutschsprachige Prosa zum I. Weltkrieg*, eds. Thomas F. Schneider and Hans Wagener, 181–193. Amsterdam: Rodopi.
Winter, Jay. 1995. *Sites of memory, sites of mourning: The Great War in European cultural history*. Cambridge: Cambridge University Press.

The Genie and the Bottle: Reflections on the Fate of the Geneva Protocol in the United States, 1918–1928

Roy MacLeod

Abstract For Americans serving in the First World War, the advent of chemical weapons made a deep impression. For chemists and soldiers, the experience of meeting—and then making—variants of "poison gas" bred both fear and determination. The wartime creation and post-war struggles of the Chemical Warfare Service reveal the deep divisions these tensions caused, both during the war and through the 1920s, when the United States extensively debated, but failed to ratify, the Geneva Protocol. By the close of the 1920s, the popular optimism that greeted postwar science and invention was clouded by visions of science as a source of new and terrible weapons. In the case of chemical weapons, professional resolve to prepare for future wars competed with a desire to protect the ideals that science represented. In ways that now seem familiar, the profession of chemistry, the chemical industry and the military became powerful allies. This paper examines a subject neglected by historians, and considers how political and professional factors combined to frustrate and delay the early ratification of the Geneva Convention by the United States. As we shall see, our knowledge of these circumstances is far from complete, and will remain so until we have a deeper understanding of the history of America's complex relationship with this toxic legacy.

R. MacLeod (✉)
History, University of Sydney, Sydney, Australia
e-mail: roy.macleod@sydney.edu.au

B. Friedrich et al. (eds.), *One Hundred Years of Chemical Warfare: Research, Deployment, Consequences*, DOI 10.1007/978-3-319-51664-6_11

1 Introduction[1]

> War is a rough and violent game. Destruction is according to its nature and must be so. Each of the belligerents finds itself in need when facing the foe. And when in need any means are permitted. "Use whatever can be used" is the first law, based on the nature of war —H. L. Gilchrist, 1928.

Thus, Colonel H. L. Gilchrist of the US Army Medical Corps, second chief of the US Chemical Warfare Service (CWS), speaking to the American Chemical Society (ACS) meeting at Chicago in August 1928, almost a decade after the Armistice. Today, the world awaits the resolution of this and many other issues that were left unresolved in 1918 and by the Treaty of Versailles a year later, which together marked what Sellar and Yeatman memorably called "the cause of nowadays" (Sellar and Yeatman 1932). Among these, we must regrettably count the legacy of chemical weapons.

This legacy is today prominent in the Middle East where, in September 2013, the Government of Syria launched a chlorine gas attack on its own population, killing 1,429 people, including 426 children. Since then, the humanitarian tragedy in the Middle East has reached epic proportions. Over 100,000 people have died and the conflict is likely to worsen. Reports of the continued use of chemical weapons reach us almost daily. For over three years, the United States has led in helping to dispose of Syria's declared chemical weapons (CW) stockpile of 1,300 metric tons and to dismantle its 23 CW production facilities, and has overseen (at this writing) the neutralization of 600 metric tons of sarin, VX, and mustard gas—what the US Senate Armed Services Committee has called "the world's worst weapons."[2] But the threat lives on. The United States, with other countries, asks what has become of the vision implicit in the Geneva Protocol of 1925 and its successor, the Chemical Weapons Convention of 1992 (entered into force in 1997), to which 190 states—including Syria—have given their assent? What is the future of the norms that these agreements once inspired, in the attempt to put the "Genie back in the Bottle"?

These questions prompt historians to reflect upon the long history of attempts to control the proliferation of chemical weapons. In so doing, however, we seldom recall the factors surrounding the failure of the United States to ratify the Protocol in 1926. Not until January 22, 1975 did President Gerald Ford, following initiatives

---

[1]I wish to express my thanks to Dieter Hoffmann and Bretislav Friedrich and to Martin Wolf and Jürgen Renn for bringing an Australian from ANZAC Day in Sydney to join this gathering. I wish also to thank the small but influential group of scholars who have studied the history of the Geneva Protocol, including Hugh Slotten, Gilbert Whittemore, Catherine Jefferson, and above all, Julian Perry Robinson and the staff of the Harvard-Sussex Program, in whose archives I have had the pleasure of working.

I dedicate this essay to the memory of Christopher Freeman, founding director of the Science Policy Research Unit at Sussex University, whose experience of war gave him a special appreciation of the Chemical and Biological Weapons Convention.

[2]US Department of State. Protocol for the Prohibition of the Use in War of Asphyxiating, Poisonous or other Gases, and of Bacteriological Methods of Warfare. Signed at Geneva, June 17, 1925, and entered into force February 8, 1928.

begun a year earlier by President Nixon, sign the instruments of ratification that brought to a close a half-century of discussion. Fifty years earlier, at the Geneva Conference for the Control of the International Trade in Arms, Munitions and Implements of War, the US played a key role in drafting a protocol that was signed by 30 nations, including the US, which prohibited the "use in war of asphyxiating, poisonous or other gases, and of all analogous liquids, materials or devices, as well as the use of bacteriological methods of warfare." But the US Senate declined to ratify the protocol, and without a formal vote.

The historian will ask, why? What were the factors that withheld US Government support for a treaty the origins of which reached back as far as the Declaration of St Petersburg in 1868, the Brussels Declaration of 1874, and the Hague Declaration concerning Asphyxiating Gases of 1899—which the United States also failed to ratify—as well as the Hague Convention on Land Warfare of 1907. These prohibited the use of "poison or poisoned weapons" in warfare, a conclusion on which the leading military powers of Europe found agreement.

Our questioning begins with the American experience of the Great War; with the conception, production, and use of chemical weapons by America; and with the implications of "science mobilized" for American industry, the military, and international relations. The subtext reveals a debate that takes place within a web of conflicting interests, in which the United States, emerging from the war as the world's principal creditor nation, declined to participate in the League of Nations. This essay highlights some questions that surround these issues, the pursuit of which unites historians of science, diplomacy, and economic history who seek to understand America's relationship with this toxic legacy of the twentieth century.

2 Science and the Great War

The beginnings of this history are well known. The war began in August 1914 as many had predicted, and soon became a scientific war, for which the Allies were ill prepared. Moral outrage, stirred by the burning of the library at Louvain, the atrocities in Belgium, and the shelling of Rheims, was quickened in October by the "Manifesto of the 93 Intellectuals" (vom Brocke 1985; Ungern-Sternberg 1996; Horne and Kramer 2001).

By November, French scientists were mobilized into munitions work, and by the spring of 1915, Britain had several new research institutions for the War Office and Admiralty (MacLeod 2000, 23–46). The advent of chemical warfare on the Western Front in April 1915 marked a new departure in the application of science to war. The same week that saw the first German use of chlorine gas at Ypres followed by the U-boat sinking of the *Lusitania* saw the beginnings of a new kind of warfare, with dimensions that have disturbed mankind ever since. By 1918, new weapons, tactics, and technologies contributed to Germany's defeat (MacLeod 2009, 37–51). In an age of modern warfare, chemical weapons had become the "new normal."

Until April 6, 1917, the United States remained officially neutral, despite the growing participation of large sections of American industry, in which the disciplines of applied science were quick to take advantage (MacLeod 2014). From this experience emerged several significant features of modernity: a powerful scientific establishment, with lobbies that found their way to Congress; the redesign of international scientific organizations along lines favoring American interests; and an increasingly close relationship with the military, in ways that hinted at the militarization of science itself.

Alongside the many wartime applications of science, chemical weapons were to dominate modern memory (MacLeod and Johnson 2006). But Americans came late to gas warfare, and were not its principal victims. US troops saw their first major engagement at Chateau Thierry only in June 1918, by which time gas weapons—and gas defenses—were already used extensively by both sides. Chlorine and phosgene damaged the unwary, but the arrival of mustard gas, new in 1917, devastated American forces on the Western Front, where inexperience and poor discipline contributed to a higher proportion of gas casualties than suffered by Britain, France, or Germany.[3]

Although the US Army struggled, American chemists had been studying and monitoring the use of gas since 1915, when the American Chemical Society mounted a campaign to enlist chemists into the war effort. As the United States began to turn itself into an arsenal for the Allies, the National Academy of Sciences created a National Research Council (NRC) to extend its wartime mandate. In 1917, the NRC responded to the call of its chairman, the astronomer George Ellery Hale: "I really believe this is the greatest chance we ever had to advance research in America" (Wright 1966, 288). America's official entry into the war in April 1917 gave chemical warfare its mandate.

In America's production of chemical weapons, the Bureau of Mines led the way and, by the end of May 1917, had mobilized 118 chemists from 21 university laboratories, three private companies, and three federal agencies to work at Camp American University (still the site of American University) in Washington, D.C. But the War Department had even bigger plans, and, in September 1917, established a Gas Service in the US Army Engineers under General William Sibert. This was modeled on the Service de Chimique of the French Army, and adopted their gas masks and drill. Following the experience of the German offensive in March 1918, the War Department created an autonomous Chemical Warfare Service (CWS), which it tasked with the full spectrum of research, production, and supply. Initially, much of its work was defensive, and an overseas branch was established under General Amos Fries, who later became the Service's director and advocate. Together, Sibert and Fries developed a substantial program of research and development which, at relatively low cost, set out to bring chemistry to America's defense (Fries and West 1921; Fries 1921). Under Sibert, the CWS began by

[3]Chemical weapons were said to cost Americans 11% of their casualties, versus 5% of the other belligerents. For debate surrounding these figures, see Baxter (2004).

Fig. 1 *Left* Major General William L. Sibert (1860–1935). Watertown Free Public Library; *Right* General Amos Alfred Fries (1873–1963). Creative Commons

developing a close relationship with American industry, and especially with the American Chemical Society, and its myriad membership (Fig. 1).

In September 1917, immediately following Germany's first use of mustard, CWS laboratories began to study its method of production; and by March 1918, the research group led by the young Lt. James B. Conant found an efficient process, sharing the credit with Sir William Pope's group at Cambridge, but producing mustard gas before the British (Whittemore 1975, 151; Jones 1980, 426–440).[4] By the Armistice, America's Edgewood Arsenal was producing 30 tons of mustard gas a day; 35% of the shells fired by American artillery in France were filled with gas; and the strategists of the United States and Britain, including Benedict Crowell at the War Department and Winston Churchill at the Ministry of Munitions, were anticipating its massive use in the great offensive planned for the Spring of 1919 (Crowell and Wilson 1921; MacLeod and Johnson 2006) (Fig. 2).

With the Armistice, the wartime relationship fostered with industrial and academic chemists continued unimpeded. Many compounds were tested, and new gases were in the offing. The CWS was particularly proud of Lewisite, while the British tested a new arsenical, code-named DA, which was capable of penetrating the most effective German gas masks (Jefferson 2014, 654). The CWS recruited

[4]Mustard Gas Warfare: Man who Makes It Tells of Science's Deadliest Weapon and How United States Army Will Use it in Quantities. *New York Times*, July 7, 1918.

Fig. 2 James Bryant Conant (1893–1978), upon becoming President of Harvard University in 1933. Creative Commons

university chemists both to study known problems and to find new problems to solve. The expanding laboratories at Edgewood and Camp American University in Washington, D.C. saw a cross-section of chemists, young and old, academic and industrial (Whittemore 1975, 151).[5] Before the war, George Burrell, chief of the Research Division at Edgewood, had gone no farther with gases than exterminating small animal pests. Lee Lewis, also at Edgewood, was better known for his work on improving the water quality of public swimming pools (Fig. 3).

Robert Williams, who studied the use of ricin from the castor oil plant, a few grains of which can kill a person, went on after the war to synthesize Vitamin B1 (Whittemore 1975, 152). Others, like Yandell Henderson, professor of physiology at Yale, who developed the first successful American gas mask, wrote on gas warfare and aviation for the *Yale Alumni Weekly*. Underlying all this effort was the theme of research as "productive scholarship"—the results of which might not appear for years.

[5]The work of Gerard Fitzgerald on the wartime Edgewood Arsenal promises fresh revelations.

Fig. 3 Winford Lee Lewis (1878–1943), who invented 'Lewisite.' Wikimedia

In April 1918, on the third anniversary of the first use of gas on the Western front, Henderson wrote:

> We must give the devil his due—the Germans have developed to a high degree the research side of science, and that is what has made it possible for the German army to make this drive [...] and it is by the use of such scholarship [that] we have got to beat those men over there and drive them back (Henderson 1918).

In September 1919, the first postwar meeting of the British Association for the Advancement of Science looked back to the carnage, and appealed to the conscience of the nation: "Science must receive from man its commission to heal the sores that it has made."[6] Similar sentiments were voiced by Woodrow Wilson, speaking at Versailles in January 1919:

[6]Science in War and After: Good Angel or Evil Genie. *The Sunday Times*, September 10, 1919.

> We must take, so far as we can, a picture of the world into our minds. Is it not a startling circumstance for one thing that the great discoveries of science, that the quiet study of men in laboratories, that the thoughtful developments which have taken place in quiet lecture rooms, have now been turned to the destruction of civilisation? [...] The enemy whom we have just overcome had at its seats of learning some of the principal centres of scientific study and discovery, and used them in order to make destruction sudden and complete; and only the watchful, continuous cooperation of men can see to it that science, as well as armed men, is kept within the harness of civilization. (Wilson 1919; Schilling 1964)[7]

In the United States, chemists working for the CWS gave the impression that the Armistice had interrupted their work (MacLeod and Johnson 2006). Their experience translated into proposals for government support and was mapped onto the platform of the Progressive Party and onto the program of "preparedness" that helped define interwar America.

3 The Coming of Geneva

The Treaty of Versailles, signed in November 1919, affirmed the prewar norm and prohibition of poison gases, in an unvarnished attempt to prevent Germany from again producing, importing, or using chemical weapons. Similar provisions were included in treaties that embraced Austria, Hungary, and Bulgaria. But the issue was too important to leave to noble sentiment. The new League of Nations set out to reach a permanent international agreement. In May 1920, the League's Permanent Advisory Commission on Military, Naval and Air Questions (PAC) ruled that the use of gas was no more cruel than the use of any other weapon and was therefore not amenable to prohibition. But the Council of the League declined to accept this argument and, prompted by reports in England that the CWS had developed aerosols to deliver gas by air, referred the issue back to the Permanent Commission. While experts considered what damage airborne attacks might do to civilian cities, the League was asked to debate an outright ban.[8]

In the early 1920s, American governments were both more and less ambitious in thinking about chemical weapons, largely favoring an international agreement, but anxious lest this entail the loss of strategic and economic advantage. In 1921, the question was brought by the Harding administration to the attention of the Washington Conference on the Limitation of Armaments, a committee of which ruled that it was impossible to limit the use of chemical weapons in war, but that it was possible to forbid their use against non-combatants. When this proposition failed to win supporters, the American delegation, led by Secretary of State Charles Evans Hughes, went further and recommended a total prohibition (Fig. 4).

---

[7]Woodrow Wilson addressing the Second Plenary Session of the Peace Conference, January 1919. US Department of State, Papers Relating to the Foreign Relations of the United States: The Peace Conference, vol. 3, 179. Washington, D.C.: USGPO, 1942–1947. In Schilling (1964).

[8]See *The Times*, March 15, 1921.

Fig. 4 *Left* Secretary of State Charles Evans Hughes (1862–1948); *Right* Chairman of the Senate Committee on Foreign Relations William Borah (1865–1940). Library of Congress

Hughes' view was supported by a national opinion poll, conducted in the United States for the American delegation, which found a majority of Americans in favor of Article V of what became the Washington Naval Treaty, prohibiting the use of chemical weapons "as justly condemned by the general opinion of the civilized world." Signed by the US, the UK, France, Italy, and Japan, this "Five-Power Treaty" was ratified by the US Senate on March 29, 1922 by a vote of 72–0, with 24 abstentions. Only one Senator—James Wadsworth, Jr. (R–NY), chairman of the Senate Committee on Military Affairs—spoke against it. Because the government of France declined to ratify certain provisions concerning submarines, the treaty failed to enter into force. But the idea had gathered momentum, and the United States was seen to have led the way. This is what Daniel Jones has called the "Lesson of 1922" (Jones, 1980, 428).

This lesson was twofold. The first turned on timing. Only seven weeks separated the signing of the Washington Treaty and its ratification by the Senate. Opponents thus had little time to mobilize. The second turned on domestic politics. In this case, public opinion had been tested, and found overwhelmingly in favor. Those who opposed had to look for support elsewhere. And this was to come.

For two years, the question slumbered. On September 27, 1924, as Catherine Jefferson has reminded us, the Washington Treaty was brought to the 5th Assembly of the League of Nations, which recommended that a draft convention restricting the use of poison gas be drawn up by its Temporary Mixed Commission on the Reduction of Armaments for submission to member states (Jefferson 2014, 647–661). This discussion, of course, excluded the United States, which remained outside the League. However, when the League convened a Conference for the Supervision of the International Trade in Arms and Ammunition and in Implements of War, President Calvin Coolidge authorized a delegation to be sent, led by Rep. Theodore Burton (1851–1929), chairman of the Foreign Affairs Committee of the House of Representatives, which included Herbert Hoover (Secretary of Commerce) and Frank Kellogg (Secretary of State). This conference met in Geneva on May 4, 1925.

Before the meeting, Theodore Burton, who personally supported a ban, persuaded Coolidge that the United States propose to the Conference a provision forbidding international trade in chemical weapons—in effect, a non-proliferation treaty. The Military and Technical Subcommittee of the Conference, to which this proposal was referred, rejected the idea as unworkable on the grounds that it would unfairly discriminate against weaker states that were unable to make weapons on their own. Given this logic, the British, Polish, and Italian delegates proposed to hold a special conference, the purpose of which would be to consider a treaty to ban all chemical and (at the suggestion of Poland) bacteriological weapons. Despite reservations from Italy and Switzerland, their proposal was accepted and a protocol was appended to the final resolution of the Conference. This became known as the Geneva Protocol, which was signed on June 17, 1925.

The Geneva Protocol restated the prohibitions laid down by the Versailles and Washington treaties, with an additional ban on bacteriological weapons. It made no provision for enforcement, nor did it limit the scope of the prohibition. It banned the "use of weapons," but not the weapons themselves. It was effectively a statement of "no first use." Nothing in the protocol specified inspections or sanctions. The protocol did not prohibit development, production, or stockpiling, nor did it provide for a means of verification. But with these limitations, it did confer a measure of legal and moral condemnation of such weapons across the world.

There were, however, many potential flaws in the provision. If, for example, the Great Powers were signatories but Germany and Japan were not, would the protocol be accepted? And what of the United States? In January 1926, President Coolidge sent the protocol to the US Senate for ratification. Coolidge supported arms control, not least because he believed that as peace was the natural ally of an expanding economy, the United States would surely benefit. In any case, the Five-Power Treaty of 1922 that banned the use of chemical weapons among its signatories had sailed through the Senate without a single negative vote. The Protocol would surely have an equally smooth passage. The administration's submission to the Senate deleted the controversial submarine clauses and expanded the reach of the ban from five to 41 nations.

This time, however, the government made three fundamental errors. The first was timing. Coolidge signed the Treaty on June 17, but William Borah (R–Idaho), chair of the Senate Foreign Relations Committee and floor manager of the protocol, failed to report the treaty out of committee until a year later, on June 26, 1926. Then, for reasons of timetabling, the debate was not scheduled until December 13, 1926—almost 18 months after the protocol was first introduced. There was ample time for opposing interests to muster (Fig. 5).

Second, the Senate was poorly briefed. When the protocol finally reached the Senate floor, most Senators remained silent, and only five spoke. The issue was not as pressing as other work before the Congress. Worse still, the White House failed to consult and win the support of the War Department and the Navy (McElroy 1991, 131), both of which were disposed towards the treaty. Without their expert backing, the Senate was obliged to look elsewhere for advice. And into this vacuum leapt Senator Wadsworth, chairman of the Senate Committee on Military Affairs, and a spokesman for the Chemical Warfare Service.

Wadsworth opposed ratification, as he had in 1922, on the grounds of "national preparedness." In 1921, Sir Edward Thorpe had told the British Association for the

Fig. 5 Senate Foreign Relations Committee, 30 April 1924. Library of Congress

Advancement of Science that the "moral sense of the civilized world is not so dulled but that, if roused, it can make its influence prevail" (Thorpe 1921). But within the next generation, a trope that Woodrow Wilson had popularized took on a new appeal.

As Frances Harbour has shown, a dozen veterans' organizations were now mobilized, sending petitions that supported "preparedness" and rejected the protocol (Harbour 1990). These factions found influential support from the American Chemical Society (ACS), which in 1925 celebrated its 50th anniversary. The ACS was then the largest chemical society in the nation and one of the largest scientific societies in the world. At least 500 of its members had been in the wartime CWS and were devoted to its survival. The future of the CWS was threatened by the coming of peace. But its leaders refused to go quietly into the night and, throughout the 1920s, marshaled commercial support for its research, with applications ranging from agriculture to perfumery (Faith 2008).

In early 1926, Edgar Fahs Smith, who had been chairman of the Chemical Weapons Subcommittee at the 1922 conference, visited Coolidge and spoke against the protocol. Meanwhile, the ACS lobbied all senators, arguing that "all history shows that any effective weapon available will be used" (Harbour 1990, 13). At its national meeting in Los Angeles in August, the ACS repeated the main arguments that framed its agenda: (1) that chemical weapons were effective; and (2) that gas was a less cruel (or, as often put, "humane") alternative to worse weapons. Echoing the defense advanced in *Callinicus* by J. B. S. Haldane, FRS, the British polymath (Haldane 1926),[9] these arguments were buttressed by the campaign for national preparedness, which resisted interference from any foreign power or the League of Nations. Ratification of the protocol, they argued, would force the United States to forego a strategic capability that it had struggled to create (McElroy 1991, 140–150). Secretary of State Kellogg met with representatives of the ACS in November 1926, but failed to secure a compromise.

For months, the question was postponed as the Senate fought over affiliation with the World Court of the League of Nations. Months passed, and it seemed that Kellogg had acquired few allies in the upper house. On December 9, 1926, the first day of debate, the only senator to speak for ratification was William Borah, who quoted General John Pershing:

> Chemical warfare should be abolished among nations as abhorrent to civilisation. It is a cruel, unfair and improper use of science. It is fraught with gravest danger to noncombatants and demoralizing to the better instincts of humanity. (McElroy 1991, 132)

---

[9]Haldane's title gave it all away: Callinicus, an eighth-century Syrian prince—his name, of Greek origin, means "beautiful victor"—used "Greek fire" to prolong the survival of the Eastern Roman Empire for 750 years.

The emollient intention of Pershing's words was lost on Senator David Reed (R–Penn), who had been an artillery officer during the War:

> Are we, then, to go against an inferior antagonist, with all the abundance of artillery that the World War has left us, to blow out of existence a lot of peasants who scarcely know what the war is about? Or are we to take advantage of this great chemical opportunity which we, as a manufacturing nation, have open to us. Would it not be more merciful, assuming that we were at war with some Central American country, to win our battles by the temporary disabling of our enemies than to blow them all over their cactus plants? (McElroy 1991, 141)

After a weekend break, debate resumed, and three other Senators called for rejection. Senator Ransdell (D–La) championed the ACS and the interests of industrial chemistry and expressed the hope that the protocol be "buried so deep it would never appear before us again" (McElroy 1991, 142). At this point, the Administration saw no hope of winning the required two thirds' vote of the Senate. After three days, the State Department withdrew the protocol without putting it to a vote. Contemporaries could have seen this as a strategic retreat—a retreat that lasted for the next 50 years. But Frederick Brown sees the conclusion as almost inevitable, for two compelling arguments. The United States "could not expect to obey agreed restraints unless they were perceived to be in the national interest." To which Senator Wadsworth added: "it is against all human nature to expect a nation to deny to itself the use of a weapon that will save it" (Brown 1968).[10] The Senate accepted the argument of the CWS that ratification would "stultify if not preclude" readiness for gas warfare; in language which the CWS might have approved, Brown adds, it would be "virtually impossible to allocate scarce resources to increase chemical warfare readiness when the use of gas in war had been prohibited." Preparedness trumped prevention (Fig. 6).

It seems clear that the Senate's de facto rejection of the Geneva Protocol accompanied a shift in American policy away from "in principle" support and towards a precautionary *realpolitik*. As Frederick Brown put it, "From enthusiastic promotion of any treaty which would reduce the possibility of gas warfare in 1921, the United States had become a rather skeptical bystander by 1931" (Brown 1968, 108–109). The protocol remained in the Senate files until 1947, when the then chairman of the Foreign Relations Committee, Arthur Vandenberg, returned it in response to a *pro forma* request from the Truman Administration to process unratified treaties. Even then, historians find no reason to suggest that the protocol's retrieval was any more than a housekeeping measure. In a Cold War of rapid movement in secret chemical and biological weapons development, no American president was likely to show an interest in reviving the subject (Harbour 1990, 20).

---

[10] Congressional Record, 69th Congress, 2nd Session LXVIII, Part I, 144–149; cited in Brown 1968, 106–107.

Fig. 6 Clockwise from *top left*: David A. Reed (1880–1953); Calvin Coolidge (1872–1933); Frank B. Kellogg (1856–1937); Theodore E. Burton (1851–1929); William Borah (1865–1940); James Walcott Wadsworth Jr. (1877–1952). Creative Commons and Library of Congress

4 A Protocol Post-mortem

Had the Geneva Protocol, in Frances Harbour's phrase, simply "slipped through the cracks"? Or were other factors at work? How can we best summarize the events of 1926?

Following the successful ratification of the Washington Treaty in 1922, Coolidge seems to have been overconfident of a similarly smooth passage in 1926. Evidently, his confidence was misplaced. More important, the Senate debate, when it finally took place, shows all the effects of intervention by the supporters of the CWS. Although the military establishment never liked the CWS, nor had it won special distinction in the field, the Service had friends in both houses of Congress and in the chemical industry, many of whom, it could be argued, had made fortunes from the war and looked forward to a profitable peace.

From the late 1920s to the early 1930s, the annual budget of the CWS averaged $1.2 million and, in 1934, fell to only $800,000 (Baxter 2004, 77). But these modest figures mask the fact that, throughout the 1920s, under Major General Amos Fries (1920–1929), the CWS opened a new chapter in military affairs, testing a range of weapons and acquiring a reputation for innovation. "Medicine and agriculture have been largely benefitted by the evil genie let loose," one sympathetic journalist commented; "From red currants to pumpkins, no fruit has been discovered that cannot be poison-gassed into extra size and nutritiousness," and CWS posters extolled the new dyes that were coming from its war-related research (Slotton 1990, 493). Public disapproval of chemical weapons was broad but not deep, and no one argued against the economic value of research. In any case, with his record of insecticides and perfumes, General Fries could boldly claim he was actually running a Chemical Peace Service (*ibid.*, 492).

A third argument against ratification came from the ACS, which was ambivalent about the implications, if not the intentions, of the protocol. Driven by its wartime experience, the ACS took the side of its chemical colleagues in the CWS, with whom it shared personal and political leanings. In 1918, it was Charles Parsons, Executive Secretary of the ACS and also Chief Chemist of the Bureau of Mines, who with approval reminded the ACS that "War, the destroyer, has been […] the incentive to marvelous chemical development with a speed of accomplishment incomprehensible in normal times" (Slotton 1990, 486).

The contributions of the war, as George Ellery Hale and the NRC foreshadowed, had already opened to Americans a "New World of Science" (Yerkes 1920). In the wake of the war, the American scientific community had acquired unprecedented recognition and public acceptance. With this also came political accountability of a kind that few had so far mastered. Not only in Germany would professional values be overtaken by patriotism when the national interest made it necessary (MacLeod and Johnson 2002, 169–179).

To the general public, the case advanced by the ACS in 1926 reflected a widely held view that chemical weapons constituted a "humane chemistry," distinguishable from the "scientific barbarism" of conventional weapons. In contrast to high explosives, which accounted for most battle casualties on the Western Front, statistics produced by the British army—widely cited but now contested—suggested that only 3% of gas casualties died of their injuries, compared with the 40% of deaths caused by all other weapons. Little was known at the time of the lasting effects of gas, which by Third Ypres in 1917 accounted for 14% of British casualties (McElroy 1991, 140).

All weapons were destructive, but some were more destructive than others, and, it was argued, applications of chemistry (and biology) might at least make war endurable. J. B. S. Haldane made what many thought an overwhelming case for gas defense—and also for preventative offence, such as the CWS advocated—on the grounds that it would serve mankind well, not as a means of preventing war, which was impossible, but as a way of making future wars end quicker. Thus, by making warfare more scientific, science would make war more efficient—shorter, simpler, and less damaging to the social order, if not to humankind.

The Geneva Protocol languished in the absence of a strong will to overcome such arguments. But perhaps there was an even deeper reason. In 1921, Will Irwin's highly popular *The Next War: An Appeal to Common Sense* told the American reading public that the German gas attack in 1915 was as significant as Columbus' discovery of America. As such, it was terrifying. The experience of the War had shown that

> Those great and little scientific minds, engaged hitherto in searching for abstract truth or in multiplying the richness of life and the wealth of nations, could be turned toward the invention of means of destruction whether they wished or no. (Irwin 1921, 28)

Irwin cited wartime rumors that a dozen bomb-loads of lewisite could destroy the population of Berlin. If so, poison gas "of a power beyond the dream of a madman, seems to be the killing weapon of the future" (Irwin 1921). The protocol contained no provisions for enforcement. Kellogg said that the United States, in ratifying the protocol, would have to depend upon the "good faith of nations." But why should the United States risk losing the future?

There was also an absence of domestic political pressure. Reportedly, there was less protest in the United States at Germany's first use of gas in France than at the loss of American life on the *Lusitania*. And when, in 1918, the US Army began to use gas of its own making, there was little domestic opposition. As Hugh Slotten has found, the *New York Times* surmised that gas had inevitably "been forced upon all the combatants by the custom of the Germans," whilst other newspapers, including the *Washington Post*, the *San Francisco Chronicle* and the *New York Tribune*, did not comment at all (Slotton 1990, 485).

If a majority of Americans were opposed to the use of chemical weapons—and surveys suggested they might be—they were also in favor of being "prepared," which ratification of the Geneva Protocol seemed to override. In the absence of infallible military expertise or domestic political pressure, the floor of the Senate was no place to rally public opinion that had found no great reason to protest.

5 Summing up the Senate

Historians may point to several possible reasons for the US Senate's failure to ratify the Geneva Protocol in 1926. They may include:

1. a lack of strategic planning by the Coolidge Administration, producing a costly delay;
2. an unenthusiastic handling of the protocol on the Senate floor, revealing no coherent strategy for securing ratification; in his annual State of the Union Address, delivered only two days before the Senate debate, President Coolidge failed even to mention the protocol;
3. a failure to mobilize public opinion in support of passage at precisely the time it was needed;
4. a failure to counteract the influence wielded by a confederation led by a Chemical Weapons Service that was fighting for its life;
5. a lack of confidence that the protocol would accomplish its goal; and
6. a failure on the part of the State Department, which neglected to consult with the Army and the Navy.

In the Senate's failure to ratify the protocol, many such factors may have been necessary, but perhaps not sufficient. What surfaces above all else from the Congressional Record is the way in which the protocol suffered from an almost seamless transition from a national narrative of disarmament to a discourse of deterrence. We now see its limitations and imperfections. We accept that the protocol was essentially a no-first-use provision. It prohibited the use of gas, but permitted states to research and reserve the right to retaliate in kind. Even when the US did ratify it in 1975, the protocol did not prohibit development, production, or stockpiling; nor did it specify means of verification. It failed to provide a mechanism for collective response. In this, it illuminated some of the security failures of both the League of Nations and the United Nations (Fig. 7, 8).

To remedy these limitations, the world had to wait for the Chemical Weapons Convention in 1993 (entering into force in 1997), which prohibited the production and use of chemical weapons, provided for the destruction of production facilities (or their monitored conversion to other functions), the destruction of all chemical weapons (including chemical weapons abandoned outside state parties' territory), an inspection regime for the production of chemicals that could be converted to chemical weapons, and international cooperation in the peaceful uses of chemistry.

SAYS AMERICA LAGS IN WAR GAS RACE

Chemical Expert Declares That Arm of Service Is Treated Like a Stepchild.

ONE MASK TO 100 SOLDIERS

Points to Activity in Germany, England, Japan and Even in Latin Republics.

Germany and other European nations, including those of Latin America, are forging ahead in chemical warfare, while the United States is falling behind, according to Dr. Harrison E. Howe, member of the American Chemical Society.

"In this country we have only one modern gas mask per hundred enlisted men in the regular army and National Guard," said Dr. Howe, quoting the reported statement that "Germany has stored, ready for use, sufficient modern gas masks to provide five for every soldier authorized by the Treaty of Versailles."

It is plain that "though the ascendency of chemical warfare is written everywhere, the United States Chemical Warfare Service is being treated as a stepchild by those in charge of army appropriations."

Should war come our troops would be required to fight the first six months inadequately protected, said Dr. Howe.

"Italy, shortly after Mussolini came into power, established a chemical warfare service modeled along the lines of our organization but with twice the number of officers," he went on. "England appears to be giving chemical warfare the same importance as its three great arms of national defense—Army, Navy and Air Service.

Fig. 7 *New York Times*, 8 March 1925, p. 26

"Chemical warfare has been placed under a committee headed alternately by a high ranking officer of the army and navy. Russia, like England, has placed chemical warfare on the same plane with other important branches of her service.

"Japan has shown great activity along chemical warfare lines, and it is stated that four divisions of troops have been abolished so that the funds thus made available can be spent on the air service and the chemical warfare service.

"Japan has been buying large quantities of chemical warfare supplies from Germany and elsewhere, and is known to be availing herself of the services of the ablest scientists from the German chemical field.

"France is perhaps as well prepared to use chemical weapons in war as any other nation, though less is known about her plans and actions than any of the other powers. Spain, Switzerland, Poland, Czechoslovakia, Sweden, all have chemical warfare service in one form or another, and it will be recalled that Spain used gas against the Moors in one of her campaigns last year. Even Mexico and certain countries in South and Central America are showing a growing interest in chemical warfare.

"Meanwhile, we spend $1,347,580 to feed the 9,230 horses in the cavalry and at the ration allowed for horses—40 cents per day—the 45,000 horses and mules in the entire army cost $6,570,000 for forage alone, not to mention housing, harness, caretakers and replacements.

"There are 122 veterinary officers in the army's latest directory of Jan. 1, 1925, and there are 83 officers in the Chemical Warfare Service. The extent to which cavalry has been superseded in modern warfare was well demonstrated by the World War. The ascendency of chemical warfare is written everywhere."

Fig. 7 (continued)

CHEMISTS DISAGREE ON FUTURE OF WAR

Inventor of 'Lewisite' Tells Illinois Gathering Gas and Planes Will Dominate.

ARMY MEN SEE LESS CHANGE

Smoke Screen Demonstration by Aircraft Precedes Statement That Science Aids Peace.

Special to The New York Times.

EVANSTON, Ill., Aug. 18.—Science is the ally of peace, declared Dr. W. Lee Lewis, inventor of "Lewisite," one of the most potent gases used in the World War, tonight at the closing conference of the American Chemical Society Institute at Northwestern University.

The statement came just after army airplanes had covered the North Shore with a smoke screen in a few minutes as a demonstration of the new technique of warfare and army men had lauded the new applications of science in developing greater military action.

"If science served warfare only, it would indeed be a human curse," said Dr. Lewis in his address that closed the national defense day program of the institute, cooperating with the National Association of Chemical Defense.

"Even as applied specifically to warfare science makes for peace because it gives the balance of power to the more intelligent and advanced race.

Air Strength Not Controllable.

"The tangible instruments of war such as are controllable by agreement will undoubtedly play a smaller part in future warfare than the less tangible. Battleships and fortresses will be less potent agents than airplanes and chemicals.

"The air strength of a nation is not controllable. It depends purely upon the extent of peace time support of aviation and its commercial development. There is less difference between a commercial airplane and a fighting airplane than there is between a merchantman and a battleship.

"A fleet of commercial planes could be converted into bombing planes in a few hours. There is little difference between a peacetime chemical plant and a wartime chemical plant so far as equipment and procedure are concerned.

"Science does make war less adventuresome, less romantic and more deadly. It is, therefore, fundamentally an ally of peace. There is little of the joust, the tournament in long range guns, submarines and poison gas."

Colonel H. L. Gilchrist of the Army Medical Corps, predicted that the next war would be one of gas attack primarily.

Changes Hard to Foresee Now.

"War at its best," he said, "is a rough and violent game. Destruction is according to its nature and must be so; otherwise, it would not be war. Each of the belligerents finds its self in need when facing the foe. And when in need any means are permitted. 'Use whatever can be used' is the first law, based on the nature of war.

"Just as the World War assumed forms which no man had foreseen in times of peace, no one can today with certainty foretell what form war in the future will take. One thing, however, does seem certain; technical skill, physics and chemistry will be used more intensively than heretofore. Toxic gases in particular will play an almost decisive part. Such being the case, our methods of defense must keep fully abreast of the rapid strides made in offense."

Dr. J. E. Mills, chief chemist of the Chemical Warfare Service of the army, was inclined to look upon the possibilities of gas attacks on civilian population less seriously.

Gas attacks against a city, it was explained, can be rendered relatively ineffective for many reasons, among which the following were cited:

"Gas cannot travel up wind; any ordinary room with doors and windows closed will shut out most of the gas for a considerable period of time."

Calls Cold Steel Still Decisive.

Major Gen. Paul B. Malone, who fought at Verdun and Chateau Thierry, and is now stationed in the Sixth Corps Area, made the point that neither air nor gas warfare will eliminate or decrease the value of other branches of the military service.

"In view of the possibility of transporting airplanes on airplane carriers across the sea, it becomes evident that the defeat of our fleet upon either ocean will place the guns of a victorious fleet within range of our harbors and bring the territory of an overseas enemy into juxtaposition with our mainland. Such a result would make the interior of our country accessible to hostile airplanes.

"Battles in the sky will assume a magnitude scarcely dreamed of during the World War, but these battles *in the sky will not be decisive.* The infantry and the field artillery, the cavalry, the engineers and all other branches of the service, but slightly affected by this character of warfare, will move forth as they always have in the past to decide the issue upon the battlefields of the world.

"Machinery will enter into and render more complex the conduct of war, but machinery will never replace man," he added.

The New York Times
Published: August 19, 1928

Fig. 8 *New York Times*, 19 August 1928, p. 23

6 Conclusion

The failure of the US Senate to ratify the Geneva Protocol reflected contrary imaginations and vested interests in the American military, industry, science, and perhaps in public opinion as well. We are accustomed to seeing the protocol as a milestone in the establishment of a diplomatic and moral norm of lasting importance. However, the circumstances of 1925–26 were significant in themselves, having wider consequences that accompanied and foreshadowed deeper conflicts. In particular, we can see—and not just between the lines—evidence of a will to power shown by the natural sciences—the "new bosses of war," as one commentator put it (Slotton 1990, 494). The war had demonstrated a new-found sense of national commitment among American scientists; in the post-war world, the question was how best to "boss" the business of war-related science. Some wanted to keep chemical weapons research under the civilian administration of the Bureau of Mines. But the advantages of a closer connection with the military were clear. The precedent set by the debate would not be lost on those planning atomic weapons research after the Second World War. The protocol debate anticipated by 35 years President Eisenhower's warnings that public policy could become the "captive of a scientific-technological elite" (Jones 1980, 439).

The early 1920s offered reason for both hope and fear—the twin legacies of the Enlightenment. The emerging relationship between science and the military weighed heavily on the conscience of those who, like J.D. Bernal, saw in the outcome of the protocol debate fresh cause to re-examine the politics of science. He and others were inspired to argue that, if science was to have a future, it must show social responsibility—a theme that gained prominence during and after the Great Depression of the 1930s. How rival nations would in future justify their use of the "worst weapon" would take on new definition, and its uses would soon occupy new spaces outside Europe and North America, in Asia, Africa, and the Middle East. But that is another story, for another time.

References

Baxter, William. 2004. *The Chemical Warfare Service of the United States Army during the Inter-War Period*. MA Thesis, Texas Tech University.

Brophy, Leo P., Wyndham D. Miles, and Rexmond C. Cochrane. 1988. *The Chemical Warfare Service: From Laboratory to Field*.In the United States Army in World War II, *The Technical Services*. Washington, D.C: Center of Military History, United States Army.

Brown, Frederick J. 1968. *Chemical Warfare: A Study in Restraints*. Princeton: Princeton University Press.

Gilchrist, H. L. 1928. Chemists Disagree on Future of War. *New York Times*, August 19.

Crowell, Benedict and Robert Forrest Wilson. 1921. *The Armies of Industry, I: Our Nation's Manufacture of Munitions for a World in Arms, 1917–1918*. New Haven: Yale University Press.

Faith, Thomas Iain. 2008. *Under a Green Sea: The US Chemical Warfare Service, 1917–1929*. Ph. D. dissertation, George Washington University.

Fries, Amos A., and Clarence J. West. 1921. *Chemical Warfare*. New York: McGraw Hill.

Fries, Amos A. 1919, Digital Edition 2013. *History of the Chemical Warfare Service in France* http://cgsc.contentdm.oclc.org/cdm/ref/collection/p4013coll7/id/886.

Fries, Amos A. 1921. The Future of Poison Gas. *Current History* 15: 419–422.

Harbour, Frances. 1990. *Chemical Arms Control: The United States and the Geneva Protocol of 1925*. New York: Carnegie Council on Ethics and International Affairs.

Haldane, J.B.S. 1926. *Callinicus: A Defense of Chemical Warfare*. London: Routledge.

Hanna, Martha. 1996. *The Mobilization of Intellect: French Scholars and Writers during the Great War*. Cambridge, MA: Harvard University Press.

Henderson, Yandell. 1918. Gas Warfare and Aviation: The Problems of Productive Scholarship Growing Out of the Present War. *Yale Alumni Weekly*.

Horne, John, and Alan Kramer. 2001. *German Atrocities: A History of Denial*. New Haven: Yale University Press.

Irwin, Will. 1921. *The Next War: An Appeal to Common Sense*. New York: E. P. Dutton.

Jefferson, Catherine. 2014. Origins of the Norm against Chemical Weapons. *International Affairs* 90 (3): 647–661.

Jones, Daniel P. 1980. American Chemists and the Geneva Protocol. *Isis* 71 (3): 426–440.

Kevles, Daniel. 1978. *The Physicists: The History of a Scientific Community in America*. New York: Knopf.

MacLeod, Roy. 2000. Sight and Sound on the Western Front: Surveyors, Scientists and the 'Battlefield Laboratory', 1915–1918. *War and Society* 18 (1): 23–46.

MacLeod, Roy, and Jeffrey Johnson. 2002. War Work and Scientific Self-Image: Pursuing Comparative Perspectives on German and Allied Scientists in the Great War. In Vom Brocke, Bernhard ed.*Wissenschaften und Wissenschaftspolitik: Bestandsaufnahmen zu Formationen, Brüchen und Kontinuitäten im Deutschland des 20 Jahrhunderts*. Berlin: Akademie-Verlag, 169–179.

MacLeod, Roy, and Jeffrey Johnson. 2006. The War the Victors Lost: The Dilemmas of Chemical Disarmament, 1919–1926. In *Frontline and Factory: Perspectives on the Chemical Industry at War, 1914–1924*, ed. Roy MacLeod, and Jeffrey Johnson. Dordrecht: Springer.

MacLeod, Roy. 2009. The Scientists Go to War: Revisiting Precept and Practice, 1914–1919. *Journal of War and Culture Studies* 2 (1): 37–51.

MacLeod, Roy. 2014. Scientists. In *Cambridge History of the First World War*, Vol. 2, ed. Jay Winter, 434–459, 704–708. Cambridge: Cambridge University Press.

McElroy, Rodney J. 1991. The Geneva Protocol of 1925. In *The Politics of Arms Control Treaty Ratification*, eds. Michael Krepon, and Dan Caldwell, 125–166. New York: St Martin's Press.

Russell, Edmund. 2001. *War and Nature: Fighting Humans and Insects with Chemicals from World War I to Silent Spring*. New York: Cambridge University Press.

Sellar, C., and R.J. Yeatman. 1932. *1066 And All That: A Memorable History of England*. London: Methuen.

Schilling, Warner R. 1964. Scientists, Foreign Policy and Politics. In *Scientists and National Policy-Making,* eds. Robert Gilpin, and Christopher Wright. New York: Columbia University Press; and *The American Political Science Review*, LVI (2): 287–300.

Slotton, Hugh R. 1990. Humane Chemistry or Scientific Barbarism? American Responses to World War I Poison Gas. *Journal of American History* 77 (2): 476–498.

Thorpe, Sir Edward. 1921. Presidential Address, *Report of the British Association for the Advancement of Science*, Edinburgh, 23.

Ungern-Sternberg, Jürgen and Wolfgang von. 1996. *Der Aufruf 'An die Kulturwelt': Das Manifest der 93 und die Anfänge der Kriegspropaganda im Ersten Weltkrieg*. Stuttgart: Frank Steiner.

Vom Brocke, Bernhard. 1985. 'Wissenschaft und Militarismus: Der Aufruf der 93 An die Kulturwelt und der Zusammenbruch der Internationalen Gelehrtenrepublik im Ersten Weltkrieg'. In *Wilamowitz nach 50 Jahren,* eds. William M. Calder, Hellmut Flashar, and Theodor Lindken. Darmstadt: Wissenschaftliche Buchgesellschaft. 647–719.

West, Clarence J. 1920. The Chemical Warfare Service. In *The New World of Science: Its Development During the War*, ed. Robert M. Yerkes. New York: Century Company.
Whittemore, Gilbert Jr. 1975. Poison Gas Research and the Ideals of American Chemists. *Social Studies of Science* 5 (2): 135–163.
Wright, Helen. 1966. *Explorer of the Universe: A Biography of George Ellery Hale Woodbury*. New York: American Institute of Physics.
Yerkes, Robert M. (ed.). 1920. *The New World of Science: Its Development during the War*. New York: Century Company.

The Soldier's Body in Gas Warfare: Trauma, Illness, *Rentennot*, 1915–1933

Wolfgang U. Eckart

Abstract The paper describes medical and psychological aspects of gas warfare 1915–1918. It is shown that exact knowledge such as lethal dosages and the type and extent of injuries had been observed in cases of accident long before the outbreak of war. Nevertheless, detailed toxicological research was carried out in the toxicological department of Fritz Haber's Kaiser Wilhelm Institute for Physical Chemistry in Berlin-Dahlem. War itself offered the opportunity for deadly field experiments. The soldiers suffered not only from physical injuries (chest pain, breathlessness, coughing, bloody sputum, multiple organ failure) but also from fear and traumatization. Given the enormous fear caused by the idea of a supposed poisoning even without symptoms, distinguishing the real and actual from the simulated in such cases must have been problematic and caused a permanent threat of being accused of malingering or even simulating. From there it was only a small step to psychic and political stigmatization as "*Rentenbetrüger*" (pension fraudsters) or being mentally ill in the late Weimar Republic and especially under National Socialism. Whereas the nation was forever grateful to the war-wounded and disabled veterans, the stigmatized were seen as being mentally ill, were sterilized, and sometimes even murdered.

1 Introduction

If the First World War may be understood as the first technical and industrial war, then this implicit metaphor not only carries with it the technical aspect of weaponry in the sense of modernity in technology, both with regard to fighting with new weapons and to their development, and further to modern industrial production under the conditions of Taylorism and streamlining of production. The soldiers themselves, too, understood the new war as a kind of industrial work, which was even reflected in their language. One "went to work" in the trenches much as in the

W.U. Eckart (✉)
Institut für Geschichte und Ethik der Medizin, Heidelberg University, Heidelberg, Germany
e-mail: wolfgang.eckart@histmed.uni-heidelberg.de

B. Friedrich et al. (eds.), *One Hundred Years of Chemical Warfare: Research, Deployment, Consequences*, DOI 10.1007/978-3-319-51664-6_12

mineshafts of the Ruhr, to change places with the exhausted men of the previous shift; there was at least as much "work" on the battlefield as in a Krupp factory hall. In sweaty cooperation and collaboration and under the pressure of an unstoppable, urgent timing cycle one became, as an artilleryman or ammunition carrier, as much a part of the mechanical processes of this huge, monstrous death machine as of those of a sheet metal factory at home. The only difference was that one created technical value as a dependent employee there, but on the front one destroyed technical mass products—and with these human lives—in great masses in paid work quite dependent on commands and orders (Eckart 2014).

The most depressing aspect, probably, of the technical and industrial modernity of the Great War was the chemical warfare, which particularly reveals the interlinkage of medical and military technology (Gradmann 2003, 131–154). Gas, as a substance of mass application and well-known toxicity, was in a great variety of forms already present at the start of the war as a waste product of the chemical industry, particularly of the I.G. Farben (Roth 2009, 6–8). Its usability was increased during the course of the war by the innovative synthesis of ever more murderous new substances. In particular, scientists had learnt, long before the war, how to liquefy gas by using high pressure, thus making it feasible to store and transport it in large quantities. As concerns the effect of chemicals on the human organism, most of this information was already known and available owing to industrial accidents.

The development and use of poison gas as a weapon definitely required medical and pharmacological knowledge, which all industrial countries involved in the war had to a greater or lesser extent; the German side had particularly good knowledge in this regard, owing to a high level of development in the chemical industry. The knowledge of the physiological effects of most poison gas types was based not only on animal experiments, but also on the industrial accidents mentioned above.

In the course of animal experiments, exact knowledge was gathered, such as the lethal dosage and the type and extent of injuries, which matched exactly what had been observed in accident cases, for example in the chemical industry. Without such knowledge, the development of new poison gases and their use would have been unthinkable. In addition, there was the fact that the use of poison gas had been anticipated, not only in the genre of technical science fiction (as we would now say) from Jules Verne to Arthur Conan Doyle. In Paris in 1911, for example, ethyl bromoacetate (more precisely: ethyl 2-bromoacetate) was used in the fight against crime, in the form of "tear gas" in cases of robbery. The reason that the use of gas and other chemicals in warfare could become the metaphor par excellence for modern, technical-industrial war cannot be established in the comparatively slight extent of direct effects. Among the roughly ten million war dead, conservative estimates placed the number of deaths through poison gas at "only" 90,000 to

100,000.[1] Among the permanently wounded and damaged soldiers, some 25 million on all sides, there were "only" about one million suffering from the effects of gas. Of considerable significance, however, was the horror of gas, to which not only the expected damage understandably contributed, but also the omnipresent possibility of its use, the ubiquitous way it spread in the area of battle, and the realization that even gas masks and protective clothing could not sufficiently guarantee the survival of man and beast.

A borderline between research serving the development of weapons and that serving to protect against them and to provide therapy for wounds could hardly be drawn. The use of poison gas against the enemy and the protection of one's own soldiers against enemy (or "friendly") poison gas were based on the same research. Fritz Haber's Kaiser Wilhelm Institute for Physical Chemistry in Berlin-Dahlem,[2] where research on poison gas was carried out, consequently had, from 1916, a rapidly growing Toxicological Department. The German army, which was the first to apply poison gas in April 1915, was far ahead of the enemy for this reason, not only as far as research into the gas weapon was concerned, but also in medical care in this area. In military tactics, poison gas extended the scenario of a possible threat and injury to soldiers to new territory, which was purposefully investigated by the medical experts: the effect of extreme terror, which was raised to a benchmark in the further development of this weapon (Kästner and Hahn 1994, 42–50; see also: Zeidler 1993). To say, however, that the effect of poison gas was "mostly a terrifying one," as the pathologist Otto Muntsch described it in 1935 (Muntsch 1935, 102), was pure cynicism. Poison gas injured the body on the outside to a limited extent at first, but it could gradually destroy a soldier's body within hours, days, or weeks. The symptoms were agonizing, terrifying, and, under some circumstances, could occur long after the actual poisoning had taken place. Above all, gas led directly to a large number of soldiers being unable to function. At the same time, gas spatialized the threat of bodily harm, and in place of a specific threat to the body from projectiles, there was that of a deadly environment, inimical to life. The hopes nourished by this, not only among the army, were quite clear. Whole sections of the front could be made to collapse, rapid gains in territory and finally the abolition of the exhausting trench war could be achieved. That these hopes would turn out to be illusory was not at all clear at the beginning of gas warfare, as there were some very quick surprise successes at the start. With the increasing use of poison gas, however, came the disillusionment, at the latest from the summer of 1915 onward. The hopes placed in the effects of this new type of warfare dissolved as fast as the substances being used. One exception was formed by chemical contact poisons and skin-damaging substances, such as mustard gas (actually a nitrogen mustard).

[1]Figures differ very much. For the Western war theatre losses (killed and severely wounded) it seems as if about 20,000 soldiers died and about 500,000 were wounded under gas attacks. For the Eastern front there are no reliable figures. All in all the number of soldiers killed under the influence of poisoned gas may be estimated at about 100,000 (Haber 1986, 243; Müller 2003, 519–522; Gaskrieg während des Ersten Weltkrieges).

[2]See particularly Szöllösi-Janze (1998).

Against this background, it is understandable that substances like these were under consideration for future wars, in order to avoid the costly stalemate of the trenches, as Fritz Haber lectured officers of the Reichswehr in 1920. Only a massive bombardment with sulphur mustard would be suitable

> to make an area with its protective constructions impossible to hold. This chemical weapon forces the abandonment of trench warfare, which the development of explosive munitions had brought with it (in Brauch 1982).

In fact, the main reason was to fill the expected gap in munitions production, as Germany had been cut off by sea blockade from the import of Chilean saltpetre—at that time still an absolute necessity for the manufacture of explosives—already a few weeks after the start of the war.[3] During the war, quite different types of gas and chemical substances were used. The sheer variety was so disturbingly great in the end that Oskar Minkowski, writing in 1921 in reminiscence, was forced to come to the conclusion that

> [t]he number of substances used by both sides in gas warfare had, with time, become a very large one. As the composition of the chemical compounds used by one side could be rapidly discovered and imitated by the other side, while on the other hand damage frequently occurred on both sides from their own gas ammunition, one cannot sensibly carry out any sort of differentiation between the substances used by any one army (Minkowski 1921, 346).[4]

In principle, of course, such a differentiation was clearly possible, at least according to the chemical groups involved. Thus, eye irritants were used, such as the tear gas ethyl bromoacetate (known technically sometimes as *White Cross*, because of the container marking), with rather limited success. More effective were the diverse irritants of the nasal membranes and the throat, known in German as Blue Cross, of which the best known are Clark I and Clark II. These had the aim of irritating the enemy's nose, throat and bronchial areas. These substances were used as "mask breakers" (Minkowski 1921; Gradmann 2003, 145). The strong irritation (hefty coughing, tearing eyes, headache, nausea, vomiting, trembling, and vertigo) would, it was rightly hoped, cause the soldiers to tear their gas masks off, thus exposing themselves completely to deadly gases. To this exact end, lung damaging and asphyxiating poison gases (Green Cross), such as chlorine, chloropicrin, diphosgene, or phosgene, were used. Their aim was the rapid killing of the enemy. While at first great hopes were placed in relying on blowing chlorine gas towards the enemy, the problem of prevailing winds in the right direction being needed, coupled with local microclimatic oddities, soon forced the military to abandon this

---

[3]See also Baumann (2011).

[4]Translated from the German original: "Die Zahl der von beiden Seiten beim Gaskampf verwendeten Stoffe war im Laufe der Zeit eine sehr große geworden. Da die Beschaffenheit der von einer Seite benutzten chemischen Verbindungen vom Gegner sehr bald ermittelt und nachgeahmt wurde, andererseits auf jeder Seite auch Schädigungen durch die eigene Gasmunition nicht selten vorkamen, kann eine Trennung der von den einzelnen Heeren verwendeten Stoffe nicht durchgeführt werden."

method. It was better to use special gas grenades, in this way shooting the gas, for instance phosgene, at the enemy (Minkowski 1921; Gradmann 2003, 141). Whereas chlorine gas disabled the soldier by means of strong irritation of the mucous membranes, thus the eyes, the nose, the nasopharyngeal zone and the respiratory tract, accompanied by lasting damage to the lung tissue, the hope with phosgene was the rapid dispatch of the enemy by lung oedema. Powerful chest pain, breathlessness, coughing, and bloody sputum were the harbingers of the soldier's death. Multiple organ failure then led to the actual death, either still on the battlefield or, at the latest, in the field hospitals near the front. Under the term Yellow Cross, finally, all those substances were subsumed that cause dermal injury, such as mustard gas and lewisite. In this case, the military goal was to disable the enemy by means of the nearly immediate pain following contact, and the long-lasting injury and extended treatment times. The substances of the Yellow Cross group were also shot at the enemy in grenades, in their oily pure form, or sprayed as aerosols. The expected effect on the body of the soldier consisted in the destruction of the dermal tissue by blistering and the formation of ulcerous areas, or in irreversible blindness in the case of eye contact. In cases of large-scale surface skin contact, or longer-lasting aerosol exposure, such substances could also prove themselves to be rapidly fatal.

It was clear from the start to the German side, and especially after phosgene and Yellow Cross (mustard gas) were ready to be used, that a very difficult situation would arise, should the enemy come into possession, for example, of phosgene, which had been used by the Germans from July 1917. For this case, and in view of the chemical war's intensification through 1918, preparations were made. In January 1918, the medical department of the Ministry of War published a 55-page white paper, entitled "Zur Kenntnis und Therapie der Gasvergiftungen" (On the knowledge and therapy of gas poisoning), which dealt in particular detail with phosgene and mustard gas poisoning. A first addition, published in May 1918, gives an impression of the problems resulting from the escalation of gas warfare, without any mention of enemy use of mustard gas. Apparently there were not enough correspondingly trained medical personnel. This is at least indicated by the instruction to keep such personnel, once trained, at the gas hospital (Eckart 2014, 79–80).

2 Perceptions in the Field

Fritz Haber had, at the beginning, relied on blowing chlorine gas out of pressurized bottles. The first mass use of this deadly poison gas was at his suggestion on 22 April 1915. In this case, success was immense, with massive enemy losses, although the exact number is still debated. The French general Henri Mordacq reported in horror on the gas attack near Ypres on 22 April 1915:

> On the banks of the canal, only some yellowish wisps of smoke could be recognised, but as we approached, via Boesinghe, some three or four hundred meters closer, we felt a hefty

tingling in our noses and throats, our ears buzzed, it became difficult to breathe; an intolerable stench of chlorine surrounded us. Soon we had to dismount, because the horses, bothered and affected by this, refused to gallop or trot [...] Near the village, the scene that we saw was more than pitiable; it was tragic. Men fleeing everywhere: infantry, Africans, riflemen, Zouaves (light infantry), artillerymen without weapons, deranged, coats off or open wide, neckbands torn off, ran like madmen into the unknown, screaming loudly for water, spitting blood; some rolled around on the ground and tried in vain to breathe. (Hanslian 1934, 44)[5]

In Germany, this success was celebrated as "Day of Ypres," in analogy to the "Day of Sedan." But the use of pressurized gas bottles was dependent on the wind, and thus very dangerous for the German troops, too. It was primarily for this reason that the chemist Walter Nernst preferred "gas bombardment" with artillery grenades (Eckart 2014, 80). In February 1916, indeed, for the first time phosgene grenades were fired from the French side, which caused the High Command to finally make new arrangements. In consequence, blowing gas out of pressurized bottles was abandoned, and replaced by bombardment with larger gas containers at shorter ranges, and smaller ones at longer ranges. Now the grenades contained the highly effective substance diphosgene. The grenades, marked in green (Green Cross), were used for the first time on 22 and 23 June near Verdun and caused massive losses on the opposing side (ibid.). Phosgene and especially diphosgene were much more destructive than chlorine gas. Both sides were affected. The physician Alfred Schroth reported in 1917 on such a phosgene gas attack:

All those cases, however, that we lose through death two or three hours after the attack on the position show a sight of the greatest horror. Breathlessness and coughing increase to asphyxiation. The sputum, at first not much and thick, is replaced by a liquid and then foamy expectoration, which slowly is coloured by blood, and finally oozes out of the nose. The appearance of the poisoned victim is wasted, and as a consequence of the lung oedema, death occurs with the victim nearly completely conscious (Brauch 1982, 70).

Simulation of illness after gas attacks became a particular problem for the German prosecution of the war. This is quite in keeping with the hysteria of the army command with regard to an increase in simulants in this area of warfare, since the army command had already been dramatically confronted with this problem in connection with war neuroses (Gaupp 1922, 71).

[5]Translated from the German original: "Man konnte am Ufer des Kanals nur noch einige gelbliche Rauchschwaden erkennen, als wir uns aber Boesinghe auf drei oder vierhundert Meter genähert hatten, fühlten wir heftiges Prickeln in der Nase und Kehle, in den Ohren sauste es, das Atmen fiel uns schwer; ein unerträglicher Chlorgeruch umgab uns. Wir mußten bald absitzen, da die dadurch belästigten und behinderten Pferde sich weigerten zu galoppieren oder zu traben. »[...] "In der Nähe des Dorfes war das Bild, das sich uns bot, mehr als bedauernswert, es war tragisch. Überall Flüchtende: Landswehrleute, Afrikaner, Schützen, Zuaven, Artilleristen ohne Waffe, verstört, mit ausgezogenen oder weit geöffneten Röcken, abgenommener Halsbinde liefen wie Wahnsinnige ins Ungewisse, verlangten laut schreiend nach Wasser, spuckten Blut, einige wälzten sich sogar am Boden und versuchten vergeblich, Luft zu schöpfen."

At the same time, there are indications of disciplinary problems caused by allegedly or actually pretended or faked gas injuries. These were described in analogy to the war neuroses:

> When the physician at the front has not himself established gas poisoning or the consequences of being buried by a shell in men who complain of this, and whom he feels himself compelled to send back to the rear, then the wounded slip or the paper must have the annotation 'allegedly'. Instead of the very certain designation 'gas poisoning' or 'nervous shock through burial', which leads in such cases to the patient having a permanent notion of serious illness, it is better to note down 'complaints (sic), allegedly following gas poisoning' or 'nervous complaints, allegedly burial'. Keeping men on the battlefield […] is [absolutely] necessary.[6]

A directive of the head of field hospitals of November 1917 was renewed, ordering that the "many soldiers reporting alleged gas poisoning, but not showing immediate signs of illness" be kept directly near the front for 24–48 h, which corresponds to the phase of acute danger in the symptoms of phosgene poisoning (ibid.).

With regard to the assumed simulation of mustard gas poisoning there were similar panicky reactions, although somewhat later. In June the High Command pointed out that the "difficulty in immediately recognizing mustard gas poisoning in those affected […] aided shirking," in that it at least enabled the supposedly poisoned soldier to receive the 24-hour observation time. In the autumn of 1918, additional reports appeared on the simulation of such poisoning, or more exactly its symptoms on the part of soldiers. Thus, in a report from a collecting point for wounded, dated October 1, 1918:

> Over the last few days, three cases of self-inflicted injury in the form of acetic acid burns have been established here without any doubt. Cloths soaked in acetic acid are placed on the skin of the lower arms and legs (ankles) and cause characteristic changes of the skin. […] The sick men claim to have been poisoned by mustard gas.[7]

The War Ministry, on October 4, 1918, pointed out "remarkable cases of self-disfigurement and the simulation of illness," among them the conjunctivitis also typical of mustard gas poisoning, created by rubbing soap in the eyes. Probably this was the reason for the directive of August 1918, later rescinded, "that no instruction about Yellow Cross is to be given for reasons of secrecy to the replacement recruits, only at the field recruit depot." The secrecy, in any case, can hardly have been directed towards the enemy, who at this point already possessed properly structured medical care (Gradmann 2003).

Poison gases displayed specifics in the area of the symptoms caused by them that could indeed awaken the suspicion of a simulation. The pathology of mustard gas poisoning, in particular, was not understood fully by contemporaries, so that the

---

[6] "1. Nachtrag" zur Dienstvorschrift "Zur Kenntnis und Therapie der Gasvergiftungen," May 1918, 57. In Gradmann (2003, 148).

[7] Bericht der Krankensammelstelle 257 vom 1.10.1918. Bayerisches Hauptstaatsarchiv, Stv. Gen. Kdo., I. AK, San A, 176. In Gradmann (2003, 152).

diagnosis was reduced to the evaluation of external symptoms, such as the skin injuries resembling burns, irritation of the eyes, and disturbances of digestion. These symptoms, however, could be reproduced using hard soap or acetic acid. Additionally, there was the problem that appearances of illness could often take a good 4–6 h to appear, with nothing appearing initially. The believable report of gas poisoning could thus—even without symptoms—attain the usual 24-hour hospital stay for observation. Because mustard gas showed a high degree of persistence, poisoning could occur without any obvious contact—for example through grenade bombardment—and could even remain unnoticed at first. When one considers the enormous fear caused by the idea of a supposed poisoning even without symptoms, distinguishing the real and actual from the simulated in such cases must be problematic, to say the least. The doctors assumed, rightly, that there were many simulators among the supposedly gas poisoned. Those who really had been poisoned "were hard to tell from the others," as Oskar Minkowski in his *Handbook of Medical Experience in the World War* pointed out. The "others" were "afraid of having inhaled poison gas, or wished only to take the opportunity to leave the battlefield" (Minkowski 1921, 370). The nerve doctor Gaupp identified a frequent source of war neuroses in imaginary or simulated gas poisoning. The enormous effect of fear, especially in the case of mustard gas, made many think that the victim must be suspected of simulation or at least was aggravating, exaggerating, his symptoms. This is clear from a whole series of measures introduced in 1918 in a situation characterized by a general threat to discipline and, at the same time, the growing efficacy of the enemy's gas warfare. The corresponding regulations or white paper (see above) emphasized the analogy to the problem of the supposed simulation in war neuroses, and advised care in the application of the designation "gas poisoning":

> Instead of the definite term 'gas poisoning' or 'nervous shock through burial', which embeds itself in such cases in the mind of the patient to become a permanent idea of his sufferings, it is better to say simply 'complaints supposedly after gas poisoning' or 'nervous complaints, allegedly burial' (ibid.).

In addition, the rule for alleged gas victims was to keep them as near to the front as possible for observation. "Keeping men on the battlefield [...] is [absolutely] necessary,"[8] was the corresponding regulation. A directive from the Ministry of War of August 1918 is also remarkable, stating that "no instruction about Yellow Cross is to be given for reasons of secrecy to the replacement recruits."[9] Secrecy towards the military enemy, who had over a year's painful experience to draw on and had long since set up effective medical facilities, was pointless. As reports of the simulation of mustard gas poisoning were already present, the point of secrecy can only have been to keep knowledge of these symptoms from the troops as far as

[8]Letter (classified): Armee-Abteilung des Bayerischen Kriegsministeriums an Generalkommandos der Armeekorps und ausbildende Stellen vom 23.8.1918. In Gradmann (2003, 152).

[9]Letter dated 27.9.1918, KM, chem. Abt, Nr. 3206/9.18 A 10. Bhsta, Stv GenKdo I. AK SanA 135. In Gradmann (2003, 152).

possible. This interpretation is also supported by the attempts made by the chemical department of the Ministry of War to counter "wrong ideas of the effect of mustard gas on the human body" (ibid.), which were circulating at the front, in such a way that one is forced to categorize these attempts as lying between euphemism and disinformation.

3 Gas and Psyche

It does not require much imagination to have an idea of the effect of the gas weapons on the psyche of the soldiers. The hoped-for effect of terror did not take long on both sides of the main front to take hold; even when only "enemy" troops were affected, soldiers knew very soon that this could reflect their own fate. The simple soldier thus did not share in the enthusiasm of the experts, the military commanders, and the politicians from the beginning. Probably many soldiers were more horrified than delighted at the new escalation of mass murder at the front, although we have few confirmations of this in the field letters. Thus, the miner F. Tholl, in a letter written from hospital on May 10, 1916, reflected on the consequences of the new kind of warfare:

> Hopefully this war of mass murder will soon come to an end. It is said that the English had to carry away their dead by the wagon after a successful German gas attack, losing thousands in one to two hours. What artillery destroys in numbers of men, is supposed to be nothing in comparison. War technology, then, is on the best road to destroying whole armies without spilling a drop of blood, choking them or putting them to sleep. What a humane way to wage war (Ulrich and Ziemann 1994, 95).

It was even worse for soldiers who had survived a gas attack, but had come into contact with some poisonous substance unknown to them, who had then dutifully reported to the doctor, only to hear their superiors suspect them of simulation and threaten them with disciplinary consequences for cowardice, instead of experiencing care and observation. This is pretty much what happened to the infantryman Birzer from the Upper Palatinate (Bavaria), who wrote to his mother, Anna Birzer, on 20 August 1917 from the trenches:

> My dearest mother! […] Last night at 2 o'clock, while I was standing at my post, the English carried out a very strong gas attack, 3 m left of me 3 gas grenades exploded. By the time I had got my gas mask on and had alarmed those below me, I had swallowed a bit of the gas. So I reported to the doctor, because every time they said, if you think you have swallowed some, go see the doctor right away (ibid., 95–96).

But Birzer had not reckoned with his company commander, who not only would not let him see the doctor, but railed at him:

> He tore a strip of me like nobody ever has, calling me a coward and a slacker, that's what he called me (ibid.).

The infantryman had observed how eight of his comrades and their company commander had died in the attack, and that otherwise a great many of his surviving comrades were in hospital already. He knew, too, that other soldiers would fall ill with delayed symptoms ("Usually it first comes over you the next day, then you become really ill"). Birzer was in despair: "Dearest mother, I cannot stand it any more, if it were not for you I would take my own life" (ibid., 280). About his company commander Birzer wrote: "Somebody like that should be shot." Birzer's letter was read by the field censor. Upon checking the incident, the infantryman's statements appeared more probable than the accusation of simulation put forward by his company commander, Reserve Lieutenant Münch. Münch was then sentenced by his regimental commander to a day's house arrest. Birzer was allowed to go to hospital for observation. Not all soldiers shared in his good luck.

The German public took up critical positions to the gas war in a very restrained manner. There were such critics, of course, for example among the pacifistic left. But reports of the special brutality of chemical warfare spread but slowly, even among pacifist circles. Thus, the anarchist and author Erich Mühsam wrote in his diary on 27 April 1915 after reports of the German chlorine gas attack near Ypres:

> […] near Ypres a victory has succeeded […] with stink bombs. They were at first only used by the Allies, in Germany they were outraged, now the whole world is outraged over Germany. I cannot deny the view that smoking out the trenches with chlorine vapours is no worse than killing the occupants with bullets and grenades. That this war is hardly a chivalrous one, is well known (Mühsam 1915).[10]

In the Reichstag (German parliament) in February 1918 a first critical debate about poison gas took place, after rumors has spread that a large, impending offensive in the West was being planned by the High Command, including a massive poison gas attack. All this had been preceded by the February call (February 8, 1918) by the International Committee of the Red Cross in Geneva to ban poison gas on the battlefield. After having been awarded the only Nobel Peace Prize of the entire war, the ICRC had finally felt morally obligated to condemn the barbaric innovations which had been introduced to warfare by the natural sciences, and to urge all concerned to keep to the Hague Articles of Land Warfare:

> Today we wish to raise our voices against a barbarous innovation which science is in the course of perfecting, that is, making it more murderous and more refined in its cruelty. We are speaking of asphyxiant and poisonous gases, the use of which, it seems, is growing to a scale hitherto unsuspected. The Regulations adopted at The Hague respecting the laws and customs of war on land contain the following: "It is especially forbidden to employ poison or poisoned weapons, and to employ arms, projectiles, or material calculated to cause unnecessary suffering." Asphyxiant or poisonous gases are without any doubt one of the

---

[10]Translated from the German original: "[…] bei Ypern ein größerer Sieg gelungen—und zwar mit Stinkbomben. Die wurden zuerst nur von den Alliierten angewandt, da entrüstete man sich in Deutschland, jetzt entrüstet man sich in aller Welt über Deutschland. Ich kann mich der Ansicht nicht verschließen, daß das Ausräuchern der Schützengräben mit Chlordämpfen nicht ärger ist als das Töten derInsassen mit Patronen und Granaten. Daß sich dieser Krieg in keinen ritterlichen Formen abspielt, weiß man ja schon."

> poisons forbidden under the Convention. Medical personnel are all unanimous in testifying to the terrible suffering caused by these gases, which is more harrowing to see than that resulting from the worst of wounds (World War I: the ICRC's appeal against the use of poisonous gases[11]).

To this day, it is difficult to understand why the ICRC was unable to issue such an appeal earlier, especially since, besides the Vatican and various Red Cross organizations in other, neutral countries, the powers involved in the war, too, had reacted quickly and positively. The President of France Raymond Poincaré let the ICRC in Geneva know that the Entente would give up the use of poison gas if the Central Powers would do likewise. The official note from the Entente of May was in the same tones, even mentioning a possible total ban of gas weapons, but placing the blame for their use entirely on the Central Powers. The German reply took a long time and was disappointing. The German Foreign Office informed Geneva on September 12, 1918, diplomatically brief and in fact untruly, only that Germany had agreed to earlier conventions against the use of poison gas; the enemy alone was responsible for the development and use of poison gas. There was no negotiating possible on this basis. In the Reichstag, however, the initiative of the Red Cross in a debate on the necessity of a "great offensive" to attain a "peace of power" in the West was certainly promptly discussed at the end of February. The only voice against the plan for such an offensive with the massive use of poison gas was that of the Berlin lawyer Oskar Cohn (1869–1934), member of the Reichstag for the USPD. It might well be, said Cohn, that the enemy could not withstand such an offensive, but then, he said—looking at the political representatives of the inner truce in the Reichstag—one "would freeze in this house from the hate of all mankind".[12] Cohn received support only from the rows of the USPD. It was a scandal that the German public, for reasons of censorship, knew nothing of the initiative of the Red Cross, although the international newspapers were full of it; in this manner, one would simply run directly into "the most horrible thing to happen in this war […], into the gas offensive in the West."[13] Gustav Stresemann, of the National Liberals, repudiated Cohn's references to the Red Cross vehemently. "In all of this," one could only see the "malicious repression of everything that Germany does" (ibid.), the attempt to "discredit our own Fatherland in the world out there," so as not to see "any wounds" in the others:

> You speak of how mankind trembles before the means with which we intend to prosecute the offensive in the West. Do you not know, then, how many thousands and thousands of German soldiers have been killed by the poison gas attacks of the enemy? […] When you speak of us having to freeze in the hate of the world, which would turn against us after this war, well—you are encouraging that hate by attacking Germany![14]

---

[11]See https://www.icrc.org/eng/resources/documents/statement/57jnqh.htm.

[12]Stenogr. Berichte d. Reichstags, 131. Sess., 22.2.1918; Bd. 311, p. 4084A.

[13]Stenogr. Berichte d. Reichstags, 131. Sess., 22.2.1918; Bd. 311, p. 4085A.

[14]Stenogr. Berichte d. Reichstags, 131. Sess., 22.2.1918; Bd. 311, p. 4088B.

At least Philipp Scheidemann (1865–1939) at the end of February 1918 for the SPD, like Cohn, again critically pointed to the press censorship, which simply prevented the public from being informed about such "great-hearted suggestions" such as that of the Red Cross—never mind allowing said public a voice. Probably the Reich Government had already sent a response to Geneva which did not reflect the general opinion in the matter, commented Hugo Haase (1863–1919) for the USPD:

> What would it have cost the German Government to respond to this suggestion by saying: Yes, we are prepared to do so, if the others also pledge themselves to so do? But no! They could not wait to see whether the others wanted this too, but right from the start they reserved this means, any means for themselves. We are not surprised; we have heard, often enough, that all means were justified in this war, if they only lead to victory, no matter how cruel such means are.[15]

The MP was wrong in this case: Berlin had not yet responded and was clearly not prepared to before the planned offensive in the West.

4 Weimar to the Nazi Period—the Need of the Traumatized

During the Weimar Republic, war trauma was basically recognized as damage incurred during military service. De facto, however, there were pension cuts and the withdrawal of state benefits in so-called "doubtful cases" already in the 1920s and early 1930s. The evidence is the exemplary pension statistics of the official pension offices in each town. Nils Löffelbein has examined this for Munich and shows that in the city of Munich and the surrounding country area alone (München-Land), some 66.4% of the benefits applications based on psychological trauma were refused from the start (Löffelbein 2013). Among these, without any doubt, were numerous soldiers whose alleged gas injuries or psychotrauma were not interpreted as real damage caused by the war, but attributed to a greedy and fraudulent attempt to obtain a pension on the basis of simulation, simulation which was insinuated and presumed. There were, in addition, general problems in providing benefits and pensions in the First German Republic.

If the Weimar democracy found itself in an emergency condition after the Reichstag elections of 1930, the year 1932 made the social catastrophe even worse. The Great Depression was worse than could have been imagined; mass unemployment and the fall into poverty took on unimaginable dimensions. The cuts in the state benefits for veterans reached a new high in the summer of 1932 with the third emergency directive released by the Papen government. Hindenburg's objections, too, who wrote to the Reich Chancellor expressing the deepest misgivings about the cuts in the veterans' pensions, or at least advocated some relief

[15] Stenogr. Berichte d. Reichstags, 135. Sess., 27.2.1918; Bd. 311, p. 4213B.

from the hardships, remained without effect. The directive came, and it gave new impetus to the feelings of revolt among the wounded veterans and their relatives.

The Nazi propaganda after 1933 was correspondingly careful to distinguish between war-wounded, disabled veterans, to whom the nation was forever grateful, and who were potentially able to place their remaining strength at the disposal of the national community, and useless "ballast existences", unable to work. The wounded of the First World War were at the top of the Nazi scale of social value, celebrated as "honorary citizens of the nation" before all other groups of the handicapped. But only physically wounded soldiers were regarded as worthy of benefits, who, in the words of Reich Health Leader Leonardo Conti, were to be classified as "highly valuable people accused of war" (Löffelbein 2013, 329). Mentally ill veterans, on the other hand, were vilified as "simulants," "hunters after pensions", "unclean elements" (*Volksschädlinge*), who damaged the reputations of the true victims of the war in public (ibid., 238–239). Soon after the National Socialist "takeover" (*Machtergreifung*), mentally ill veterans (ca. 16,000 between 1934 and 1938) were deprived of all pensions (Neuner 2011, 198). Not only that a great number of them were sterilized, some of them became also victims of the so-called "euthanasia" and were killed between 1939 and 1945 (ibid., 315–324). We don't know how many of those mentally ill (most of them patients with "war neurosis") had been traumatized by gas attacks. However, it must have been a considerable number because gas attacks rated high among the causes of "war neurosis."

In order to answer the rather obvious question as to why, then, since the war there had been several thousand mentally ill front-line soldiers, an ideological maneuver was thought up, the construction of a direct connection between traumatization and the Weimar "system period" of 1918 until 1933. The war, according to this, was not the cause of the mental suffering of the veterans. Rather it was the Weimar welfare system, which had supposedly produced large numbers of pension neuroses, anti-social and psychopathic elements. "Anti-social" behavior was supposedly directly furthered by the climate of the Weimar welfare state.

5 Summary

The most depressing aspect, probably, of the technical and industrial modernity of the Great War was chemical warfare, which particularly reveals the interlinkage of medical and military technology. Seen from the military aspect, the use of poison gas was not very effective and by no means decisive for ending the war. Seen, on the other hand, from the psychological and humanitarian point of view, it was a disaster for the soldiers' minds and bodies wherever it was put to use. The fear of "gas" was paralyzing, and the wounds caused by most of the poisonous substances were terrible. Chemical warfare must be looked upon as the first failure of science and technology in the twentieth century. Scientists completely submitted themselves to the murderous necessities of war and not only provided their knowledge but participated totally in the perfidious creativity of mass murder on the battlefields. The paper outlines this subjugation of science to the military and then changes its perspective to the soldiers' perception of chemical war on the battlefield and after the war, which was shaped by dread and long-lasting traumatization. The paper's last part describes the political and psychic stigmatization of the mentally traumatized in Germany. Many of them had been physically wounded and mentally shocked by poison gas. Whereas to the war-wounded and disabled veterans the nation was forever grateful, the mentally ill were stigmatized, sterilized, and some of them even murdered. Thus, WWI chemical warfare continued its terrible destruction long after the armistice on the battlefields.

References

Baumann, Timo. 2011. *Giftgas und Salpeter. Chemische Industrie, Naturwissenschaft und Militär*. Dissertation, Universität Düsseldorf.

Brauch, Hans Günter. 1982. *Der chemische Alptraum oder gibt es einen C-Waffen-Krieg in Europa?*. Berlin: Dietz.

Eckart, Wolfgang U. 2014. *Medizin und Krieg: Deutschland 1914–1924*. Paderborn: Schöningh.

Gaskrieg während des Ersten Weltkrieges. Wikipedia. https://de.wikipedia.org/wiki/Gaskrieg_während_des_Ersten_Weltkrieges. Accessed 10 Jan 2016.

Gaupp, Robert. 1922. Schreckneurosen und Neurasthenie. In *Handbuch der Ärztlichen Erfahrungen im Weltkriege*: *Geistes- und Nervenkrankheiten*, vol. 4, ed. Karl Bonhoeffer, 68–101. Leipzig: Barth.

Gradmann, Christoph. 2003. "Vornehmlich beängstigend." Medizin, Gesundheit und chemische Kriegführung im deutschen Heer 1914–1918. In *Die Medizin und der Erste Weltkrieg*, eds. Wolfgang U. Eckart and Christoph Gradmann, 131–154, 2nd ed. Herbolzheim: Centaurus.

Haber, Ludwig F. 1986. *The poisonous cloud. Chemical warfare in the First World War*. Oxford: Clarendon.

Hanslian, Rudolf. 1934. *Der deutsche Gasangriff bei Ypern am 22. April 1915. Eine kriegsge-schichtliche Studie*. Berlin: Gasschutz und Luftschutz.

Kästner, Ingrid and Susanne Hahn. 1994. Der Toxikologe Otto Muntsch (1890–1945) und die deutsche Kampfstoffforschung. *1999. Zeitschrift für Sozialgeschichte des 20. und 21. Jahrhunderts* 4: 42–50.

Löffelbein, Nils. 2013. *Ehrenbürger der Nation. Die Kriegsbeschädigten des Ersten Weltkriegs in Politik und Propaganda des Nationalsozialismus*. Essen: Klartext.

Minkowski, Oskar. 1921. Die Erkrankung durch Einwirkung giftiger Gase. In *Handbuch der Ärztlichen Erfahrungen im Weltkriege 1914/18*: *Innere Medizin*, vol. 3, ed. Ludolf von Krehl, 340–383. Leipzig: Barth.

Mühsam, Erich. Tagebücher. Heft 13. http://www.muehsam-tagebuch.de/tb/diaries.php#d_1915_04_27. Accessed 1 Dec 2014.

Müller, Rolf-Dieter. 2003. Gaskrieg. In *Enzyklopädie Erster Weltkrieg*, ed. Gerhard Hirschfeld, Gerd Krumeich, and Irina Renz, 519–522. Paderborn: Schöningh.

Muntsch, Otto. 1935. *Leitfaden der Pathologie und Therapie der Kampfgaserkrankungen*. Leipzig: Thieme.

Neuner, Stephanie. 2011. *Politik und Psychiatrie. Die staatliche Versorgung psychisch Kriegsbeschädigter in Deutschland 1920–1939*. Göttingen: Vandenhoeck & Ruprecht.

Roth, Karl-Heinz. 2009. Die Geschichte der I.G. Farbenindustrie AG von der Gründung bis zum Ende der Weimarer Republik. http://www.wollheim-memorial.de/de/ig_farben. Accessed 12 July 2013.

Szöllösi-Janze, Margit. 1998. *Fritz Haber 1868–1934. Eine Biographie*. München: Beck.
Ulrich, Bernd, and Benjamin Ziemann (eds.). 1994. *Frontalltag im Ersten Weltkrieg: Wahn und Wirklichkeit. Quellen und Dokumente*. Frankfurt am Main: Fischer.
World War I: the ICRC's appeal against the use of poisonous gases. Full text of the appeal to the belligerents, issued by the International Committee on 8 February 1918. http://www.icrc.org/eng/resources/documents/statement/57jnqh.htm. Accessed 13 July 2013.
Zeidler, Manfred. 1993. *Reichswehr und Rote Armee. Wege und Stationen einer ungewöhnlichen Zusammenarbeit*. München: Oldenbourg.

Chemical Weapons Research on Soldiers and Concentration Camp Inmates in Nazi Germany

Florian Schmaltz

Abstract In 1944 and 1945 scientists and physicians in the Allied military intelligence gathered evidence on the criminal human experiments with chemical weapons conducted on inmates of the Nazi concentration camps in Sachsenhausen, Natzweiler, and Neuengamme during World War II. Some of the experiments were judged during the Nuremberg Medical Trial (Case I) and French military tribunals at Metz and Lyon after liberation. Based on this evidence and on further archival sources, this paper will examine the preconditions and settings of these experiments, the perpetrators involved, and what is known about their purpose and outcome. Furthermore, the paper will raise the question if and how the experiments in the concentration camps were linked to other experiments conducted in Nazi Germany for the Wehrmacht at military research establishments such as the Gas Protection Laboratory (Heeresgasschutzlaboratorium) in Spandau, the Militärärztliche Akademie, the Heeresversuchsstelle Raubkammer, or by universities. The paper will focus on experiments with chemical agents in German concentration camps and analyze how rivalry and division of labor between the military and the SS in human experimentation with chemical agents went hand in hand.

1 Organizational Structures of Chemical Warfare Research in Germany

Chemical warfare research in military and academic contexts is generally an issue of secrecy. It encompasses screening, identification of potential chemical agents suitable for use as weapons, means and methods for their large-scale industrial production, storage and deployment, as well as defensive research in toxicology on animals and humans. It also includes possible medical prophylaxis and treatments, as well as measures and technologies for detecting chemical agents and protecting soldiers and civilians against the severe injuries and health risks involved. In contrast to other fields of scientific research, most of the results on chemical warfare

F. Schmaltz (✉)
Max Planck Institute for the History of Science, Berlin, Germany
e-mail: schmaltz@mpiwg-berlin.mpg.de

B. Friedrich et al. (eds.), *One Hundred Years of Chemical Warfare: Research, Deployment, Consequences*, DOI 10.1007/978-3-319-51664-6_13

issues have remained unpublished. In the case of Nazi Germany, military agencies and private companies involved in the research, development, and production of chemical weapons systematically destroyed their records from Fall 1944. Since the intention was to avoid written evidence, orders were given only orally so that events remained undocumented. This holds even truer for the human experiments conducted in concentration camps. The determination to cover up and destroy evidence of war crimes and crimes against humanity by the perpetrators in Nazi Germany makes it difficult to establish the historical facts. Uncertainties will remain.

The utilization of hydrogen cyanide (known under the trademark Zyklon B) or carbon monoxide as gasses used in the destruction of the European Jews, Sinti, and Roma in the extermination camps at Auschwitz, Majdanek, Sobibor, Treblinka, and other places, or the murder of invalids and handicapped persons at these killing centers will not be discussed in this paper. Both compounds were not suitable for extra mural deployment as war gas because of their fugacity.[1]

In the Weimar Republic, research and development of chemical weapons had been organized in a covert network of smaller dislocated working groups and laboratories in Germany and abroad by means of a secret collaboration between the Reichswehr and the Red Army (Müller 1985; Brauch and Müller 1985; Groehler 1992; Krause and Mallroy 1993; Schmaltz 2005).

After the Nazis had seized power in January 1933, the research and development of chemical weapons became high priority in the context of the armament policy of the new regime. Efforts were made to establish a greater research infrastructure with more than 1000 employees working in three army-operated chemical warfare research centers. The largest institution by far became the testing ground and laboratories of the Wehrmacht at Raubkammer near Munster with an average of 500 employees, peaking at 800 in 1944 (Mills 1945, 9). The second in size was the so-called Gas Protection Laboratory (Heeresgasschutzlaboratorium) installed by the Army Ordnance Office (Heereswaffenamt, Wa Prüf 9) in the Citadel Spandau, which insulated and shielded its 450 employees from the public. Despite its name, the Gas Protection Laboratory also conducted offensive research on new chemical agents. In addition, the Gas Protection Office of the Army Ordnance Office (Wa Prüf 9) was established in Berlin with an average of 143 employees, peaking at 200 in 1944 (Mills 1945, 9) (Table 1).

The Military Medical Academy (Militärärztliche Akademie) in Berlin reopened in October 1934 in the building of the Kaiser Wilhelm Academy which had been closed consistent with the Treaty of Versailles (Neumann 2005, 70). There, the Department of Pharmacology and Military Toxicology (Institut für Pharmakologie und Wehrtoxikologie), headed by Otto Muntsch and Wolfgang Wirth, played an

[1]For the history of Zyklon B and the gas chambers in German concentration camps, see Szöllösi-Janze (1994), Ebbinghaus (1998), Kalthoff and Werner (1999), Joly (2000), Hayes (2004, 272–300), Morsch and Perz (2011), Trunk (2011).

Table 1 Organization of German chemical warfare research and development[a]

Heeresversuchsstelle (Army experimental station), Raubkammer (Munster)	Heereswaffenamt (Army ordnance office)—Wa Prüf 9, Berlin (Group 1–12)	Heeresgasschutzlaboratorium (Army gas protection laboratory), Spandau[b]	
		Dept.	Research field
R I: Field trials	1: Organization and direction	F1	Chemical synthesis
R II: Chemical analysis	2: Gas defense	F2	Analytical chemical
R III: Decontamination and detection	3: Decontamination and gas protection		Lab. 1 Polarimetry
R IV: Vehicles and workshops	4: Mechanical		Lab. 2 Sarin and tabun
5: Extra mural research	5: Extra mural research		Lab. 3 Examination of loot, HCN determination, ampoules and storage
6: Gas and smoke weapons	6: Gas and smoke weapons		Lab. 4 Mustard (derivates)
7: Testing field trials	7: Testing field trials		Lab. 5 Chamber trials
8: CW Manufacture (technology)	8: CW Manufacture (technology)		Lab. 6 Sarin research
9: Finance	9: Finance	F 3	Microchemistry
10: Incendiaries	10: Incendiaries		Microchemical and physicochemical
11: Patents	11: Patents		Physical measurements
12: Veterinary	12: Veterinary		Smoke
13: Intelligence	13: Intelligence		Adsorption and desorption of charcoal
			Molecular weight measurements
		F 4	Measurements of concentrations on films
		F 5	Library
		IIL	Individual
		IIIL	Collective protection
		IIIaL	Filter units
		IIIbL	Detection and recognition
		IIIcL	Decontamination/CW munitions
		IVL	Mechanical
		VWL	Physiological chemical
		VIaL	Ground contamination
			Lab. 1: Dust
			Lab. 2: Organic chemicals
			Lab. 3: Organic chemicals
			Lab. 4: Mustard
		VIbL	Smoke and thermal generators

(continued)

Table 1 (continued)

Heeresversuchsstelle (Army experimental station), Raubkammer (Munster)	Heereswaffenamt (Army ordnance office)—Wa Prüf 9, Berlin (Group 1–12)	Heeresgasschutzlaboratorium (Army gas protection laboratory), Spandau[b]	
		Dept.	Research field
		VIcL	Small scale development of shells—HE tests for shells
		VIdL	Meteorology
		VIIL	Toxicological Institute (animal- and human experiments)
		VIIIL	Semi-technical scale plant chemical storage

[a]A. K. Mills, Investigations of Chemical Warfare Installations in the Munsterlager area, including Raubkammer. *CIOS File No.* XXXI-86, (London SHAEF Combined Intelligence Objectives Sub-Committee—G-2 Division: 1945), Tables I-III on pp. 27–33
[b]Ibid., Table II, pp. 29–31

important role in chemical warfare research.[2] The Military Medical Academy and the Army Ordnance Office (Wa Prüf 9) established a network of outposts at universities in Marburg, Munster, Giessen, Würzburg, Greifswald, and at the Academy of Medicine at Danzig (Oehler-Klein and Neumann 2004; Schmaltz 2006b; Eberle 2015, 505–524). Furthermore, several institutes of the Kaiser Wilhelm Society were also included in this network after 1933 (Schmaltz 2005, 2009).

While the hegemony of military institutions in chemical warfare research remained dominant until the end of World War II, from 1942 on the SS tried to assert itself against the Wehrmacht through its research organization, the SS-Ahnenerbe. Initially founded as a registered association in 1935, the SS-Ahnenerbe was incorporated between the end of March and April 1942 into the Personal Staff of the Reich leader of the SS, Heinrich Himmler (Kater 1997, 11, 302, 463; Schleiermacher 1988, 79–83; Reitzenstein 2014, 34). In July 1942, the SS-Ahnenerbe established the Institute for Applied Military Research (Institut für wehrwissenschaftliche Zweckforschung), with a special branch headed by the anatomist August Hirt at the "Reichsuniversität Straßburg" to foster chemical warfare research (Schmaltz 2005, 530). With its direct link to Himmler, the SS-Ahnenerbe had privileged access to concentration camp inmates as subjects for human experimentation. From May 1944 all SS and police agencies had to apply to Himmler for personal authorization to conduct human experiments in concentration camps. Applications had to be submitted to the Reichsarzt-SS, substantiating scientific objectives as well as the required number of prisoners and duration of experiments.[3] There is no doubt that military experts were informed about specific human experiments with chemical

[2]For Muntsch's career, see Kästner and Hahn (1994), Neumann (2005, 83).

[3]Himmler (Reichsführer SS), copy, May 15, 1944, Bundesarchiv Berlin (BArch), R 26 III/729, fol. 36; Schmaltz (2005, 176–177), Hahn (2008, 480).

agents in concentration camps. Some of the military experts were also involved in their preparation and evaluation.

In addition to military laboratories and academic research institutions, chemical warfare research was also undertaken in the laboratories of private chemical companies. However, chemical warfare research in private companies such as I.G. Farbenindustrie on nerve agents apparently only involved animals and self-experiments on humans. Regarding the military, there are no documents available that give evidence of forced human experiments in industrial research laboratories (Schmaltz 2005, 455–459).

2 Chemical Weapons Research on Humans in Military and Academic Institutions

New compounds suitable for chemical warfare were regularly first tested on animals and humans at the Military Medical Academy in several gas chambers with a volume of 2–3 cubic meters before toxicological studies were conducted in the 10, 30, and 100 m^3 gas chambers of department VII L, or the larger gas chambers of the Gas Protection Laboratory at Spandau with 250 and up to 1000 m^3 (Mills 1945, 9–10). The human experiments were conducted in self-experiments by the scientists and on soldiers, officer cadets, members of student companies (*Studentenkompanien*), and convalescent companies (*Genesungskompanien*). These experiments covered toxicological evaluations, defensive protection technologies (gas masks and protection gear), and the treatment of injuries caused by chemical agents (Schultz 2001; Kopke and Schultz 2001, 242–246; Baader 2002; Neumann 2005, 288–298; Woelk 2003, 283). The participants from the military knew that the experiments implied health risks. Officially, military test persons participated voluntarily, but we can assume that peer pressure as well as compensation offered of between 5 and 100 Reichsmarks may have been an incentive (Kopke and Schultz 2001, 243–244; Neumann 2005, 289–290). Apart from the pain experienced during the actual experiments, the long-term health problems and consequential suffering are well documented for a number of cases (Spiegelberg et al. 1961). The publicist Ernst Klee claimed that on several occasions, death row inmates at Plötzensee Prison were transferred to the Gas Protection Laboratory and subjected without their consent to experiments with chemical agents (Klee 1997, 272–273).[4] This statement is based on only one testimony of a hearsay witness, who did not accompany the prisoners to the alleged experimentation.[5]

[4]Affidavit of the former juridical officer Affidavit Walter Strelow, November 27, 1946, NG-405, in Dörner et al. (2000b, microfiche 4/7764-7766).

[5]For a critical review of Klee's (mis)interpretation, see Kopke and Schultz (2001, 245–246), Neumann (2005, 295–296).

Nonetheless, the experiments in the military institutions remained ethically and legally dubious. The "Regulations Concerning New Therapy and Human Experimentation," issued by the Reich Ministry of Interior in 1931 prohibited experimentation "in all cases where consent has not been given."[6] To date, no sources on any internal discussions concerning ethical frameworks or the implementation of regulations for human experiments in military institutions during the Nazi era have been found. As the historian Ulf Schmidt has emphasized, military researchers either ignored the 1931 guidelines or were unaware of their existence (Schmidt 2013, 236; Roelcke 2017). In many cases, it is unclear if sufficient animal testing had taken place prior to the human experiments with chemical agents. In contrast to this complete lack of any institutionalized regulation of ethical issues concerning human experiments, the Nazi regime established such regulations for animal experiments in line with the animal protection law of 1933. In spring 1939, the medical service of the Wehrmacht (*Sanitätsinspektion*) restricted animal experiments to scientific laboratories, and a number of military institutes established frequent expert inspections.[7] The German attack against Poland further weakened the limitations set by medical ethics. German soldiers suffering from battle wounds and infections and civilians affected psychologically by Allied air raids, along with chemical warfare experts and physicians all radicalized their approach to exploiting vulnerable concentration camp inmates as subjects of human experiments. During World War II, human experiments dealing with agents suitable for chemical warfare were conducted in the concentration camps at Sachsenhausen in 1939, at Natzweiler from 1942 to 1944, and at Neuengamme in 1944.

Rumors about another series of human experiments with war gasses on inmates of a sub-camp of the concentration camp Groß-Rosen, who were forced to work in the nerve gas factory at Dyhernfurth near Breslau where tabun was produced and filled in shells from 1942 onward, are not confirmed by available sources. There is no doubt, however, that camp inmates were forced to work at Dyhernfurth in the extremely dangerous tabun production and filling stations with only insufficient protection, and consequently suffered severe damage to their health (Czernik 1974; Groehler 1989, 245–248; Ebbinghaus 1999, 185–186). Accidents—some of them fatal—occurred frequently, even among the German workers (Jones 1945, 10). While eyewitness accounts confirm that emergency treatments with atropine were used, no evidence has been established so far that camp inmates were subjected to standardized human experiments in a controlled manner.[8]

[6]Rundschreiben des Reichsministers des Inneren vom 28.2.1931: Richtlinien für die neuartige Heilbehandlung und für die Vornahme wissenschaftlicher Versuche am Menschen 1931, see Schmidt (2013, 236), Sass (1983), Grodin (1992, 129–132).

[7]Waldmann (OKW B 49 OKH/AHA S In II) to Militärärztliche Akademie (copy), April 29, 1939; Müller (OKW B 49 OKH/AHA S.In II) to Militärärztliche Akademie (addendum), May 31, 1939, Bundesarchiv-Militärarchiv (BA-MA) Freiburg, RH 12-23/1740.

[8]The former prisoner Tadeusz Karol, who survived the Dyhernfurth concentration camp, testified that he was ordered to enter the filling station without gas protection gear in order to examine a possible contamination with tabun. Karol collapsed after being injured by the nerve agent. An

3 Experiments in Concentration Camps

3.1 Sachsenhausen

On September 8, 1939, one week after the invasion of the Wehrmacht in Poland, Polish troops who were withdrawing accidently used sulphur mustard mines instead of regular explosives to blow up a bridge at Jaslo. This incidence caused mustard gas injuries to 14 German soldiers, two of which were fatal. The incident immediately led to an investigation by German chemical warfare experts.[9] In direct response to this incident, the Military Medical Academy and the SS initiated several series of tests to evaluate possible treatments of skin wounds caused by sulphur mustard gas. At least two series were conducted in the concentration camp at Sachsenhausen on a total of 31 prisoners. The wounds were treated with different drugs: (1) Freskan (code name F 1000 and F 1001), a powder produced by the company Dr. Fresenius (Bad Homburg) to cure skin burns; (2) the Holzmannsche-Lost-Heilmittel; and (3) probably Thiosept, an ointment based on sulphurous shale oil (Figs. 1 and 2, Table 2).

For the first series of experiments, Reichsarzt SS Dr. Ernst Grawitz ordered SS physician Dr. Hugo-Heinz Schmick, then in charge of the surgical ward at Sachsenhausen concentration camp, to conduct the experiments.[10] Schmick worked together with camp physician Dr. Walter Sonntag. On October 13, 1939 sulphur mustard was applied to the upper arms of 23 inmates.[11] According to an account by the former political prisoner Hans Kargl,[12] he and four other inmates from his barrack (Theuer, Steinmeyer, Hahn, and Grunert) were treated with a "yellow liquid" which was smeared in a radius of about 3 cm on both upper arms causing blistering,

(Footnote 8 continued)

atropine injection saved his life. See OK Wroclaw, Ds 1/68, pp. 225–227 and 244–247, eyewitness testimony by Tadeusz Karol, cited after Witkowski and Rudy (1987, 135–136). I am grateful to Esther Chen (Max Planck Institute for the History of Science) for her helpful advice and explanation concerning the Polish publication.

[9]Wolfgang Wirth (Heereswaffenamt Gasschutzabteilung): Bericht über die Verwendung von Lostminen durch die Polen bei Jaslo am 8.9.1939, September 16, 1939, BA-MA Freiburg, Bestandsergänzungsfilm WF-01/20871 (National Archives Washington, DC, Microfilm, T-77, reel 876, frames 5624376-5624396); Martinetz (1996, 167–168), Gellermann (1986, 135–136).

[10]Hugo Heinz Schmick (1909–1982) became a member of the NSDAP in August 1933 (No. 3681138) and of the SS (No. 84693) in May 1933. From June 1939 he was assigned to the concentration camp Sachsenhausen where he was ordered to establish the surgical ward. SSO Akte Hugo-Heinz Schmick, geb. 30.3.1909, BArch, VBS 286/6400039545; NSDAP-Ortsgruppenkartei Hugo-Heinz Schmick, geb. 30.3.1909, BArch, VBS 250. Vernehmungsprotokoll von Dr. Hugo Schmick durch den Untersuchungsrichter beim Landgericht Duisburg, July 21, 1951, Archives of the Memorial and Museum Sachsenhausen (AMMS), JD 1/22, pp. 19–22; Kopke and Schultz (2001, 247).

[11]Dr. Sonntag, Abschlußbericht über die mit L. am 13. Oktober 1939 geimpften 23 Fälle, December 22, 1939 (=NO-198), BArch, NS 19/1582, fol. 2.

[12]For the biography of Hans Kargl (1884–1960) see Ley and Morsch (2007, 335–337).

Fig. 1 Walter Sonntag (Courtesy of the Bundearchiv R 9361-III/195957)

open wounds.[13] These were then treated with an ointment (probably Thiosept) and Freskan. According to Kragl, the treatment caused violent pain.[14] Assisted by the orderly Fritz Langheinrich, the wounds and the healing process were documented in medical records, on film, and photographs.[15] In the second series of experiments, eight prisoners were treated with mustard gas on both arms. After three days, the blisters were opened to infect the wounds of two prisoners with a mixed flora of streptococcus, staphylococcus, and pneumococcus bacteria. Another two of the eight

[13]Hans Kargl: Erlebnisniederschrift über die Zeit der Verfolgung und Inhaftierung während des Naziregimes. (Typoskript), undated, AMMS, P3 (Stadtarchiv Hanau, 103/85), fol. 55–56; Kopke and Schultz (2001, 116–117).

[14]Ibid. For the name of the ointment, see Landgerichtsrat Peterek (Untersuchungsgericht Duisburg): Vernehmung von Mathias Mai, December 6, 1950, AMMS, JD 1/22, fol. 6–8.

[15]Wissner (Kriminalpolizei Düsseldorf): Vernehmung von Fritz Langheinrich, November 18, 1949, Archives of the Memorial and Museum Sachsenhausen JD 1/22, fol. 2–5.

Fig. 2 Hugo Heinz Schmick (Courtesy of the Bundesarchiv NSDAP Zentralkartei)

Table 2 Concentration camp Sachsenhausen: mustard gas experiments

Time period	Researcher	Objective	Subject
13 Oct.–Dec. 1939	Walter Sonntag Hugo Heinz Schmick	Therapeutic effect of Freskan powder	23 concentration camp inmates[a]
	Fritz Langheinrich	Holzman remedy	
Dec. 1939	Walter Sonntag Hugo Heinz Schmick	Ditto and antiseptic effect on bacterial infections	8 inmates

[a]The numbers of victims vary. While Sonntag noted in his report that 23 inmates were used, former prisoner Frank Cyranek estimated a number of 28 to 30 prisoners. See Vernehmung von Franz Cyranek, copy, (undated), Archives of the Memorial and Museum Sachsenhausen (hereafter cited as AMMS), JD 1/22, fol. 16

prisoners received the same treatment on the fourth day.[16] Some of the wounds reached a size of 7 × 18 cm. The infected prisoners developed sepsis with high temperatures, shivering, swelling of the glands, and enlarged spleens.[17] The prisoners' suffering led to the insight that neither Holzmann's remedy nor the Freskan powder had any healing effect on the mustard gas wounds or the infections.[18] In January 1940 Reichsführer SS Heinrich Himmler was informed about the negative results.[19] While the experiments of the SS were taking place at the Sachsenhausen concentration camp, the Wehrmacht had also started a series of human experiments investigating the efficacy of Freskan powder F 1000 and F 1001 for the decontamination and therapy of skin lesions caused by mustard gas.[20] The two chemical war experts who conducted those experiments were Ludwig Lendle and Wolfgang Wirth. Lendle was one of the leading German pharmacologists at the time and since 1936 director of the Institute for Pharmacology at the University Münster. In August 1939 Lendle was ordered on duty to the Institute for Pharmacology and Military Toxicology at the Military Medical Academy.[21] There he collaborated with the head of this institute, Wolfgang Wirth.[22]

[16]SS-Untersturmführer Dr. Sonntag, Vorläufiger Bericht über 8 Fälle von 'Öl-O'-Verätzungen und deren Behandlung mit dem Mittel 'H' bzw. 'F 1001' unter Setzung einer Infektion in 4 der Fälle, December 22, 1939, BArch, NS 19/1582, fol. 4.

[17]Ibid., fol. 5–6.

[18]Ibid., fol. 6.

[19]Grawitz to Personal Staff RFSS, January 5, 1940, BArch, NS 19/1582, fol. 3. Himmler received the letter on January 16, 1940.

[20]Bericht über die Eignung der Freskanpuder F 1000 und F 1001 zur Entgiftung und Behandlung von Lostschäden der Haut, December 21, 1939, secret, signed Lendle and Wirth, pp. 1–49 and 1 Table, BA-MA Freiburg, RH 12–23/1728.

[21]For Lendle's biography and his activities concerning chemical warfare research, see Schmaltz (2005, 468–470); Lendle, Lebenslauf, January 29, 1945, BArch, R 9361 II, Parteikorrespondenz, Ludwig Lendle, born February 2, 1899; for his postwar career, see Schmidt (1985).

[22]Wolfgang Wirth (1898–1996), studied studied chemistry in Munich and Würzburg and medicine in Berlin. He worked as assistant to Gauss at the University Clinic and as assistant to Werner Haase at the Laboratory for physiological Zoology of the Biologische Reichsanstalt für Land- und Forstwirtschaft. Wirth conducted research on chemical agents during the Weimar Republic in Germany and in the context of the secret collaboration with the USSR on behalf of the Army Ordnance Office. Wirth became a member of the SA in 1933 and the NS-Lehrerbund, the NSV, and the Reichskolonialbund in 1936. From April 1936 until January1938 he worked for the Heereswaffenamt Wa Prüf 9 (Army Ordnance Office, Chemical Warfare Department) and changed in February 1938 to the Militärärztliche Akademie (Military Medical Academy), where he headed the Pharmacological Institute. From June 1941 Wirth was the provisional commander and from August 1942 to August 1943 the commander of the Lehrgruppe C—Forschungsgruppe (Teaching Group C/Research Group) at the Military Medical Academy. Wirth was arrested in June 1945 and interned in Nuremberg during the Doctors Trial between December 1946 and August 1947. In 1948 he entered the Pharmacological Department of the Farbenfabriken Bayer A.G. in Wuppertal which he headed from 1954 until his retirement in 1963. See Military Government of Germany. Fragebogen Wolfgang Wirth, June, 14, 1946, pp. 19–28; Lebenslauf Wolfgang Wirth, 30.8.1944, and Lebenslauf 1.6.1944, US Army—Freedom of Information/Privacy Office—Fort G. Meade, Investigative Records Repository, file Wolfang Wirth. For Wirth's biography, see Kopke and

Lendle and Wirth conducted a series of human experiments on 23 officer cadets (*Fähnriche*) at the Military Medical Academy. They began by smearing one gram of LOST on two parts of the underarm of the soldiers. No decontamination measures followed. The dose applied was strong enough to cause deep skin lesions leading to necrosis, which only healed after 6–8 weeks.[23] They came to the conclusion that the capacity of Freskan powder F 1001 to detoxify was insufficient and that Losantin led to better results.[24] Although the healing process with these powders had been rather ineffectual, both Lendle and Wirth recommended the extension of the experiments with Freskan powder to "patients with more expanded and less penetrative LOST injuries as preferable."[25]

It has been debated in historiography whether Lendle and Wirth conducted their experiments in direct cooperation with the SS physicians who had been in charge of the mustard gas experiments in Sachsenhausen. Christoph Kopke and Gebhard Schultz interpreted the relations of the army chemical warfare experts with the SS physicians as cooperation. Refuting this claim, historian Alexander Neumann emphasized that there was no hint or even a covert allusion to the Sachsenhausen experiments in the report by Lendle and Wirth (Neumann 2005, 291). However, this is not the case since Lendle and Wirth reported in two experiments about "a round blister frequently emerging around an anemic corona." They tried to "expose the base of the lesion by brushing it with a sterile steel brush, as had been done with the cases treated at Sachsenhausen."[26] The explicit reference to "cases treated at Sachsenhausen" clearly indicates that Lendle and Wirth had knowledge about the medical treatment of the wounds of concentration camp inmates who were subjected to the mustard gas experiments at the time they conducted their own experiments on cadets from the Military Medical Academy. This contemporary source gives clear evidence that Lendle and Wirth, as army physicians, had established relations with the SS, which included an exchange of expert knowledge and experience from human experimentation with chemical agents and the therapeutic treatment of mustard gas injuries. After the defeat of Germany, when being interrogated in 1947 in a statutory declaration, Wirth denied any knowledge of experiments undertaken in German concentration camps.[27] It was not before 1951, when the public

(Footnote 22 continued)

Schultz (2001, 247–249), Klee (2001, 298–303), Woelk (2003, 271–276), Neumann (2005, 275–276, 278–285), Kopke and Schulz (2006).

[23]Bericht über die Eignung der Freskanpuder F 1000 und F 1001 zur Entgiftung und Behandlung von Lostschäden der Haut, December 21, 1939, secret, signed Lendle and Wirth, p. 3, BA-MA Freiburg, RH 12-23/1728.

[24]Ibid., pp. 4–5, BA-MA Freiburg, RH 12-23/1728.

[25]Ibid., p. 6, BA-MA Freiburg, RH 12-23/1728.

[26]Bericht über die Eignung der Freskanpuder F 1000 und F 1001 zur Entgiftung und Behandlung von Lostschäden der Haut, December 21, 1939, secret, signed Lendle and Wirth, pp. 3–4, BA-MA Freiburg, RH 12-23/1728.

[27]Interrogation of Wolfgang Wirth. Office of U.S. Chief of Council for War Crimes. Vernehmung No. 799 Dr. Wirth, February 12, 1947, p. 16, National Archives, Washington, RG 282, Microfilm M1019, roll 90; Klee (1997, 302); Kopke and Schultz (2001, 246, fn. 43).

prosecutor conducted a preliminary investigation of Schmick, including the human experiments on camp inmates with mustard gas, that Wirth admitted he had visited the camp and seen the victims in person. According to his statement from 1951, he had received an order by Generaloberstabsarzt Anton Waldmann in October or November 1939 to observe the experiments conducted in the concentration camp of Sachsenhausen. During his visit to Sachsenhausen, Wirth met with physicians who presented to him about "6, perhaps also 10 persons who had injuries approximately the size of the palm of the hand."[28] Wirth also remembered seeing a film screening at Sachsenhausen that documented the course of the disease on photographs. From what he had observed, Wirth drew the conclusion that he "could not determine a difference between persons who had been treated with the antitoxin and those who remained untreated" (Kopke and Schultz 2001, 249).

3.2 *Natzweiler*

3.2.1 The Sulphur Mustard Experiments of August Hirt

From 1942 to 1944 human experiments with sulphur mustard (aka LOST, named after their inventors Wilhelm LOmmel and Wilhelm STeinhaus) were conducted at the concentration camp Natzweiler on the initiative of August Hirt, SS-Sturmbannführer and director of the Anatomical Institute at the Reichsuniversität Straßburg (Mitscherlich and Mielke 1947, 92–98; Kater 1997, 248; Ebbinghaus 2000, 42–43; Steegmann 2005, 392–395; Schmaltz 2005, 531–535; Reitzenstein 2014, 131–149). In doing so, Hirt received support from the SS-Ahnenerbe. So as to gain more influence in the natural sciences, the SS-Ahnenerbe established in July 1942 the Institute for Military Scientific Research with the department "H" at the Reichsuniversität Straßburg—"H" as in Hirt.[29] Commissioned by the Wehrmacht in 1939, Hirt had already studied whether the intake or injection of vitamins or their application with an ointment offered a suitable therapy for treating severe skin lesions caused by LOST.[30] Supported by the SS-Ahnenerbe's General Secretary Wolfram Sievers, he succeeded in winning Himmler's approval for the LOST experiments.[31] In mid-July 1942 Himmler

[28]Regional court councilor Meyer (Der Untersuchungsrichter des Landgerichts Duisburg), Vernehmungsprotokoll von Wolfgang Wirth, August, 28, 1951, AMMS, JD 1/22, fol. 23–25; Kopke and Schultz (2001, 248–249), Woelk (2003, 277–278).

[29]Himmler to Sievers, July 9, 1942, BArch, R 26 III/729, fol. 195. For Hirt's biography, see Kasten (1991), Lachmann (1977), Wojak (1999), Lang (2004, 123 ff.), Uhlmann (2011), Uhlmann and Winkelmann (2015), Reitzenstein (2014, 105 ff.).

[30]Hirt, Bericht über Lost-Untersuchungen im Auftrag der Wehrmacht (copy), undated, BArch, NS 19/1582, fol. 46–49; Wolfgang Wirth, Re: Hirt: interrogation of 21.9.1945 (Major Tilley), September 22, 1945, The National Archives (Kew), FO 1041/104.

[31]Aktenvermerk zu den Forschungen von Hirt, June 26, 1942, BArch, NS 19/1209, pp. 5–8.

decreed that Hirt was to conduct his research assignments in connection with the concentration camp Natzweiler.[32] Following a visit by Hirt and Sievers to the concentration camp on August 31, 1942[33] the SS-Ahnenerbe administration began preparations for animal testing with LOST at Natzweiler in late September. Stables were set up, fodder provided, and stockbreeding developed.[34] The experimental toxicological und pharmacological research methodology used for chemical agents during World War I, whereby human experiments were preceded with a series of animal testing and their mutual findings correlated, was also applied at Natzweiler. In late October 1942 Sievers first ordered 20 g of LOST for Hirt from the Waffen-SS.[35] In mid-November, Hirt's assistant Karl Wimmer established a laboratory at Natzweiler and began to select inmates as test objects for the experiments.[36] The first LOST experiment conducted on November 25, 1942 on 15 inmates failed because the agent provided by the Waffen-SS proved to be ineffective.[37] In early December 1942 Hirt continued the experiments with a second delivery of LOST,[38] which did not, however, proceed as expected.[39] The results of the animal testing were not applicable to humans: Unlike the experiments on rats, the human experiments conducted on inmates showed that the Vitamin A treatment obviously did not induce protection, but quite the opposite, that is, hypersensitivity.[40] In late January 1943, Sievers and Hirt discussed at Natzweiler and Dachau the extension of the LOST experiments in both concentration camps. So as to elaborate valid regulations for the troop's treatment, a "major rat experiment" was to be conducted on 1000 animals. Subsequently, the therapeutic effect of four vitamins for the treatment of LOST injuries was to be examined on 240 KZ inmates.[41] Gerit Hendrik Nales, a former Dutch inmate who worked as an orderly at the Natzweiler sick bay from November 1942, testified during the Nuremberg Doctors' Trial that between April and May 1943 a blistering substance had been smeared on the

[32]Rudolf Brandt an Glücks, July, 13, 1942, BArch, NS 21/904 and Sievers, Aktenvermerk, November 3, 1942; BArch, NS 21/905.

[33]Aktenvermerk Sievers, September 17, 1942, BArch, R 26 III/729, fol. 133.

[34]SS-Obersturmbannführer Vogel to SS-Ahnenerbe, September 23, 1942 and Aktenvermerk Wolff, September 28, 1942, BArch, NS 21/904.

[35]Chef des Amtes Ahnenerbe to SS-Hauptsanitätslager der Waffen-SS, October 22, 1942, BArch, NS 21/905.

[36]Hirt to Sievers, November 13, 1942, BArch NS 21/905.

[37]Hirt: Versuchsbericht, November 30, 1942 and Hirt to Sievers, November 26, 1942, BArch, NS 21/905.

[38]Handschriftlicher Vermerk, December 4, 1942 concerning Hirt's "Bericht über die mit dem übersandten L-Stoff angestellten Versuche," November 30, 1942; SS-Ahnenerbe to SS-Hauptsanitätslagers der Waffen-SS, December 1, 1942; Vermerk über ein fernmündliches Gespräch mit Stabsarzt Dr. Wimmer, December 4, 1942; Wimmer, Empfangsbestätigung über 20 g Lost, December 4, 1942; Hirt to Sievers, December 22, 1942, BArch, NS 21/905.

[39]Hirt to Sievers, December 31, 1942, BArch, NS 21/905.

[40]Hirt an Sievers, January 6, 1943, BArch, NS 21/906.

[41]Sievers to Hirt, Bezug: Besprechungen am 25.1.1943 in Natzweiler, undated, BArch, NS 21/906.

forearms of 15 German inmates, inflicting "terrible, festering wounds" on the skin that spread to the whole body and caused some inmates to go blind.[42] According to Nales, three inmates died in horrible pain within a couple of days.[43] The symptoms described indicate LOST experiments. The names of the victims who died of edema of the lungs or pneumonia are known: on December 21, 1942 Karl Kirn; on December 28, 1942 Friedrich Karl Tries; and on December 31, 1942, Wilhelm Müssgen (Steegmann 2010, 425; Reitzenstein 2014, 141–142). In 1944, Hirt submitted a report summarizing the results of his LOST experiments in form of a proposal for a therapy of mustard gas wounds. His report did not mention the circumstances of the experiments conducted on concentration camp inmates or the suffering of the victims. He concluded that a mix of vitamins (A, B-complex, C) given orally, or Vitamin B-1 injected with glucose would give the best results (Fig. 3).[44]

3.2.2 The Phosgene Experiments of Otto Bickenbach

On 17 March 1943 the Institute for Military Scientific Research, mentioned above, invited selected scientists from the Reichsuniversität Strasburg's medical faculty to a conference.[45] One of the speakers was the physicist Otto Bickenbach—like Hirt an avid member of the NSDAP. Since 1939 Bickenbach had been researching possible treatments for the effects of the poison gas phosgene ($COCl_2$), which was used in combat during World War I. He had tested on animals the possible therapeutic and prophylactic effects of hexamethylenetetramine against pulmonary edema caused by phosgene poisoning (Schmaltz 2005, 521–562; 2006a). Schering AG marketed this medicine under the brand name Urotropin to treat cystitis and meningitis (Schmaltz 2005, 524). Due to the results of the animal testing, Bickenbach considered Urotropin "a very efficient protectant against the suffocation symptoms caused by the phosgene poisons."[46] At the conference, hosted by the SS-Ahnenerbe, Bickenbach screened a film he had shot himself to document the phosgene experiments conducted on cats and apes up to 1940.[47] In consequence, Sievers suggested that Bickenbach continue his experiments "in connection" with Hirt in Natzweiler.[48] Bickenbach agreed to the cooperation with

[42]Affidavit Gerrit Hendrik Nales, NO-1063, in Dörner et al. (2000b, microfiche 3/01640f).

[43]Interrogation of Gerrit Hendrik Nales on June 30, 1947. Trial transcript, Dörner et al. (2000b, microfiche 2/10586 ff. and 2/10594).

[44]Hirt and Wimmer, Behandlungsvorschlag für Kampfstoffverletzungen mit Lost, 1944 (=NO-99), BArch, NS 19/1582, fol. 74–76.

[45]Sievers, Aktenvermerk zu der Konferenz vom 17.3.1943, April 5, 1943, BArch, NS 21/906.

[46]Französische Republik. Ständiger Militärgerichtshof des sechsten Bezirks in Strassburg. Trial proceedings, May 6, 1947, NO-3848, Dörner et al. (2000a, microfiche 3/2529).

[47]Aktenvermerk von Sievers zu einer Unterredung mit Bickenbach am 17.3.1943 über Kampfstoff-Forschung, April 5, 1943, BArch, NS 21/906.

[48]Ibid.

Fig. 3 August Hirt "Reichsuniversität Straßburg" (© Hans-Joachim Lang)

the SS because it gave him access to a large number of KZ inmates as test objects for his experiments. In early April 1943 Sievers asked camp commander Josef Kramer about the exact spatial volume of the gas chamber under construction there so that Bickenbach could calculate the gas concentration and thus the phosgene dose required for the human experiments.[49] On April 12, 1943 Kramer reported that the gas chamber was now "completed" and had "a spatial volume of 20 cubic meters."[50] In mid-September 1942 Bickenbach agreed to cooperate with a working group at the Institute for Military Scientific Research.[51] Two days after receiving the news concerning the operative gas chamber in Natzweiler, Sievers reported to Himmler in person the results of Hirt's LOST experiments so far. Bickenbach was

[49]Sievers to Commander of the concentration camp Natzweiler, April 5, 1943, BArch, NS 21/906.

[50]Kramer to SS-Ahnenerbe, Betr.: G-Zelle im KL Natzweiler, April 12, 1943, BArch, NS 21/906. Kramers specifications are not precise. In 1946, the gas chamber of the camp had been examined technically by a French commission of experts. The floor plan measured 2.40 × 3.50 m with a ceiling height of 2.60 m. The volume therefore was 21.84 m^3. See Camp de Concentration du Struthof. Rapport d'expertise de MM. les professeurs et docteurs Simonin (Strasbourg), Piédelièvre (Paris) Docteur Fourcade (Strasbourg), January 15, 1946, BArch Ludwigsburg, B 162/335, fol. 66.

[51]Sievers, Aktenvermerk, September 17, 1942, BArch, R 26 III/729, fol. 122.

then also asked to "deliver a short report about the resistance to, or the rejection of his phosgene experiments and defense proposals by the Wehrmachtdienststellen."[52] Two weeks later, Hirt informed Bickenbach that the experiments under his responsibility could now begin. Consistent with the statement of Ferdinand Holl, a political prisoner who served as *kapo* (prisoner functionary) in the Natzweiler barrack reserved for the SS-Ahnenerbe, the first phosgene experiments took place in June 1943. According to his estimate, approximately 90 to 150 inmates were subjected to phosgene—50 to 60 of whom suffocated in agony.[53] Contradictory statements by Holl regarding the number of subjects involved and the number of victims who died, as well as the question of whether this early series had actually taken place, are still being discussed among historians today.[54]

3.2.3 New Series of Phosgene Experiments in June and August 1944

In 1944 the Natzweiler gas chamber was used again for several test series with phosgene.[55] Helmut Rühl, Bickenbach's assistant, was responsible for the measurement of phosgene concentration in correlation to the humidity of the gas chamber.[56] Rühl began to work on the construction apparatus for the measurements in January 1944 but had difficulties with the calibration of the instruments.[57] The measuring method used by Rühl had been developed by Wolfgang Wirth, head of the Institute for Pharmacology and Military Toxicology of the Military Medical Academy in Berlin (Wirth 1936). Wirth visited Rühl in Strasburg and gave him advice on the final adjustment of the instruments before the last series of phosgene experiments began at Natzweiler.[58] Although we do not know how much Wirth learned about the experiments conducted in the concentration camp at Natzweiler, his technical support may be seen as further

[52]Sievers to Hirt, April 14, 1943, BArch, NS 21/906.

[53]The number of test victims can no longer be clearly established. In the Nuremburg Doctor's Trial, Ferdinand Holl first gave the number as 150 victims. During cross-examination, he stated that he had witnessed about four series of experiments with LOST, each involving 30 inmates (i.e., a total of 120 inmates). According to Holl, each test series led to 7–8 casualties. Regarding the phosgene experiments, he confirmed three series, each with 30 inmates (i.e. altogether 90 victims). Cf. interrogation of Ferdinand Holl on January 6, 1947, Wortprotokoll, in Dörner et al. (2000a, microfiche 2/01092-01096).

[54]The testimonies of Holl have been reviewed and analyzed, Reitzenstein (2014, 134, 141–142, 168–169, 358 fn. 1007). See also Schmaltz (2005, 535 and 561), Steegmann (2005, 394–395).

[55]The implementation of the experimental series was delayed due to a conflict between Hirt and Bickenbach, see Schmaltz (2005, 538–543).

[56]Dr. Helmut Rühl an Karl Brandt, 2. Bericht: Untersuchungen über den Konzentrationsabfall des Phosgens in der Verwendeten Kammer und seine Hydrolyse unter Einfluss der Luftfeuchtigkeit, undated, NO-1852, Dörner et al. (2000a microfiche 3/02775-02777).

[57]Bericht von Dr. Helmut Rühl über seine Tätigkeit an dem Forschungsinstitut der Medizinischen Fakultät Straßburg, [1950], p. 2, The National Archives (Kew), FO 1060/570.

[58]Helmut Rühl to Karl Brandt, 2. Bericht: Untersuchungen über den Konzentrationsabfall des Phosgens in der Verwendeten Kammer und seine Hydrolyse unter Einfluss der Luftfeuchtigkeit, undated, NO-1852, Dörner et al. (2000a, microfiche 3/02775).

evidence of scientific networks linking the human experiments of the Wehrmacht to criminal human experiments.[59] On June 14, 1944 Bickenbach's assistants Helmut Rühl and Fritz Letz went to Natzweiler to equip the gas chamber with the measuring apparatus. Hirt and Bickenbach followed the next day and began with the phosgene experiments which ended on August 8, 1944.[60] Twelve of the 40 inmates involved in the experiments were forced to take Urotropin orally; 20 inmates received injections and a "control group" of eight inmates remained "unprotected."[61] Apart from some "preventive detained" German inmates, most of the test victims had been transferred by the SS from the "Gipsy camp" Auschwitz-Birkenau to Natzweiler. The inmates had to report in groups of four to the experiments. The phosgene dose was gradually increased from experiment to experiment, while the dose of Urotropin was simultaneously reduced. Willy Herzberg, one of the survivors, told how Bickenbach himself led the inmates into the gas chamber, where he smashed vials filled with phosgene on the ground. Bickenbach then left the gas chamber and subsequently the doors were locked. After ten minutes in the gas chamber, Herzberg heard a "muffled splashing" caused by the "bursting lungs" of his fellow prisoners, who broke down with foam in their mouths, noses, and ears.[62] His own breathing became distressed and he had the feeling as if "someone was sticking needles into his lungs." On his chest he sensed "a pressure, as if hundreds of kilos were put upon it," and he "already thought that he would not survive this."[63] According to Bickenbach's final report, 14 inmates sustained pulmonary edema of varying degrees during the test series. In the final series, the established lethal dose of phosgene was considerably exceeded (Fig. 4).

All four inmates (Zirko Rebstock, 37; Adalbert Eckstein, 20; Andreas Hodosy, 32 and Josef Reinhardt, 38), who died at the end of the last test series, were German Sinti—thus indicating a systematic selection of victims based on racist criteria for the most perilous experiment. In his final report to Karl Brandt in 1944, Bickenbach explained in detail the degree to which the limit values of the lethal effects of phosgene poisoning could be reduced with Urotropin.[64] The phosgene experiments at Natzweiler show that human experiments, which were unethical and without doubt a medical war crime, could still produce new scientific insights. The transgression of ethical boundaries, making the death of the test subjects an integral part of the epistemology of the

[59]When interrogated in 1947, Wirth claimed that he had not heard of the phosgene experiments in Natzweiler before the Nuremberg War Crime Trials (Woelk 2003, 282). Eidesstattliche Erklärung von Wolfgang Wirth, February 2, 1947, VDB Nachtrag 1 (Hanloser), Dörner et al. (2000a, microfiche 4/4171-4173).

[60]Staatsanwaltschaft Bochum: Vernehmungsprotokoll von Otto Bickenbach, November 4, 1955, BArch Ludwigsburg, B 162/4206, fol. 1093f.

[61]Bickenbach to Karl Brandt: 7. Bericht. Die schützende Wirkung einer Inhalation von Hexamethylentetramin-Aerosol auf die Phosgenvergiftung, undated, p. 15, NO-1852, Dörner et al. (2000a, microfiche 3/02791).

[62]Staatsanwaltschaft Holtfort: Zeugenvernehmung von Willy Herzberg, July, 1, 1981, BArch Ludwigsburg, B 162/19282, fol. 80.

[63]Ibid., 81.

[64]Bickenbach to Brandt, 7. Bericht, undated, NO-1852, (Dörner et al. 2000a, microfiche 3/02794).

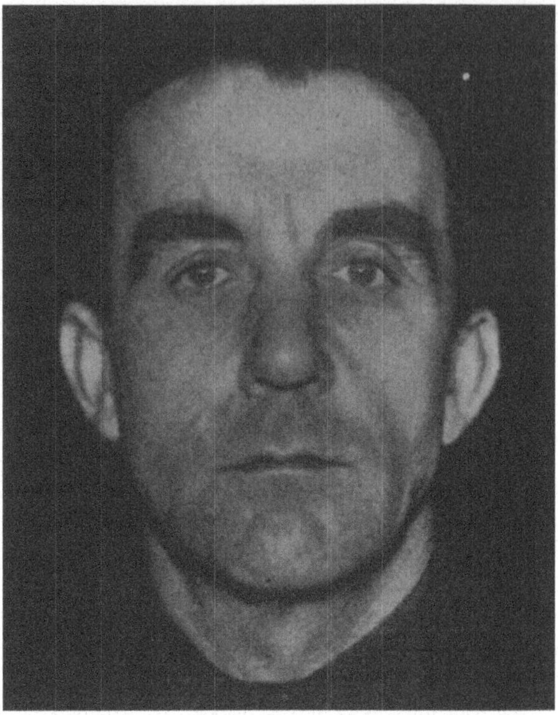

Fig. 4 Otto Bickenbach in French imprisonment (Bayle 1950, 925)

experiments in the context of brutal and dehumanized medicinal practice, delivered empirical data that could not have been obtained under any other circumstances.

3.3 Neuengamme

The last series of experiments with chemical warfare agents in a concentration camp took place between December 1944 and March 1945 at Neuengamme (Groehler 1978, 277–279; 1989, 240–242; Klee 1997, 177–179; Kalthoff and Werner 1999, 193–196). They were initiated by the head of the Reichsanstalt für Wasser- und Luftgüte, Professor Karl Ludwig Werner Haase (Fig. 5).[65]

[65]Karl Ludwig Werner Haase (1903–1980) studied chemistry in Berlin, where he also worked on his dissertation at the Kaiser Wilhelm Institute für physical Chemistry and Electro Chemistry. He completed this in 1924 at the Institut for Plant Nutrition (Institut für Pflanzenernährung) in Hohenheim. In 1925 he received his doctorate in Berlin and started to work for the Preußische Landesanstalt für Wasser-, Boden- und Lufthygiene (in 1942 renamed Reichsanstalt für Wasser- und Luftgüte). Haase joined the NSDAP in April 1940. During World War II he was a member of a commission of the Speer Ministry (Reichsministerium für Rüstung- und Kriegsproduktion) that dealt with development of mobile drinking water devices. See Affidavit Werner Haase, January 27,

Fig. 5 Ludwig Werner Haase (Werkstoffe und Korrosion, 6. Jg., 1955, 2. Innenseite: "Die Vortragenden der Korrosionstagung/DECHEMA-Jahrestagung—11. und 12. November 1954 in Frankfurt/Main")

Following the destruction of the large water dam "Möhne Reservoir" by a Royal Air Force air raid in May 1943, Haase was assigned as a consultant for its restoration. Once he had alerted the president of the Reichsanstalt to the possible risk of the Allied Forces contaminating the water with bacteria, glass dust, viruses, or chemical agents, Haase was authorized to explore new decontamination methods for chemical agents in laboratory research.[66] In spring 1944 Haase advanced a new method for the decontamination of water that had been poisoned with the blister agent Lewisite [$C_2H_2AsCl_3$; dichloro(2-chlorovinyl)arsine]. This involved the

(Footnote 65 continued)

1947, VDB 1, Dörner et al. (2000a, microfiche 3/4996); BArch, NSDAP Ortsgruppenkartei, Haase, Ludwig Werner, born May 2, 1903. See also the entries on Haase in Kürschners Gelehrten-Lexikon, Oestreich (1954, 761), Schuder (1961, 633–634; 1970, 945).

[66]Aussage von Ludwig Werner Haase, November 21, 1963, BArch Ludwigsburg, B 162/1001, fol. 365–366.

application of hypochlorous acid.[67] The organic arsenic agent Lewisite, produced as a chemical weapon during World War I, causes severe blistering and burns, resulting when ingested in great pain, nausea, vomiting, and tissue damage (Pechura and Rall 1993; Bey and Walter 2003). The Wehrmacht immediately tested Haase's method at Raubkammer/Munsterlager.[68] Within the Military Medical Academy, Haase's method was controversial because the first results of the decontamination experiments were ambiguous.[69] In summer 1944 the Institute for Pharmacology and Military Toxicology headed by Wolfgang Wirth investigated the toxicity of Lewisite decomposition products.[70] By the end of August 1944, the apparatus for the decontamination process was available.[71] In late September 1943 the president of the Reichsanstalt für Wasser- und Luftgüte sent a copy of Haase's preliminary report to the head of the Ministry of Interior's health department, Fritz Cropp, who in turn forwarded it to Reich Health Leader Leonardo Conti.[72] SS-Obergruppenführer Conti immediately submitted the report to Himmler asking for support to further develop the new method of decontamination since traditional procedures would fail due to the insufficient availability of active charcoal for such large amounts of water.[73] Himmler authorized experiments in a concentration camp and Neuengamme was chosen as the location.[74]

Haase and his assistant Dr. Jaeger had been preparing the experiments since June 1944 and planned to install the decontamination apparatus by the end of July at the Neuengamme concentration camp. On August 5 and September 1, 1944 they visited

---

[67]Ibid.

[68]Ibid.

[69]While the head of the Institut für Wehrpharmazie und angewandte Chemie, Konrad Gemeinhardt, pled against follow-up studies, the head of the Institut für Pharmakologie und Wehrtoxikologie, Wolfgang Wirth, recommended further investigations. See Hemmrich (Der Sonderbeauftragte des Reichsministers für Rüstung und Kriegsproduktion für die Entseuchung und Entgiftung von Trink- und Brauchwasser: Niederschrift über die Besprechung am 28.4.1944, May 5, 1944, p. 3, BA-MA Freiburg, RH 12-23/1707.

[70]OKH Chef H Rüst u SAN (Nr 93601 44 g S In II G III) to Ministerialrat Georg Hemmrich, Betr.: Genußfähigkeit arsenhaltigen Trinkwassers, August, 28, 1944, BA-MA Freiburg, RH 12-23/1707 and Wirth to Zettel (Generalkommissar des Führers für das Sanitäts- und Gesundheitswesen), Betr.: Kampfstoffentgiftung von Wasser (draft), secret, August 24, 1944, ibid.

[71]Zettel to Wirth, August 19, 1944, BA-MA Freiburg, RH 12-23/1707 and Wirth to Zettel, August 24, 1944, ibid.

[72]Reichsanstalt für Wasser- und Luftgüte to Ministerialdirektor Dr. Cropp, September 27, 1943, BArch, NS 19/3819, fol. 2 and Rudolf Brandt (Personal Staff RFSS) to Reichsarzt-SS und Polizei Dr. Grawitz, October 8, 1943, ibid., fol. 3.

[73]Conti to Himmler, October 1, 1943, BArch, NS 19/3819, fol. 1. In 1944, the shortage of active charcoal was so severe that the army could not meet its demands for gas mask filters, see Schreiben der Heeres-Rohstoffabteilung unter Az. 66 b 91.30 H Ro Va zu Nr. 99/44 v. 22.8.44 to San In./Org II, Betr.: Trinkwasser und Entgiftung (Geheime Kommandosache), BA-MA Freiburg, RH 12-23/1707.

[74]Rudolf Brandt to Reichsarzt-SS und Polizei Dr. Grawitz, October 8, 1943, BArch, NS 19/3819, p. 3; Rudolf Brandt to Conti, October 3, 1943, ibid., p. 4 and Grawitz to Rudolf Brandt, March 30, 1944, ibid., p. 6.

Neuengamme.[75] Probably on one of these days, hypochlorous acid was added to the drinking-water supply at Neuengamme to perform a large-scale test on approximately 10,000 inmates to see if the water with the added decontamination compound would lead to health problems.[76] According to Haase and Regierungsbauinspektor Kumpfert, no observed health problems were reported after consumption of the water with the decontamination compound.[77] On October 10, 1944 Haase continued testing with Wolfgang Wirth from the Military Medical Academy at the army's proofing ground at Raubkammer.[78]

In November the Personal Staff of Reichsführer SS asked for the promised results.[79] By the end of November, Haase announced that he and his research assistant Dr. Jaeger would prepare and conduct the experiments at Neuengamme. Haase attributed the delay of the experiments to the difficulties presented by the required physiological pre-examinations conducted by the director of the Pharmacological Institute Felix Haffner at the University of Tübingen, as well as the results of other institutes at Dresden, Hamburg, Raubkammer, and Wuppertal regarding a possible effect of poor water quality on the results.[80] Haffner's research project was authorized by the Heereswaffenamt and furthered by the Reich Research Council (Fig. 6).[81]

By the end of November Helmut Poppendick, Chief of Personal Staff of the Reichsführer SS and Police, acting as designated principal investigator at Neuengamme, characterized the human experiments at Neuengamme as a "control experiment on a large scale for a final assurance" of the decontamination method,

---

[75]See the Tagesjournal der Reichsanstalt für Wasser- und Luftgüte, daily journal entries dating June 7, July 28.7. August 5, and September 1, 1944, Archiv des Umweltbundesamtes, Tagesjournale. I am grateful to Dr. Karsten Linne who informed me about these documents.

[76]Grawitz telexed to Rudolf Brandt on September 8, 1944 that the hypochlorous acid had proven harmless ("WIRKSAEURE IN DEN NOTWENDIGEN KONZENTRATIONEN KEINE GESUNDHEITSCHAEDIGUNG VERUSACHT") and that new experiments with chemical agents would start now (Telex Grawitz to Rudolf Brandt, September 8, 1944, BArch, NS 19/3819, fol. 8). For the number of concentration camp inmates exposed to the detoxification acid, see Friedrich Konrich (Präsident der Reichsanstalt für Wasser- und Luftgüte). Berichterstatter Haase und Regierungsbauinspektor Kumpfert: 13. Bericht über die Wirkung hoher W-Säurekonzentrationen im Wasser auf Menschen und Tiere, October 26, 1944, Staatsarchiv Nürnberg, KV-Anklage, Dokumente, NO-153.

[77]13. Bericht über die Wirkung hoher W-Säurekonzentrationen im Wasser auf Menschen und Tiere, October 26, 1944, Staatsarchiv Nürnberg, KV-Anklage, Dokumente, NO-153.

[78]Wirth (Wi G III) to Org II, Betr.: Entgiftung von Trinkwasser, (draft), October 31, 1944, BA-MA Freiburg, RH 12-23/1707 and Wirth to Karl Brandt, Betr.: W-Säureverfahren von Prof. Haase, December 13, 1944, BA-MA Freiburg, RH 12-23/1740.

[79]Telex from Rudolf Brandt to Poppendick, November 3, 1944, BArch, NS 19/3819, fol. 9.

[80]Haase spoke of the "Außenstellen in Dresden, Hamburg, Raubkammer und Wuppertal." Except for the Heeresversuchsanstalt at Raubkammer, it is not clear which laboratories he meant, see Haase: Aktenvermerk, November 22, 1944, BArch, NS 19/3819, fol. 11.

[81]For Haffner's research project, see Felix Haffner, Wasserentgiftungsmethoden (Kampfstoffe). Pharmakologisches Institut Universität Tübingen. OH-09/0012—DE-009/752/43 Gkdos. Oberregierungsrat Dr. Wagner. approved September 1943, BArch, R 26 III/12.

Fig. 6 Helmut Poppendick, November 5, 1947 at Nuermberg (Courtesy of the United States Holocaust Memorial Museum, Photograph #07322)

since Haase and his collaborators had allegedly already continuously drunk the decontaminated water without suffering any health damages. By that time, 1200 units of the decontamination apparatus were already in production.[82] In early December 1944 Karl Brandt asked Wirth for a statement on the Haase method.[83] Wirth was not able to provide an evaluation report since he had not received the relevant report from Haase. Complaining about an insufficient supply of research data, Wirth conducted comprehensive toxicological and pathohistological experiments to establish an empirical basis for the evaluation of Haase's methods. Wirth promised to submit a report to the Reichsführer SS by mid-December 1944 (Table 3).[84]

Between December 3 and 15, 1944 approximately 150 inmates of a so-called "Schonbaracke (recovery barrack) at Neuengamme were subjected to the drinking water experiments. According to Haase's report, the water had first been poisoned with Lewisite for 15 days, with doses increasing in rates of up to a maximum of

[82]Helmut Poppendick to Rudolf Brandt, November 23, 1944, BArch, NS 19/3819, fol. 10.

[83]Karl Brandt to Wirth, December 6, 1944, BA-MA Freiburg, RH 12-23/1740.

[84]Wirth to Karl Brandt, Betr.: W-Säureverfahren von Prof. Haase, December 13, 1944, BA-MA Freiburg, RH 12-23/1740.

Table 3 Lewisite and mustard gas experiments at the concentration camp Neuengamme

Date	Agent	Number of test persons	Involved
5 Aug. and 1 Sept.1944	Hypochlorous acid (compound for decontamination)	Approx. 10,000 concentration camp inmates	Haase Jaeger
3 Dec.– 12 Dec.1944	Lewisite	Approx. 150	
	Dora (Lewisite dry)		
	Nitrogen mustard		
Planned in Feb.1945	8 CW agents	16 planned	Poppendick
	Nitrogen mustard		

"approximately 100-fold" the amount the Military Medical Academy considered noxious.[85] Purportedly, the SS camp physicians did not observe any health damage in the camp prisoners.[86] The experiments included two different agents: Lewisite and Dora (a dry form of Lewisite) were used. SS-Oberführer Helmut Poppendick concluded from Haase's report that further experiments in January were necessary to establish possible damage caused by long-term consumption of the water since the amount of arsenic ingested was still considered "significant."[87] However, Himmler did not consider further experiments necessary since the dosage tested had been high enough.[88] Haase and SS-Sturmbannführer Hermann Friese, who acted as an expert consultant for the SS in issues of chemical warfare, favored further experiments after hearing about Himmler's skeptical appraisal. He argued that additional experiments with nitrogen mustard gas were necessary because this agent had a different chemical composition and reaction than Lewsiste, the arsenic compound tested so far. The experiments with nitrogen mustard were conducted in January 1945.[89] In February 1945 Poppendick reported that the nitrogen mustard experiments had been completed and achieved the "same favorable result."[90]

As stated in a final report by the Reichsanstalt für Wasser- und Lufthygiene in March 1945, this series of tests with nitrogen mustard gas had actually been initiated by Wolfgang Wirth during a meeting with Karl Brandt as early as December 4, 1944, one day after the experiments on concentration camp inmates had begun.[91]

[85]The report written by Haase is quoted in the letter from SS-Oberführer Poppendick (Reichsführer-SS—Reichsarzt-SS und Polizei) to Rudolf Brandt, December 20, 1944, BArch, NS 19/3819, fol. 12.

[86]Ibid.

[87]SS-Oberführer Poppendick to Rudolf Brandt, December 20, 1944, BArch, NS 19/3819, fol. 12.

[88]Rudolf Brandt to Poppendick, December 31, 1944, BArch, NS 19/3819, fol. 13.

[89]For the academic career of Friese, see also Albrecht et al. (1991, 74).

[90]Poppendick to Rudolf Brandt, February 8, 1945, BArch, NS 19/3819, 15.

[91]Konrich (Präsident der Reichsanstalt für Wasser- und Luftgüte). Berichterstatter: Dr. Jaegers und Regierungs-Bauinspektor Kumpfert. Bericht Nr. 25 über die in Hamburg-Neuengamme durchgeführten Versuche vom 30.3.1945, in Staatsarchiv Nürnberg, KV-Anklage, Dokumente, NO-154.

In an affidavit for the defense in the Nuremberg Doctors' Trial, Wirth confirmed that he probably attended a meeting in Brandt's office on December 12, 1944 when two devices for the decontamination of water poisoned with chemical agents were discussed. Wirth denied having given advice to extend the experiments with nitrogen mustard.[92] Though the report by Konrich gives no evidence that Wirth personally visited Neuengamme, or that he participated directly in the experiments, his advice, however, led to its implementation in January 1945 when inmates were forced to drink decontaminated water that had been previously poisoned with nitrogen mustard gas. This shows that the chemical warfare researchers from army agencies were not only aware of the unethical experiments undertaken in German concentration camps, but were actually involved as expert consultants in the specific division of labor for the SS physicians in the camps.

After the nitrogen mustard experiments were completed in early February 1945, the scientists proposed another series of experiments with a far more radical approach to forced human participation. The experimental design now not only included intentional health damage and disabilities, but even the death of camp inmates. Poppendick urged that the ingestion of chemical agents be examined in more detail since all available data on harmful doses were "nothing but a pure guess."[93] According to Poppendick's letter, Haase wanted to force eight concentration camp inmates to ingest eight important chemical agents in harmful doses, and another eight inmates to ingest a lower dose that he considered harmless, in order to determine the threshold: "Since damage or cases of death do not have to be taken into account for the first eight test persons, prisoners facing death sentences should be used."[94]

On February 16, 1945 Himmler withdrew his approval "in consideration of the current situation."[95] This was, as far as we know, the first time ever that Himmler refused to allow human experiments to be conducted in a concentration camp. Even with the advance of the Allied Forces and Germany's final defeat unavoidable, the scientists still tried to make use of the last opportunity to exploit the lives of the concentration camp inmates at their disposal and ruthlessly subjected them to lethal

---

[92]Eidesstattliche Erklärung von Wolfgang Wirth, Feb. 2, 1947, VDB Nachtrag 1 (Hanloser), Dörner et al. (2000a, microfiche 4/4171-4173). Woelk (2003, 282) does not discuss the reliability of Wirth's affidavit. As Kopke and Schultz have shown, Wirth committed perjure when he denied under oath that he had absolutely no knowledge of human experiments in German concentration camps, see Kopke and Schultz (2001, 247–249), Schmidt (2007, 295). Karl Brandt and Siegfried Hanloser denied any specific knowledge of the involvement of Wirth in the water decontamination experiments at Neuengamme: Kreuzverhör von Karl Brandt (34. Verhandlungstag), February 2, 1947, Microfiche-Edition Ärzteprozess, Dörner et al. (2000a, microfiche 2/02654-2655); Kreuzverhör Siegfried Hanloser (39. Verhandlungstag), February 18, 1947, ibid., microfiche 2/03057-3059.

[93]Poppendick to Rudolf Brandt, February 8, 1945, BArch, NS 19/3819, fol. 15.

[94]Ibid.

[95]Handwritten note by Rudolf Brandt from February 16, 1945, see Poppendick to Rudolf Brandt, February 8, 1945, BArch, NS 19/3819, fol. 15.

human experiments. In March 1945 Dr. Jaeger visited Neuengamme again.[96] Whether his intention was to retrieve the decontamination apparatus or to conduct the final experiments remains unclear.

4 Conclusion

Concerning the issue of the informed consent of subjects on whom experiments with chemical agents were performed, the crucial question of voluntary participation and informed consent mark an important difference between those experiments conducted under the auspices of military institutions, or those that took place in the concentration camps. While the former provided some room for manoeuvre, this was not the case in the concentration camps. For those experiments, there was no informed consent and no attempts were made to avoid the unnecessary suffering. The experiment designs of August Hirt, Otto Bickenbach, and Ludwig Haase took the death of involuntary test persons into account. In the case of Haase, it was only the impending military defeat of Nazi Germany that fortunately inhibited the implementation of the last deadly series of experiments.

Cooperation, competition, and division of labor went hand in hand. As the examples presented here have shown, competition and rivalry between chemical warfare experts from the military and the SS led to a specific division of labor in human experimentation with chemical agents. Plans to conduct the human experiments in concentration camps were not always initiated by the SS, August Hirt, or Helmut Poppendick. As the cases involving airforce officers Bickenbach and Haase from the Reichsanstalt für Wasser- und Luftgüte show, scientists from other groups and institutions, in addition to the SS, were also driving forces in conducting criminal experiments on decontamination methods for drinking water poisoned with chemical agents. Sources indicate that it is also likely that Wolfgang Wirth, head of the Institute for Pharmacology and Military Toxicology of the Military Medical Academy, proposed to expand those experiments on concentration camp inmates at Neuengamme to investigate nitrogen mustard.

For a better understanding of human experimentation during the Nazi regime, it is crucial to analyze historically the epistemology of human experimentation and to take seriously the research motivation and aims of the scientists involved. Knowledge production and dissemination of human experimentation with chemical agents was not restricted to SS doctors, but included a much broader group of scientists in the army and airforce, at universities, in research organizations such as the renowned Deutsche Forschungsgemeinschaft and the Reich Research Council, or at the Reichsanstalt für Wasser und Luftgüte.

[96]Eintrag Nr. 35/2 zu den Reisekosten Jaegers nach Hamburg-Neuengamme vom 6.3.1945 im Tagesjournal 1945 der Reichsanstalt für Wasser- und Luftgüte, Archiv des Umweltbundesamtes, Tagesjournale. I would like to thank Dr. Karsten Linne, who found this document, for generously supplying a copy.

References

Albrecht, Helmuth, Bettina Gundler, and Technische Universität Braunschweig. 1991. *Catalogus professorum der Technischen Universität Carolo-Wilhelmina zu Braunschweig*. Beiträge zur Geschichte der Carolo-Wilhelmina 9. Braunschweig.

Baader, Gerhard. 2002. Lost-Levisit-Kampfstoffversuche in der deutschen Militärmedizin, 1939–1945. *Virus. Beiträge zur Sozialgeschichte der Medizin* 3: 12–28.

Bayle, François. 1950. *Croix gammée contre caducée. Les expériences humaines en Allemagne pendant la deuxième guerre mondiale*. Neustadt: Centre de L'Imprimerie Nationale.

Bey, Tareg, and Frank G. Walter. 2003. Senfgas, Stickstofflost, Lewisit und Phosgenoxim. Hautschädigende Militärkampfstoffe und deren Bedeutung für Rettungsdienste. *Notfall & Rettungsmedizin* 6(5): 327–336.

Brauch, Hans Günter, and Rolf Dieter Müller. 1985. *Chemische Kriegführung—Chemische Abrüstung. Dokumente und Kommentare*. Berlin: Arno Spitz Verlag.

Czernik, Hendryk. 1974. Filie obozu koncentracyjnego Gross-Rosen w Brzegu Dolnym (Dyhernfurth I i II). *Studia nad faszyzmem i zbrodniami hitlerowskimi*, 191–248.

Dörner, Klaus, Angelika Ebbinghaus, Karsten Linne, Karl Heinz Roth, and Paul Weindling. 2000a. *Der Nürnberger Ärzteprozeß 1946/47. Wortprotokolle, Anklage- und Verteidigungsmaterial, Quellen zum Umfeld*. München: K.G. Saur.

Dörner, Klaus, Angelika Ebbinghaus, Karsten Linne, Karl Heinz Roth, and Paul Weindling. 2000b. *Der Nürnberger Ärzteprozeß 1946/47. Wortprotokolle, Anklage- und Verteidigungsmaterial, Quellen zum Umfeld*. Erschließungsband zur Mikrofiche-Edition: Bearbeitet von Karsten Linne. München: K.G. Saur.

Ebbinghaus, Angelika. 1998. Der Prozeß gegen Tesch & Stabenow. Von der Schädlingsbekämpfung zum Holocaust. *1999. Zeitschrift für Sozialgeschichte des 20. und 21. Jahrhunderts* 13(2): 16–71.

Ebbinghaus, Angelika. 1999. Chemische Kampfstoffe in der deutschen Rüstungs- und Kriegswirtschaft. In *Krieg und Wirtschaft. Studien zur deutschen Wirtschaftsgeschichte 1939–1945*, ed. Dietrich Eichholtz, 171–194. Nationalsozialistische Besatzungpolitik in Europa 1939–1945. Berlin: Metropol-Verlag.

Ebbinghaus, Angelika. 2000. Einleitung. In *Der Nürnberger Ärzteprozeß 1946/47. Wortprotokolle, Anklage- und Verteidigungsmaterial, Quellen zum Umfeld*. Erschließungsband zur Mikrofiche-Edition, eds. Klaus Dörner, Angelika Ebbinghaus, Karsten Linne, Karl Heinz Roth, and Paul Weindling. Bearbeitet von Karsten Linne, 11–69. München: K.G. Saur.

Eberle, Henrik. 2015. *Ein wertvolles Instrument. Die Universität Greifswald im Nationalsozialismus*. Köln: Böhlau.

Gellermann, Günther W. 1986. *Der Krieg, der nicht stattfand. Möglichkeiten, Überlegungen und Entscheidungen der deutschen Obersten Führung zur Verwendung chemischer Kampfstoffe im zweiten Weltkrieg*. Koblenz: Bernhard & Graefe Verlag.

Grodin, Michael A. 1992. Historical Origins of the Nuermberg Code. In *The Nazi Doctors and the Nuremberg Code. Human Rights in Human Experimentation*, eds. George J. Annas, and Michael A. Grodin, 121–144. New York: Oxford University Press.

Groehler, Olaf. 1978. *Der lautlose Tod*. Berlin (DDR): Verlag der Nation.

Groehler, Olaf. 1989. *Der lautlose Tod. Einsatz und Entwicklung deutscher Giftgase von 1914 bis 1945*. Reinbek: Rowohlt.

Groehler, Olaf. 1992. *Selbstmörderische Allianz. Deutsch-russische Militärbeziehungen 1920–1941*. Berlin: Vision Verlag.

Hahn, Judith. 2008. *Grawitz, Genzken, Gebhardt. Drei Karrieren im Sanitätsdienst der SS*. Münster: Klemm & Oelschläger.

Hayes, Peter Francis. 2004. *From cooperation to complicity. Degussa in the Third Reich*. Cambridge: Cambridge University Press.

Joly, Hervé. 2000. L'implication de l'industrie chimique allemande dans le Shoah: Le cas de Zyklon B. *Revue d'histoire moderne et contemporaire* 47 (April–Juni): 368–400.

Jones, Irving H. 1945. *I.G. Farbenindustrie A.G. Frankfurt/Main*. CIOS File No. XXX-19 (Item 30), ed. SHAEF Combined Intelligence Objectives Sub-Committee—G-2 Division.

Kalthoff, Jürgen, and Martin Werner. 1999. *Die Händler des Zyklon B*. Hamburg: VSA Verlag.

Kasten, Frederick H. 1991. Unethical Nazi Medicine in Annexed Alsace-Lorraine: The Strange Case of Nazi Anatomist Professor Dr. August Hirt. In *Historians and archivists. Essays in modern German history and archival policy*, ed. G.O. Kent, 173–208. Fairfax, Va.: George Mason University Press.

Kästner, Ingrid, and Susanne Hahn. 1994. Der Toxikologe Otto Muntsch (1890–1945) und die deutsche Kampfstofforschung. *1999. Zeitschrift für Sozialgeschichte des 20. und 21. Jahrhunderts* 9(3): 42–50.

Kater, Michael H. 1997. *Das "Ahnenerbe" der SS 1935–1945. Ein Beitrag zur Kulturpolitik des Dritten Reiches*, 2nd ed. Studien zur Zeitgeschichte 6. München: R. Oldenbourg.

Klee, Ernst. 1997. *Auschwitz, die NS-Medizin und ihre Opfer*. Frankfurt am Main: S. Fischer Verlag.

Klee, Ernst. 2001. *Deutsche Medizin im Dritten Reich. Karrieren vor und nach 1945*. Frankfurt am Main: Fischer Verlag.

Kopke, Christoph, and Gebhard Schultz. 2001. Menschenversuche mit chemischen Kampfstoffen bei der Wehrmacht und SS. Ein Forschungsbericht. In *Medizin und Verbrechen. Festschrift für Walter Wuttke*, ed. Christoph Kopke, 239–257. Ulm: Klemm & Oelschläger.

Kopke, Christoph, and Gerhard Schulz. 2006. Die Menschenversuche mit dem Kampfstoff Lost im KZ Sachsenhausen (1939) und die Debatte über die Rolle des Wehrmachtstoxikologen Wolfgang Wirth. In *Medizin im Zweiten Weltkrieg*, ed. Wolfgang U. Eckart, and Alexander Neumann, 113–129. Paderborn: Schönigh.

Krause, Joachim, and Charles K. Mallroy. 1993. *Chemische Waffen in der Militärdoktrin der Sowjetunion. Historische Erfahrungen und militärische Lehren 1919–1991*. Baden-Baden: Nomos-Verlag.

Lachmann, Ernest. 1977. Anatomist of infamy: August Hirt. *Bulletin of the History of Medicine* 51: 594–602.

Lang, Hans-Joachim. 2004. *Die Namen der Nummern. Wie es gelang, die 86 Opfer eines NS-Verbrechens zu identifizieren*. Hamburg: Hoffmann und Campe.

Ley, Astrid, and Günter Morsch. 2007. *Medizin und Verbrechen. Das Krankenrevier des KZ Sachsenhausen 1936–1945*. Schriftenreihe der Stiftung Brandenburgische Gedenkstätten 21. Berlin: Metropol-Verlag.

Martinetz, Dieter. 1996. *Vom Giftpfeil zum Chemiewaffenverbot. Zur Geschichte der chemischen Kampfmittel*. Frankfurt am Main: Verlag Harri Deutsch.

Mills, A.K. 1945. *Investigations of Chemical Warfare Installations in the Munsterlager area, including Raubkammer*. CIOS File No. XXXI-86, ed. SHAEF Combined Intelligence Objectives Sub-Committee—G-2 Division, London.

Mitscherlich, Alexander, and Fred Mielke. 1947. *Das Diktat der Menschenverachtung. Eine Dokumentation*. Heidelberg: Verlag Lambert Schneider.

Morsch, Günter, and Bertrand Perz, eds. 2011. *Neue Studien zu nationalsozialistischen Massentötungen durch Giftgas: Historische Bedeutung, technische Entwicklung, revisionistische Leugnung*. Schriftenreihe der Siftung Brandenburgische Gedenkstätten. Berlin: Metropol-Verlag.

Müller, Rolf Dieter. 1985. World power status through the use of poison gas? German preparations for chemical warfare, 1919–1945. In *The German military in the age of total war*, ed. Wilhelm Deist, 171–209. Warwickshire: Berg Publishers.

Neumann, Alexander. 2005. *"Arzttum ist immer Kämpfertum". Die Heeressanitätsinspektion und das Amt "Chef des Wehrmachtssanitätswesens" im Zweiten Weltkrieg*. Düsseldorf: Droste.

Oehler-Klein, Sigrid, and Alexander Neumann. 2004. Die Militarisierung der Medizin an der Universität Gießen und ihre Beziehungen zu den Sanitätsinspektionen von Heer und Luftwaffe im Zweiten Weltkrieg. *Mitteilungen des Oberhessischen Geschichtsvereins Gießen* 89: 95–188.

Oestreich, Gerhard. 1954. *Kürschners deutscher Gelehrten-Kalender 1954*, 8th ed. Berlin: de Gruyter.

Pechura, Constance M., and David P. Rall. 1993. *Veterans at risk: The health effects of mustard gas and lewisite*. Washington, D.C.: National Adademy Press.

Reitzenstein, Julien. 2014. *Himmlers Forscher: Wehrwissenschaft und Medizinverbrechen im "Ahnenerbe" der SS*. Paderborn: Schöningh.

Roelcke, Volker. 2017. The use and abuse of medical research ethics: The German Richtlinien/guidelines for human subject research as an instrument for the protection of research subjects—and of medical science, ca. 1931–1961/64. In *From clinic to concentration camp: Reassessing Nazi medical and racial research, 1933–1945*, ed. Paul Weindling, 33–56. Milton Park: Routledge.

Rundschreiben des Reichsministers des Inneren vom 28.2.1931: Richtlinien für die neuartige Heilbehandlung und für die Vornahme wissenschaftlicher Versuche am Menschen. 1931. *Reichsgesundheitsblatt* 55: 174–175.

Sass, Hans-Martin. 1983. Reichsrundschreiben 1931: Pre-nuremberg German regulations concerning new therapy and human experimentation. *Journal of Medicine and Philosophy* 8 (99): 99–111.

Schleiermacher, Sabine. 1988. Die SS-Stiftung "Ahnenerbe." Menschen als Material für "exakte" Wissenschaft. In *Menschenversuche. Wahnsinn und Wirklichkeit*, ed. Rainer Osnowski, 70–87. Köln: Volksblatt Verlag.

Schmaltz, Florian. 2005. *Kampfstoff-Forschung im Nationalsozialismus. Zur Kooperation von Kaiser-Wilhelm-Instituten, Militär und Industrie*. Geschichte der Kaiser-Wilhelm-Gesellschaft im Nationalsozialismus 11. Göttingen: Wallstein Verlag.

Schmaltz, Florian. 2006a. Otto Bickenbach's human experiments with chemical warfare agents at the concentration camp Natzweiler in the context of the SS-Ahnenerbe and the Reichsforschungsrat. In *Man, medicine and the state. The human body as an object of government sponsored research in the 20th century*, ed. Wolfgang U. Eckart, 139–156. Beiträge zur Geschichte der Deutschen Forschungsgemeinschaft 2. Stuttgart: Steiner.

Schmaltz, Florian. 2006b. Pharmakologische Nervengasforschung an der Militärärztlichen Akademie und an den Universitäten Marburg, Danzig und Leipzig im Zweiten Weltkrieg. In *Medizin im Zweiten Weltkrieg*, ed. Wolfgang U. Eckart, and Alexander Neumann, 171–194. Paderborn: Schöningh.

Schmaltz, Florian. 2009. Chemical weapons research in National Socialism: The collaboration of the Kaiser Wilhelm institutes with the military and industry. In *The Kaiser Wilhelm Society under National Socialism*, ed. Susanne Heim, Carola Sachse, and Mark Walker, 312–338. Cambridge: Cambridge University Press.

Schmidt, Gerald. 1985. Ludwig Lendle (1899–1969). In *Neue Deutsche Biographie*, ed. Historische Kommission bei der Bayrischen Akademie der Wissenschaften, 201–202. Berlin: Duncker & Humblot.

Schmidt, Ulf. 2007. *Karl Brandt: The Nazi doctor. Medicine and power in the Third Reich*. London: Continuum.

Schmidt, Ulf. 2013. Accidents and experiments: Nazi chemical warfare research and medical ethics during the Second World War. In *Military medical ethics for the 21st century*, ed. Michael L. Gross, and Don Carrick, 225–244. Farnham, Surrey: Ashgate.

Schuder, Werner. 1961. *Kürschners deutscher Gelehrten-Kalender 1961. A-N*, 9th ed. Berlin: de Gruyter.

Schuder, Werner. 1970. *Kürschners Deutscher Gelehrten-Kalender 1970*. Berlin: Walter de Gruyter.

Schultz, Gebhard. 2001. Militärisches Sperrgebiet. Die Zitadelle Spandau im Nationalsozialismus. *Berlinerische Monatsschrift* 7: 51–59.

Spiegelberg, Ulrich, Karl Dierkes, and Hans Bürger-Prinz. 1961. *Psychopathologisch-neurologische Schäden nach Einwirkung synthetischer Gifte*. Wehrdienst und Gesundheit: Abhandlungen aus Wehrmedizin, Wehrpharmazie und Wehrveterinärwesen. Darmstadt: Wehr und Wissen Verlagsgesellschaft.

Steegmann, Robert. 2005. *Struthof. Le KL-Natzweiler et ses kommandos: une nébuleuse concentrationnaire des deux côtés du Rhin, 1941–1945*. Strasbourg: Nuée bleue.

Steegmann, Robert. 2010. *Das KZ Natzweiler-Struthof und seine Außenkommandos an Rhein und Neckar 1941–1945*. Berlin: Metropol.

Szöllösi-Janze, Margit. 1994. Von der Mehlmotte zum Holocaust. Fritz Haber und die chemische Schädlingsbekämpfung während und nach dem Ersten Weltkrieg. In *Von der Arbeiterbewegung zum modernen Sozialstaat. Festschrift für Gerhard A. Ritter zum 65. Geburtstag*, ed. Jürgen Kocka, 658–682. München: Saur.

Trunk, Achim. 2011. Die todbringenden Gase. In *Neue Studien zu nationalsozialistischen Massentötungen durch Giftgas. Historische Bedeutung, technische Entwicklung, revisionistische Leugnung*, eds. Günter Morsch, and Bertrand Perz, 23–49. Berlin: Metropol Verlag.

Uhlmann, Angelika. 2011. August Hirt und seine Mitarbeiter Kiesselbach, Wimmer und Mayer. Die Karrieren vor der Reichsuniversität Straßburg. *Revue d'Allemagne et des Pays de Langue Allemande* 43(3): 333–340.

Uhlmann, Angelika, and Andreas Winkelmann. 2015. The science prior to the crime. August Hirt's career before 1941. *Annals of Anatomy* 204: 118–126.

Wirth, Wolfgang. 1936. Über die Wirkung kleinster Phosgenmengen. *Naunyn-Schmiedebergs Archiv für Pathologie und experimentelle Pharmakologie* 181: 198–206.

Witkowski, Józef, and Aleksandra Rudy. 1987. Gaz bojowy tabun a podobóz Dyhernfurth. *Przegląd Lekarski* 34 (1): 131–142.

Woelk, Wolfgang. 2003. Der Pharmakologe und Toxikologe Wolfgang Wirth (1898–1996) und die Giftgasforschung im Nationalsozialismus. In *Nach der Diktatur. Die Medizinische Akademie Düsseldorf nach 1945*, eds. Wolfgang Woelk, Frank Sparing, Karen Bayer, and Michael G. Esch, 269–287. Essen: Klartext-Verlag.

Wojak, Irmtrud. 1999. Das "irrende Gewissen" der NS-Verbrecher und die deutsche Rechtssprechung. Die "jüdische Skelettsammlung" am Anatomischen Institut der "Reichsuniversität Straßburg." In *"Beseitigung des jüdischen Einflusses..." Antisemitische Forschung, Eliten und Karrieren im Nationalsozialismus*, ed. Fritz-Bauer-Institut, 101–130. Jahrbuch 1998/99 zur Geschichte und Wirkung des Holocaust. Frankfurt, New York: Campus Verlag.

No Retaliation in Kind: Japanese Chemical Warfare Policy in World War II

Walter E. Grunden

Abstract This essay examines Japan's Chemical Warfare (CW) policy in World War II as revealed in interrogations of high-ranking military officers conducted by United States military intelligence after the war. Based upon these interrogations and an examination of recorded incidents of chemical weapons use, it may be concluded that Japanese CW policy permitted use of chemical weapons in China where the enemy did not possess the capacity to retaliate in kind, but largely prohibited their use in the Pacific against the Allies, whom they feared could respond in kind with overwhelming force. Thus, the threat of retaliation in kind served as a successful deterrent to CW employment in the Pacific Theater. For its part, the US refrained from using poison gas largely due to President Franklin D. Roosevelt's moral abhorrence of chemical weapons, but also because it was not in a position logistically to engage in CW on a large scale until late in the conflict, at which time the use of nuclear weapons made the issue moot.

From 1937 to 1945, the military services of Japan used chemical weapons on over 2000 occasions, primarily in the China Theater of Operations. In contrast, there were only a few occasions of use against Allied forces in the Pacific. The primary reason for this great disparity in incidents of use was Japan's fear of retaliation in kind. While engaged in combat against military forces in China, the Imperial Japanese Army used a variety of chemical weapons without concern of retaliation in kind by the technologically inferior Chinese military, which was utterly lacking in chemical weapons and whose soldiers often lacked even basic protective gear such as gas masks. In China, Japanese military forces often found themselves at a numerical disadvantage and used chemical warfare (CW) as a means to compensate. In most instances, the Japanese used tear gas and smoke candles, but there are numerous recorded incidents of more debilitating and lethal gases also being deployed. Combat in the Pacific Theater, however, was a different matter. Japanese military forces tended to use CW while on the offensive in open terrain, such as in

W.E. Grunden (✉)
Bowling Green State University, Bowling Green, OH, USA
e-mail: wgrund@bgsu.edu

B. Friedrich et al. (eds.), *One Hundred Years of Chemical Warfare: Research, Deployment, Consequences*, DOI 10.1007/978-3-319-51664-6_14

China; but, when on the defensive in more restrictive environments, such as in close combat on the various Pacific islands against Allied forces, CW was not a viable option. More important was the fact that in the Pacific the Japanese were primarily up against the armed forces of the United States, which not only had the ability to respond in kind, but—it was thought—were backed by a national industrial capacity that could utterly annihilate Japan with chemical weapons should the Japanese initiate this type of warfare.

This essay examines Japan's CW policy in China and the Pacific Theater during World War II and argues that the perceived ability of the enemy to retaliate in kind was the primary factor in determining the use of chemical weapons by the Japanese. Legal prohibitions against using poison gases, such as those set forth in the Geneva Protocol of 1925, which Japan did not ratify in any case, were a secondary consideration (Robinson 1971, 289). Moreover, there was a significant disconnect between the stated official policy governing the use of chemical weapons and their actual employment upon the battlefield. Interrogations of high-ranking army officers, such as General Tojo Hideki,[1] conducted by US military intelligence after the war revealed that responsibility for approving use of lethal chemical weapons remained intentionally hazy, even though the actual chain of command was clear. This ambiguity in policy implementation served the interests of the top brass, who, after the war, attempted to avoid prosecution by exploiting the opacity of Japan's CW policy in China and the Pacific.

In Japan, military interest in chemical weapons originated with reports of the use of poison gas at Ypres on April 22, 1915. The Army Technology Review Board, which was responsible for monitoring innovations in weaponry, began to investigate the potential of developing an array of chemical weapons, poison gas launchers, and gas masks. One of the first Japanese scientists to pursue an interest in chemical weapons was Koizumi Chikahiko, a physician assigned to the School of Hygiene at the Army Medical College (*Rikugun Gun'i Gakkō*) who specialized in the study of industrial toxins. By the end of 1915, Koizumi emerged as the lead researcher in CW for the army and ultimately earned the moniker of "father of chemical warfare in Japan." In September 1917, the Army Medical College constructed a new Chemical Weapons Laboratory to support his research, and in the following year, Koizumi was named the laboratory's "Chief of Research on Protective Devices." He would go on to develop one of the first gas masks to be adopted for general use by the army (Tsuneishi 1984, 100–103; Tsuneishi and Asano 1982, 51–56). Research and development of chemical weapons was not given a high priority at this time, however, as Japan was not a principal belligerent in the war and lacked the relative urgency of the other participants. As a result, Japan's foray into chemical weapons developed more slowly than in Europe and the United States (Grunden 2005, 165–196; Murata et al. 1996, 16–31; Robinson 1971, 287–289).

[1]Japanese names appear here with family name first and given name second, as is the custom in Japan.

Japan's CW program grew sporadically throughout the 1920s and 1930s. In April 1919, the Army Institute of Scientific Research (*Rikugun Kagaku Kenkyūjo*) was established as the central R&D facility of the Army Technical Headquarters (*Rikugun Gijutsu Honbu*), at which time it assumed jurisdiction over the army's CW program. In 1923, disaster struck. The Great Kantō Earthquake devastated much of Tokyo and leveled several buildings housing the army's R&D facilities for its CW program, including some on the campus of the Army Medical College and some belonging to the Army Institute of Scientific Research. The army exploited the disaster as an opportunity to upgrade its program, allocating a sum of ¥1.2 million for the construction of new laboratories. The Imperial Japanese Navy followed by initiating its own CW program that year at the Navy Technical Research Institute (Tsuneishi 1984, 102–105; Tanaka 1988, 11).[2] In the war that was to come, three military institutions would account for nearly all of the CW research being conducted in Japan: the Sixth Army Technical Institute under the Army Institute of Scientific Research, the Imperial Japanese Army Air Service's Third Laboratory, and the Imperial Japanese Navy's Sagami Naval Research Department (Robinson 1971, 287–289).[3]

In 1927, the army appropriated the island of Okunoshima and established its central chemical weapons production facility there. Located in the Inland Sea in Hiroshima prefecture, this small island, only four kilometers in circumference, provided secrecy and a measure of safety, being removed some three kilometers from the nearest city of Tadanoumi. In May 1929, the Okunoshima facility began production of tear and mustard gases, but would later produce an array of other lethal gases as well (Tanaka 1988, 12–14). At its peak capacity, Okunoshima produced some 200 tons of mustard gas (H), 50 tons of lewisite (L), 80 tons of diphenylcyanoarsine (DC), 50 tons of hydrocyanic (Prussic) acid (HCN), and 2.5 tons of chloroacetophenone (CN) per month.[4]

A Chemical Warfare School was established at Narashino, Chiba prefecture, in 1933, under the auspices of the Inspector General of Military Education, and served as the principal training facility for both the Imperial Japanese Army and Navy. According to a US military intelligence report, Narashino was "splendidly equipped, well staffed, and effective in the fulfillment of its mission until the end of the

[2]On the navy CW program, see General Headquarters, United States Army Forces, Pacific, Scientific and Technical Advisory Section, "Report on Scientific Intelligence Survey in Japan: September and October 1945," November 1, 1945, vol. IV, Chemical Warfare, RG 457, Entry 9032, Box 765, US National Archives, College Park Md., pp. 19, 22, and appendices CW-3-1, CW-4-1, and CW-8-5. [Hereafter cited as GHQ, "Report on Scientific Intelligence Survey."].

[3]Ibid., pp. 3, 39, 46–47, appendix CW-3-1.

[4]Ibid., p. 42, appendix CW-2-2, CW-6-1, and CW-6-2.

war," and graduated 3074 officers between 1939 and 1945.[5] The training of Japanese soldiers in defense against gas warfare was well organized and well executed, and all Japanese troops and a large number of reservists received CW training.[6] One overall significant shortcoming of Japan's CW program, however, was that no separate, independent chemical warfare division with oversight of all CW activities in the military services was ever established, which ultimately resulted in a "failure to develop an integrated, balanced and coordinated program."[7] As a result, the Japanese military forces never achieved more than "a limited tactical capability" with chemical weapons (Robinson 1971, 289).

As "the curtain opened" on the war in China with the Marco Polo Bridge Incident on July 7, 1937, so began Japan's foray into chemical warfare (Murata et al. 1996, 10). The Japanese army began to use "gas" weapons against the Chinese almost immediately, with the first incident reported as early as July 18, 1937.[8] Reports of Japanese use of smoke, tear gas, and poison gas steadily increased as the conflict in China dragged on and the war situation worsened (Yoshimi 2004, 49–68; Wakabayashi 1994, 3–8). Reports received through "official sources" stated there was "no proof that lethal or toxic chemicals were used prior to the fall of 1939." However, "lethal gases definitely appeared in the summer of 1941," though their use was "confined to restricted areas where the Chinese were exerting pressure," and they were used in such cases "generally to support Japanese counter-attacks."[9] In the "Ichang Incident," for example, a three-day battle that began on October 8, 1941, Japanese forces barraged the Chinese with gas shells for over four hours and dropped more than 300 gas-laden bombs on Chinese positions. An investigation conducted by a US Chemical Warfare Service (CWS) officer assigned to General Joseph Stillwell in the China-Burma-India Theater confirmed that mustard and CN gases had been used and that there was evidence suggesting lewisite may have been deployed as well. In this incident, there were 1600 confirmed casualties, 600 of which were killed in action as the result of Japan's use of poison gases.[10] During the war in China, Imperial Japanese Army forces are alleged to have used chemical

[5]Office of the Chief Chemical Officer, GHQ, AFPAC, Tokyo, Japan, "Intelligence Report on Japanese Chemical Warfare," vol. I, "General Organization, Policies and Intentions, Tactics," May 15, 1946, RG 319, Entry 82, Box 2097, File: "Japanese Chemical Warfare Policies and Intentions —US Army Forces, Pacific," US National Archives, College Park, Md., p. 19. [Hereafter cited as Chief Chemical Officer, "Intelligence Report on Japanese Chemical Warfare."].

[6]United States Military Intelligence Service, *Enemy Capabilities for Chemical Warfare*. Washington, DC: US War Department, 1943, 87–88.

[7]Chief Chemical Officer, "Intelligence Report on Japanese Chemical Warfare," p. 21.

[8]Ibid., p. 8.

[9]"Condensed Statement of Information Available Concerning Japanese Use of War Gas," RG 319, Publications File, Entry 82, Box 2098, US National Archives, College Park, Md., p. 1. [Hereafter cited as "Condensed Statement."].

[10]Ibid., pp. 1–2.

weapons on as many as 2091 separate occasions, with estimates of casualties ranging from 36,968 to 80,000, including both military personnel and civilians.[11]

In the aftermath of World War II, the United States Army Chief of Staff, together with Army Intelligence (G-2) and officers of the CWS, conducted a thorough investigation of Japanese activities in chemical warfare to assess Japan's capacity to wage large-scale war using chemical weapons and "to ascertain whether the Japanese possessed knowledge, techniques, materiel, or procedures superior to our own and worthy of adoption."[12] Because of the general order to destroy evidence upon Japan's surrender on August 15, 1945, documentation for the investigators was lacking (Drea et al. 2006, 9–11). As a result, interrogations of high-ranking military personnel served as one of the most important sources of information and played a key role in the investigation. Through the interrogations, US intelligence personnel attempted to discern what constituted chemical warfare policy within the Imperial Japanese Army and who was responsible for its implementation.

Among the first to be questioned was General Tojo Hideki, who had served as Vice-Minister of War from July to December 1938, and then as Minister of War from July 1940 to July 1944. He served concurrently as Prime Minister from October 17, 1941 to July 22, 1944. As such, he was in a position not only to have influenced the formation of CW policy, but to have overseen its implementation as well. Interrogated on April 2, 1946, just a few weeks before the International Military Tribunal for the Far East, or "Tokyo Trials," were to begin, Tojo was understandably very cagey with his answers. He admitted CW *research* was conducted in Japan, but emphasized that it was done only in a "defensive sense" and that "precautions" were taken "in the same spirit."[13] He was adamant that the use of chemical weapons was forbidden because, had they been used, "it would have been disastrous for Japan." On this point, he articulated three specific reasons why he personally opposed the use of chemical weapons. First, it was against international law, which, he stated, "Japan had to follow." Secondly, he cited the industrial superiority of the United States. Finally, he stated that Japan is "an island country and if it were used, it would be very unfortunate for her." For these reasons, he stated, "I made a tremendous fuss about this and absolutely forbade its use, so I

[11]The figures given for incidents of poison gas attacks range from 886 to 2091 separate occasions. See "Condensed Statement", p. 1; Awaya (1992, 3–6). Documentation of such numbers remains problematic even after the release of numerous seminal documents concerning Japanese CW in China. See Awaya and Yoshimi (1989), Drea et al. (2006).

[12]Chief Chemical Officer, "Intelligence Report on Japanese Chemical Warfare," p. 1.

[13]Geoffrey Marshall, Colonel, CWS, Chief Chemical Officer, General Headquarters, United States Army Forces, Pacific, Office of the Chief Chemical Officer, "Japanese Chemical Warfare Policies and Intentions," April 13, 1946, "Interrogation of General Hideki Tojo," conducted April 2, 1946, RG 319, Entry 85A, MIS#: 261223, US National Archives, College Park, Md., p. 1. [Hereafter cited as Marshall Interrogations.].

prohibited it, both from the standpoint of policy and strategy." He added, "as War Minister, I had enough voice in this sphere to see that it was not used as a military policy."[14]

When asked about specific orders being issued to field commanders concerning chemical weapons at the beginning of the war, he flatly stated, "During the time of the China Incident, it was forbidden and gas could not be used without my consent." But his chief interrogator, Lieutenant Colonel John E. Beebe, Jr. of the US Chemical Warfare Service, persisted, knowing that chemical weapons had been used on numerous occasions in the China Theater. Tojo began to prevaricate and attempted to make a distinction between simple "harassing agents" such as smoke and tear gas and "casualty agents" such as poison gases that could be debilitating or fatal. Tojo argued that "casualty agents" were "absolutely forbidden," but that harassing agents—those that result in coughing, sneezing, and tear production—were "used to a certain extent." Tojo was adamant that he refused requests from the Supreme Commander in the China Theater to use casualty agents, though he did approve use of harassing agents.[15]

Beebe now had Tojo in a corner. He asked, "You mentioned earlier that Japan was obligated by international law and treaties not to use gas warfare. How, then, can you explain the use of harassing agents against the Chinese?" Tojo replied, "The police all over the world use tear gas and sneezing gas. They are used even in your country." Beebe followed, "Was not the use of these harassing agents also prohibited by international law?" Tojo flatly stated, "In fact, they were in use by the police throughout the world. How about the atom bomb?"[16] In this exchange, Beebe established that Japan did not adhere to any strict interpretation of international law concerning the use of chemical weapons and that Tojo himself had approved their use in the China Theater, though Tojo was adamant that he had approved only the use of harassing agents.

Next, Beebe interrogated General Kawabe Masakazu, who had an extensive service record in the China Theater. Kawabe was a Major-General and had served as a commander in the Permanent China Brigade during the Marco Polo Bridge Incident in July 1937. From August 1937, he served as the Deputy Chief of Staff of the North China Army; from February 1938 to January 1939, he served as the Chief of Staff of the Central China Expeditionary Army.[17] With this service record, Kawabe was certainly in a position to know of Japanese employment of CW in China. In a previous interrogation, conducted one month prior, Kawabe had

---

[14]Ibid., pp. 1–2.

[15]Ibid., pp. 2, 4–5.

[16]Ibid., p. 6.

[17]In January 1939, Kawabe returned to Japan to assume the post of Deputy Inspector-General of Military Education, then became Inspector-General himself in September. After serving a short stint as the C-in-C of the IJA 3rd Army, in August 1942 he was promoted to Chief of Staff of the China Expeditionary Army, in which post he served until March 1943. See: Kawabe Masakazu (1886–1965). *The Pacific War Online Encyclopedia*. http://pwencycl.kgbudge.com/K/a/Kawabe_Masakazu.htm. Accessed April 11, 2015.

disavowed any knowledge of chemical weapons having been used in China, nor would he subsequently admit to having requested approval for their use. But now, Kawabe made exceptions for "special smoke," which in this cased referred to a smoke candle the Japanese called *aka-to* ("red candle"), which was actually diphenylcyanarsine (DC), a sneezing gas frequently used by Japanese forces in China. Beebe questioned Kawabe concerning various battles in which such "special smoke" had been deployed, including one occasion where some six or seven thousand special smoke candles had been used. When pressed for information on these incidents, Kawabe consistently replied, "I do not remember," a refrain he repeated often.[18]

Beebe then questioned Kawabe about a document that bore his name, entitled, "Lessons from the China Incident," which was published by the Inspectorate General of Military Education, Chemical Warfare Section, on April 15, 1939.[19] Beebe confronted Kawabe with a copy of "Lessons" and pointed to the entry for chapter six, entitled, "The Chinese Army as Seen from the Point of View of Chemical Warfare." This chapter enumerated the many deficiencies of the Chinese military forces in defending against chemical weapons, including evaluations of the poor quality of Chinese gas masks, the fact that they were not issued to all Chinese soldiers, and the observation that those who did have them often lacked proper training in their use. One entry in this chapter noted that some Chinese soldiers had "died by asphyxiation, sticking their noses and mouths into the ground" to avoid breathing the smoke.[20]

When confronted with such damning evidence of Japan's CW in China, Kawabe again disavowed any knowledge of poison gases and stated these must have been instances of use of "special smoke," which he insisted was a non-toxic gas. Beebe pressed Kawabe further about policy concerning the use of more lethal gases, including mustard, lewisite, and phosgene. To this, Kawabe replied, "The use of these gases was not thought about. No one had it. It was forbidden." Kawabe then attempted to differentiate "special smoke" from poison gases, but claimed that this was his "private opinion" and that he could not speak for the army. Beebe followed, "What about international law and treaties on the use of gas?" Kawabe answered, "In my opinion treaties did not cover special smoke."[21] Thus, in such a manner, the top brass of the Imperial Japanese Army parsed the language of international law that prohibited chemical weapons and rationalized the use of "special smoke" as a non-lethal gas.

---

[18]Marshall Interrogations, "Continued Interrogation of General Masakazu Kawabe," April 10, 1946, pp. 1–4.

[19]Ibid., pp. 1, 6–8. A copy of "Lessons" may be found in Chief Chemical Officer, "Intelligence Report on Japanese Chemical Warfare," pp. 87–124.

[20]Marshall Interrogations, "Continued Interrogation of General Masakazu Kawabe," pp. 6–7.

[21]Ibid., pp. 8–9.

Beebe encountered this rhetoric yet again when interrogating General Hata Shunroku. Like Kawabe, Hata also began his long and infamous career in the China Theater. Hata assumed command of the Central China Expeditionary Forces in February 1938 and held that post until December, when he became a member of the Supreme War Council. In September 1939, Hata was appointed Minister of War and served under two successive prime ministers until being replaced by Tojo in July 1940. In 1941, Hata was appointed Commander-in-Chief of all armies in the China Theater, including the North, Central, and South China Armies.[22] Clearly, Hata should have been quite well-informed about CW policy in China. In his interrogation of Hata, conducted on April 11, 1946, Beebe had him articulate a clear chain of command from the Cabinet level on down through the top ranks of the army. Then, Beebe sought to identify the level at which the employment of "gas warfare" could be authorized. Hata was firm in his assertion that approval for use of poison gas could only come from the Imperial General Headquarters, and that it was expressly forbidden while he served as commander of the Central China Expeditionary Force in 1938. Beebe next asked Hata *when* he received permission to employ gas weapons. Hata replied, "In each case we were authorized by Imperial General Headquarters. There was no blanket authority." Beebe followed with "When did you first receive authority to employ gas?" Surprisingly, Beebe then started to get from Hata the sort of information he was seeking.

Although it is difficult to determine tone from a transcript, Hata's answer appears rather matter-of-fact. He stated, "Tear gas could be used at any time. No specific permission was necessary. Sneezing gas (*aka-to*) could also be used at any time."[23] He also admitted that they had achieved "very good results" using tear and sneezing gases and experienced few to no casualties when these were deployed in combat against the Chinese, most of whom did not have gas masks and would "break and run" or flee in disorder "the minute sneezing gas was used." When pressed to make a distinction between these agents and poison gas, Hata stated, "Poison gas is one which kills or has a permanent disabling effect. I think mustard, lewisite and phosgene are poison gases, but tear and sneezing gases are not. Where avoidable, we did not use tear and sneezing gas, nor did we use it recklessly. Only where we expected great loss to ourselves or the enemy did we use it."[24]

Hata had revealed perhaps more than he intended, for he had actually confirmed for Beebe that, in fact, no special authorization was needed to use such gases, that they could be used "at any time," and that they were a common weapon frequently used by Japanese soldiers in China. Indeed, when asked if he had ever ordered his

---

[22]His authority did not extend to the Kwantung Army in Manchuria, which remained under a separate command. Marshall Interrogations, "Interrogation of General Shunroku Hata," April 11, 1946, pp. 1–2. See also: Hata Shunroku (1879–1962). *The Pacific War Online Encyclopedia*. http://pwencycl.kgbudge.com/H/a/Hata_Shunroku.htm. Accessed April 11, 2015.

[23]Marshall Interrogations, "Interrogation of General Shunroku Hata," p. 3.

[24]Ibid., p. 5.

troops to discontinue using these gases in China, Hata replied, "No. They could use sneezing gas without stopping. That was the policy of Imperial General Headquarters."[25] Hata may have been truthful on this point, but on so many others, Tojo, Kawabe, and Hata were obviously lying. Japanese forces in China clearly had used gases other than the tear and sneezing varieties. Of 65 incidents of CW reported in a document dated October 6, 1944, the vast majority indicated use of vomit-inducing gases, blistering agents, and in one instance, the possible use of a nerve toxin.[26] In any case, it was now clear that CW policy in the China Theater permitted the widespread and common use of harassing agents such as tear and sneezing gases, and that other, more lethal types of poison gases were also used, although on a more limited basis.

From 1942 through 1943, Japan's use of poison gases such as vomiting and blistering agents as well as mustard gas and lewisite actually increased on the China front.[27] Although US President Franklin D. Roosevelt had denounced Japan's employment of CW in China as early as 1938, there was little he could do to stop it. With the US now in the war, however, and with more incidents of poison gas use being reported, Roosevelt attempted to take a more aggressive stand. On June 5, 1942, he publicly stated, "I desire to make it unmistakedly [sic] clear that if Japan persists in this inhuman form of warfare against China or against any other of the United Nations, such action will be regarded by this Government as though taken against the United States and retaliation in kind and in full measure will be meted out" (Rosenman 1950, 258). Any such threats to retaliate in kind at that time, however, were largely hollow and not likely to be realized as the US did not then possess sufficient quantities of chemical weapons in the Pacific to respond on a large scale (Spiers 1986, 73–75; Moon 1984, 12–14). But it was important, perhaps, for the US to clearly articulate its own CW policy in order to attempt to deter further Japanese use of chemical weapons in China.

As the Allies began to advance further across the Pacific, Japanese military forces' use of chemical weapons diminished significantly. In the Pacific Theater, chemical weapons were to be used only on the defensive and *only if* the Allies used them first. This policy was strictly observed with very few exceptions. For example, during the Battle of Guadalcanal, which typified the intensity of combat that was all too common in the jungles of the Southwestern Pacific, on two occasions, January 23 and 28, 1943, Japanese soldiers resorted to using toxic smoke against US troops. Such incidents could be looked upon as the actions of a few desperate men and

---

[25]Ibid., pp. 5–9.

[26]"Reports of Incidents of Use of Gas by Japanese," October 6, 1944, RG 319, Publications File, Entry 82, Box 2098, US National Archives, College Park, Md., pp. 1–5.

[27]Ibid.

were not likely pre-planned.[28] Nonetheless, in April 1943, President Roosevelt issued yet another stern warning stating that if any of the Axis Powers used gas against any of the Allies, the US would "retaliate with overwhelming force." But Roosevelt also made clear that the US would not be the first to initiate chemical warfare.[29]

Roosevelt's warnings and his articulation of US CW policy appear to have had an impact. Tojo acknowledged as much during his interrogation, stating, "I thought, as I had from the beginning, that the use of gas would be very disadvantageous for Japan because of America's tremendous industrial capacity and this statement of the President strengthened my own ideas."[30] This response was echoed by Major General Akiyama Kinsei, who served as the director of the army's CW training school at Narashino from 1935 to 1940. During his interrogation, Akiyama confirmed that Roosevelt's threat of massive retaliation likely prevented the spread of CW attacks throughout the Pacific.[31] Roosevelt's declaration may also have led to a wider de-escalation of CW in China, as well as the actual termination of large-scale industrial production of poison gases in Japan.[32] As the war in Europe turned decisively against Germany following the D-Day invasion in June 1944, Japan's military leaders began to worry that Germany might resort to using chemical weapons to stop the Allied advance on Berlin. In the summer of 1944, the Japanese army ordered the recall of all stocks of gas munitions in the field to depots in rear echelon positions. This order was given as a precaution "against irresponsible use by isolated units in desperate situations which might provoke full scale retaliation."[33] Not only were the Japanese concerned that the US would retaliate against Japan if Germany initiated CW in Europe, they now sought to minimize the possibility that any chemical weapons would be used by their own forces in the Pacific Theater.[34]

The Japanese apparently trusted Roosevelt's pledge not to initiate a first strike, but they also took precautions not to precipitate one by the US late in the war. Theoretically, at that point, the US could have hit Japan with a CW attack without

---

[28]The report indicates it was a "choking gas," but does not elaborate on the means of dissemination, whether by smoke candle, grenade, mortar, or otherwise. See "Reports of Incidents of Use of Gas by Japanese," October 6, 1944, RG 319, Publications File, Entry 82, Box 2098, US National Archives, College Park, Md., p. 3; "Condensed Statement," p. 1.

[29]Marshall Interrogations, "Interrogation of General Hideki Tojo," p. 4.

[30]Ibid.

[31]Major H. Skipper, interview with Major General K. Akiyama, October 16, 1945, in GHQ, "Report on Scientific Intelligence Survey," appendix CW-13-3.

[32]In September 1944, the army's primary facility for chemical weapons production at Okunoshima was converted to the manufacture of conventional explosives. See Target No. 635, "Manufacture of Poison Gases," RG 319, Entry 85A, MIS#: 235950-1, US National Archives, College Park, Md., p. 1.

[33]Chief Chemical Officer, "Intelligence Report on Japanese Chemical Warfare," pp. 7–8.

[34]According to the report of the Chief Chemical Officer, "The Japanese were even prepared to overlook small scale local tactical use by the Allies to avoid general gas warfare [… although] Retaliation would have been attempted in the event of large-scale attacks." Ibid., pp. 7–8.

violating the Geneva Protocols' prohibitions against first use. Allied CW policy, as stated in the Combined Chiefs of Staff document 106/2, and as clearly articulated in numerous declarations by Roosevelt and Prime Minister Winston Churchill, considered an attack on any United Nations ally to be an attack on the US or Great Britain. Thus, because Japan had already initiated CW attacks upon China, the US would have been justified in retaliating in kind (Moon 1984, 12–13, 1996, 501). But it did not. Historian John Ellis van Courtland Moon provides several reasons for US restraint. First and foremost was the "widespread moral revulsion against chemical weapons" engendered by their use in the First World War. Secondly, Allied CW policy was limited to deterrence and retaliation. Third, chemical weapons "offered limited military advantages and carried serious liabilities." Finally, Moon argues, the US was "unprepared throughout the war to wage chemical warfare in the Pacific Theater." In short, "preparations always lagged behind policy" (Moon 1989a, 40–42; b, 317).

Although deploying chemical weapons on a large scale presented certain logistical difficulties in the Pacific Theater, they were not insurmountable. Military and government leaders in the US began to call for their use as early as the summer of 1943. The "Island Hopping" campaign in the Pacific had resulted in the accumulation of excessive casualties, particularly among the US Marines, to whom the duty of being first to land and establish beachheads usually fell. Following the battle of "Bloody Tarawa" in November 1943, the Chief of the US Chemical Warfare Service, General William N. Porter, argued for the employment of poison gas against the remaining Japanese forces in the Pacific. He argued, "the tactical advantages of using gas against entrenched enemy positions were undeniable." Moreover, they were justified, as the Japanese had already used poison gas in the China Theater. But his request was denied. At that time, Roosevelt and the Chiefs of Staff were concerned that such use would proliferate to Europe and provide Hitler a rationale for using chemical weapons against Allied forces in any attempted cross-channel invasion. At least while Germany remained in the war, the Allies could not risk any proliferation. (Moon 1984, 17) Yet, even after the defeat of Germany, when US Army Chief of Staff General George C. Marshall proposed CW use in a pending invasion of the home islands of Japan, his request was denied on "moral and policy grounds" (Moon 1989a, 42). Once Germany was defeated, numerous "liabilities" of CW use disappeared, but the tactical advantages remained.

Others argued that the military advantages of using CW against Japan in the Pacific far outweighed any liabilities once Germany was out of the war. Experiments with mustard gas in late 1943 suggested it would be a highly effective weapon in the Pacific, especially in tropical jungles with high humidity. Experiments conducted at the Dugway Proving Ground in Utah suggested that various gas weapons—in combination with more conventional weapons—might be effective in attacking defensive positions in caves, which could be critical should the Allies have to fight all the way to Japan and invade the home islands proper. But another study undertaken jointly by a team of Americans and Canadians in July 1944 concluded, "comparisons of CW requirements with actual HE [high explosive] expenditures in specific operations do not add materially to the picture"

(Freeman 1991, 32–37; Moon 1989b). The stigma attached to CW was a significant factor in delaying its initiation in the Pacific, although the US continued to stockpile a variety of chemical weapons in the summer of 1945 in preparation for the pending invasion of Japan. (Allen and Polmar 1997; Moon 1989b) Ultimately, however, the US did not employ CW in the Pacific Theater. That decision was obviated by the use of an even more devastating weapon—the atomic bomb.

References

Allen, Thomas B., and Norman Polmar. 1997. Gassing Japan. *MHQ: The Quarterly Journal of Military History* 10(1): 38–43.

Awaya, Kentarō. 1992. Japanese mustard gas in China: Then and now. *Sino-Japanese Studies* 4: 3–6.

Awaya Kentarō, and Yoshimi Yoshiaki, eds. 1989. *Jyūgonen sensō gokuhi shiryōshū 18: Dokugasu sen kankei shiryō* [Collection of Top-Secret Documents on the Fifteen-Year War: Documents Concerning Poison Gas Warfare, vol. 18]. Tokyo: Fuji Shuppan.

Drea, Edward, et al. 2006. *Researching Japanese war crimes records: Introductory essays*. Washington, D.C.: Nazi War Crimes and Japanese Imperial Government Records Interagency Working Group.

Freeman, Karen. 1991. The unfought chemical war. *The Bulletin of the Atomic Scientists* 47 (10): 30–39.

Grunden, Walter E. 2005. *Secret weapons & World War II: Japan in the shadow of Big Science*. Lawrence: University Press of Kansas.

Hata Shunroku (1879–1962). 2012. *The Pacific War online encyclopedia*. http://pwencycl.kgbudge.com/H/a/Hata_Shunroku.htm. Accessed 20 Oct 2016.

Kawabe Masakazu (1886–1965). 2012. *The Pacific War online encyclopedia*. http://pwencycl.kgbudge.com/K/a/Kawabe_Masakazu.htm. Accessed 20 Oct 2016.

Moon, John Ellis van Courtland. 1984. Chemical weapons and deterrence: The World War II experience. *International Security* 8(4): 3–35.

Moon, John Ellis van Courtland. 1989a. Chemical warfare: A forgotten lesson. *Bulletin of the Atomic Scientists* 45(6): 40–43.

Moon, John Ellis van Courtland. 1989b. Project SPHINX: The question of the use of gas in the planned invasion of Japan. *The Journal of Strategic Studies* 12(3): 303–323.

Moon, John Ellis van Courtland. 1996. United States chemical warfare policy in World War II: A captive of coalition policy? *The Journal of Military History* 60(3): 495–511.

Murata, Tadayoshi, et al. 1996. *Nihongun no Kagakusen [Chemical Warfare of the Japanese Army]*. Tokyo: Otsuki Shoten.

Robinson, Julian Perry. 1971. *The Rise of CB Weapons: The Problem of Chemical and Biological Warfare: A Study of the Historical, Technical, Military, Legal, and Political Aspects of CBW, and Possible Disarmament Measures*, vol. 1. Atlantic Highlands: Stockholm International Institute for Peace and Conflict Research.

Rosenman, Samuel I. (ed.). 1950. *The public papers and addresses of Franklin D. Roosevelt, 1942: Humanity on the defensive*. New York: Harper and Brothers.

Spiers, Edward M. 1986. *Chemical warfare*. Urbana, IL: University of Illinois Press.

Tanaka, Yuki. 1988. Poison gas: The story Japan would like to forget. *Bulletin of the Atomic Scientists* 44 (8): 10–19.

Tsuneishi, Keiichi, and Asano Tomizō. 1982. *Saikin-sen butai to jiketsu shita futari no igakusha [The Bacteriological warfare unit and the suicide of two physicians]*. Tokyo: Shinchōsha.

Tsuneishi, Keiichi. 1984. C. Koizumi: As a promoter of the ministry of health and welfare and an originator of the BCW research program. *Historia Scientiarum* 26: 95–113.

Wakabayashi, Bob Tadashi. 1994. Documents on Japanese poison gas warfare in China. *Sino-Japanese Studies* 7: 3–33.

Yoshimi Yoshiaki. 2004. *Doku gasusen to Nihongun [Poison gas warfare and the Japanese army]*. Tokyo: Iwanami Shoten.

The 1925 Geneva Protocol: China's CBW Charges Against Japan at the Tokyo War Crimes Tribunal

Jeanne Guillemin

Abstract The 1925 Geneva Protocol, which bans the wartime use of chemical and also biological weapons, was an emphatic reaction to the use of chemical weapons in World War I, but legal institutions that would sanction violations of the treaty have evolved only with difficulty. An important example of a legal failure to support the protocol occurred at the 1946–1948 International Military Tribunal for the Far East (IMTFE), just when it might be expected that Imperial Japan would be charged for its chemical and biological warfare (CBW) waged against China from the late 1930s into World War II. In 1937, the Chinese officially presented its first complaints to the League of Nations about Japan's battlefield use of chemical weapons (mustard gas, phosgene and tear gases) against defenseless Chinese troops and civilians. In addition, in early 1941 and after, China accused Japan of launching plague attacks against key Chinese cities, killing hundreds and terrorizing thousands. None of these accusations, although supported by evidence, brought about serious international recriminations for Japan. Once World War II ended, China expected to revive these charges at the IMTFE in Tokyo. Instead, under the influence of a few key figures in US military intelligence, the trial's International Prosecution Section (IPS) deleted the Chinese charges and for decades Japan's infraction were lost to history. Analysis of this legal failure points to the obstacles posed by growing Cold War antagonisms between the United States and the Soviet Union, which prompted a general American retreat from prosecuting Japan, its new democratic ally in East Asia, as well as the internal processes at the IPS that favored more blatant incidents of Japanese wartime aggression—such as the well-documented 1937 "Rape of Nanjing" and abuses of Allied prisoners of war. After the silence imposed at the IMTFE, chemical and biological weapons proliferated with few restraints until the Cold War ended in 1992. At the same time, the international framework for war crimes prosecution greatly changed with greater attention put on crimes against civilians. Yet, lacking precedent, international readiness to legally sanction violations of the Geneva Protocol—as with the

J. Guillemin (✉)
Security Studies Program, MIT Center for International Studies Cambridge, Cambridge, MA, USA
e-mail: jeg@mit.edu

B. Friedrich et al. (eds.), *One Hundred Years of Chemical Warfare: Research, Deployment, Consequences*, DOI 10.1007/978-3-319-51664-6_15

2013 and 2017 murders of Syrian civilians with nerve gas—remains nearly as ambiguous as it was in 1946.

1 Introduction

Much of the history of war in the nineteenth and twentieth centuries was shaped by the great tension between advanced mechanized warfare and the idealistic ventures that arose to stop its mounting catastrophic impact. The nineteenth-century rise of national armies and industrial advances in weaponry began a pattern of increasingly destructive conflicts—the mass slaughter of troops and ruination of terrain—followed by valiant attempts to restrict the conduct of war, which were then followed by worse wars with more dangerously powerful and efficient weapons. The battlefield carnage and economic disruption caused by the Crimean War (1853–1856) and the American Civil War (1861–1865) were deplored. Yet in their aftermath, major state militaries plunged ahead to acquire more efficiently designed rifles, mobile heavy artillery, and machine guns.

The 1899 and 1907 Hague Conventions were a breakthrough in articulating new norms to reinforce the concept that the "right of belligerents to adopt means of injuring the enemy is not unlimited" (Boserup 1973, 152). Along with promoting the rights of prisoners of war and the protection of hospital ships, the conventions addressed specific weapons. The use of armed balloons on civilian populations and of expanding "Dum Dum" bullets (named after the British factory in India that produced them) were prohibited. The use of "smoke and noxious gas" in war was also prohibited, no small matter at the time (Hudson 1931). The role of chemicals in the production of weapons was well established and accelerating, with a growing impact on battlefield combat (Smart 2004). Gunpowder produced sickening sulfur fumes; sulfuric acid was used to make nitric acid, which was then used to make mercury fulminate for percussion caps, which meant more toxic clouds from explosives. New ideas to make weapons of chloroform, hydrochloric acid, cyanide, arsenic, and nauseating smokes and stink bombs in war started circulating in the 1860s. Other toxic substances poised for military use were chlorine, hydrogen cyanide, cyanogen chloride, phosgene, and mustard agent, all discovered or synthesized in the late eighteenth and early nineteenth centuries (Sartori 1943).

The 1907 Hague Conventions extended the 1899 provisions beyond their original five-year limit. The major Western powers appeared committed to the conventions, as did Japan and China. In 1915, though, Europe blundered into World War I and a new epoch of weapons innovations began. This time, long-range mortar, tanks, submarines with torpedoes, and the introduction of fighter airplanes expanded the dimensions of battle.

A major innovation in World War I was the introduction of chemical weapons, invented to overcome the conventional boundaries of trench warfare. In April 1915 at Ypres in Belgium, the German military released 150 tons of chlorine gas from a lineup of 5700 canisters. Carried by the wind, the gas passed in minutes to the French and British trenches and quickly killed 1000 soldiers and injured another 4000. This surprise attack provoked competition to discover more potent chemical weapons, with the Germans, British, Italians, Russians, and later the Americans engaging in an unprecedented arms race (Lepick 1998). Soon phosgene, mustard gas, and an assortment of tear gases and blistering agents (vesicants) were in battlefield use, causing death, burns, blindness, other injuries, and terror among surviving combatants. The race was also on to develop masks, suits, and blankets that could protect troops from gas attacks.

World War I brought a scale of devastation in terms of deaths and economic and political upheaval that outstripped all previous wars (MacMillan 2013). When it ended in 1918, international cooperation among nations, backed by law, appeared necessary to prevent future chaos. Supported by a range of visionaries, the creation of the League of Nations in 1920 heralded a new, institutional approach to peace centered on the political resolution of state conflicts before they escalated (Kennedy 1987). Members would, in theory, submit to arbitration rather than take up arms and they would act in each other's defense in the event of unprovoked aggression. The 1922 establishment of the Permanent Court of International Justice (PCIJ) at The Hague offered the option for member nations to settle their disputes through legal hearings, which they did, at a rate of five cases per year for the next decade (ICC 2012).

The League also promised a radically new era of arms control. By Article 8 of its Covenant, members affirmed that "the maintenance of peace requires the reduction of national armaments to the lowest point consistent with national safety and the enforcement by common action of international obligations" (Ames 1922, 306). Also in Article 8, member states agreed to "full and frank information as to the scale of their armaments, their military, naval and air programs and the condition of such of their industries as are adaptable to warlike purposes." Although the United States, reverting to an isolationist posture, refused to join the League, its representatives maintained an active presence in its deliberations and arms control initiatives. The new Soviet Union was distinctly an outsider, but its emissaries did attend important debates.

In addition to the League's active agenda, the international community began building on the 1907 Hague Conventions. In the spirit of arms control, Americans took the lead in assembling the 1922 Washington Conference on the Limitation of Armaments, which resulted in a treaty to restrict poison gas and submarine warfare (SIPRI 1971, 46–47). The United States, the United Kingdom, France, Italy, and Japan, recognized as the premier military power in the Far East, signed the treaty. Because of French objections to the provisions about submarines, the Washington Treaty of 1922 never came into force; yet it paved the way for the 1925 Geneva Protocol which forbade the use of chemical weapons in war. The text of the Protocol banned the use of "asphyxiating, poisonous or other gases, and of all analogous liquids, materials or devices" already "justly condemned by the general opinion of the civilized world" and "universally accepted as part of International Law, binding alike the conscience and the practice of nations" (SIPRI 1971, 341). To this provision, the treaty added that the prohibition would be extended to a looming innovation, "the use of bacteriological methods of warfare."

Despite its strong resolve, the Protocol raised an uncomfortable issue: what would be the international political repercussions if it were violated? In principle, the reaction of the League of Nations should have been two-fold: that the member states would act as restraints on each other and that, as a bloc, they would come to the aid of any member aggressed upon by another state. The implication of the treaty's text was that even states opting not to join would be beholden to established laws of war.

The Geneva Protocol was widely embraced in the interwar years; 43 nations became parties, with the British Empire, France, Italy, Germany, and the Soviet Union leading the way. From the outset, however, it was no guarantee against chemical or biological weapons proliferation. Instead, it allowed a provision for retaliation in kind, which meant that, based on the perception of an imminent or even a long-range threat, each state on its own could decide the level of research, development, and munitions production it needed for adequate defense.

Two important nations, the United States and Japan, refused to join the treaty. In an isolationist phase, the US Senate refused to ratify the Protocol, although sentiments against chemical warfare were strong in America and the US Army had little inclination to reenact the excesses of World War I (Brown 2006, 49–96). In principle, the United States adhered to the treaty's ban on use and limited itself to retaliation in kind. Its Chemical Warfare Service, invented late in World War I, languished and it had little inclination to investigate germ weapons (Guillemin 2005, 27).

Japan refused to become a party to the Geneva Protocol for different reasons. Having shown its military strength in its 1905 war with Russia, Japan was considered Asia's most advanced nation by Western powers, especially the British and Americans. With its population and industries growing and its natural resources limited, Japan had already annexed Korea, acquired Taiwan, and established a profitable lease for trade in Manchuria (Northeastern China). As Japan's ambitions to solidify its empire grew, so did its interest in modern, science-based weapons, including chemicals. During World War I, the Japanese military was more intrigued than repelled by Europe's chemical warfare and, following the war, its experts moved ahead to develop chemical agents with techniques learned in Germany and the United States. In 1928, Japan built a production facility on Okunoshima Island, near Hiroshima, mainly for mustard and tear gases. By that time, its Army Institute of Scientific Research included forty chemical weapons laboratories, twenty affiliated workshops, and an annex staffed with hundreds of scientists and technicians (Grunden 2005, 181–182).

2 World War II and the Post-war IMTFE

Even as the League advocated peace, the concept of total war, with scenarios of air attacks on enemy civilians, cities, and industrial centers behind the lines, was taking hold among state powers (Buckley 1999). As early as 1921, the Italian General Giulio Douhet, a pioneer of strategic bombing, predicted that aeronautics would open up a revolutionary new way to make war. "Air power makes it possible," he wrote, "not only to make high explosive bombing raids over any sector of the enemy's territory, but also to ravage his whole country by chemical and bacteriological warfare" (Douhet 1942, 6–7).

In the early 1930s, the legal means to lasting peace began unraveling. One of the first aggressive actions came from Japan, where moderates and militarists struggled over how to achieve Japan's destiny as a great power (Jansen 2002, 576–585; Gordon 2014, 181–199). In 1931, feeling the impact of the great world depression —its foreign trade had been halved since 1929—Japan made a move that violated the international "Open Door" policy for China that it had agreed to in 1922, in affirmation of China's right to sovereignty, territorial integrity, and control of its national destiny. Claiming Chinese troops had attacked the Japanese Kwantung Army at a railway junction near Mukden (Shenyang), Japanese troops soon conquered Manchuria. At a meeting in Geneva in February 1933, League members voted for Japan's withdrawal. Japan's reaction was to quit the League. That year, Nazi-controlled Germany also quit, claiming that the League's World Disarmament Conference had acted with prejudice by denying it arms parity with France.

Fascist Italy was the next to leave the League. Its 1934–1936 war with Ethiopia was patently aimed at increasing Italy's colonial holdings in Africa. The League attempted reconciliation between the two members, but to no avail. Starting in December 1935, Italy violated the Geneva Protocol by using asphyxiating and poisonous gases on undefended Ethiopian troops and civilians (SIPRI 1971, 175–189). The response of League members was uneven and tepid, establishing a precedent of non-intervention. Despite protests from Ethiopia's Emperor Haile Selassie and documented proof of gas casualties from Red Cross physicians, the League failed to implement effective military and trade sanctions against Italy or deter further gas attacks. The conflict ended in May 1936 with Ethiopia's defeat and its incorporation into Italian East Africa. Italy exited the League soon after.

In July 1937, the Japanese used a conflict on a rail line near Beijing as an excuse to instigate war against China to expand their territory. After quickly gaining control of Beijing, the Japanese captured Shanghai and then China's capital Nanjing. League members protested, but none rushed to openly confront Japan in China's defense. In October of that year, China's delegates to the League protested Japan's use of chemical weapons (identified as mustard gas) on defenseless Chinese troops on the path from Shanghai to Nanjing (SIPRI 1971, 189–192). In 1938, the Chinese returned with documented accusations of more Japanese chemical attacks. But as in the case of Italy's attacks on Ethiopia, League members did not rise to China's defense; it could only remind its members that "the civilized world" had rejected chemical weapons.

On September 1, 1939, Germany invaded Poland and another catastrophic world war began. By then, Germany, Italy, and Japan had found common ground in their alliance as Axis powers, which was sealed by the 1940 Tripartite Pact. Japan's surprise attack on Pearl Harbor in December 1941 brought the United States into the war and extended the Allied-Axis conflict to the Pacific and Asia. In 1942, the Chinese reported to the Allies that Japan had attacked four of its cities with plague, killing hundreds; in the tumult of the war, BW experts in the United Kingdom and the United States dismissed this complaint as not credible (Guillemin 2017, 15).

In a short time, World War II demonstrated that, armed with advanced weapons, especially long-range bombers, major industrial nations at war could wreck the globe (Buruma 2013). From 1939 to 1945, some 60 million people were killed, once-thriving cities and natural environments were laid waste, major states went bankrupt or nearly so, masses of people were forced to migrate and died of epidemics, exposure, or starvation, and political instability became a norm. In Germany, Poland, and the Soviet Union, entire populations were murdered in Nazi concentration camps and in mass executions on the eastern front. Allied troops died by the hundreds of thousands on the western front, German and Soviet soldiers died by the millions in the East. In the Pacific-Asia region, the estimated body counts, difficult to reckon in poor and disadvantaged countries, matched those in Europe.

At the war's end, the Allied response was to create the United Nations as a fortified reincarnation of the League of Nations—this time with a provision for armed peace-keeping intervention and a new court for conflict resolution. A complementary Allied response was to invent the international military tribunal (IMT), staging one in Nuremberg, Germany, and the other in Tokyo, to prosecute the high Axis leaders individually for their war crimes. Three major criminal charges were agreed on: first, for the waging of aggressive war (including conspiracy in its promotion and planning); second, for the violation of the customary rules of war; and, third, for crimes against humanity, pertaining especially to the torture and killing of civilians (Taylor 1992, 56–74).

Compared to the spirit of the immediate post-World War I years, the fervor for arms control in 1945 was singularly lacking. To the contrary, advanced weaponry was competitively sought. The Allies in Europe (the Americans, British, Soviets, and the French) sought to appropriate as much as they could of Nazi Germany's arsenals, from V-2 rockets to nerve gases (Jacobsen 2014). In recruiting expert German physicists, chemists, and biologists, the unsavory and even criminal aspects of their careers were suppressed. Similarly, in Japan, the US Army and its G-2 intelligence division in Tokyo diligently sought to ascertain what, if any advances in weaponry could be acquired as "war booty" (Home and Low 1993). The mobilization of American industry and science for military goals had succeeded in winning the war; the result was a validation of strategic weapons, from incendiaries to conventional bomb, culminating in the atom bomb which immediately became "a symbol of industrial might, scientific accomplishment, and national prestige" (Cirincione 2007, 17).

In principle, the second category of IMT war crimes, concerning customary rules of war, included violations of the Hague Conventions and the 1925 Geneva Protocol, along with treaties on the treatment of POWs and non-combatants. Encouraged, China decided to revive its wartime chemical and biological warfare charges against Imperial Japan. The Japanese had signed the Hague Conventions and, although not a party to the Geneva Protocol, the ban on chemical and bacteriological weapons use in war was internationally accepted. With that intent, along with charges of unprovoked war and mass killings, Chinese prosecutors arrived early at the International Military Tribunal of the Far East (IMTFE), which began organizing in December 1945 and opened in May 1946.

Using basically the same charter as the Nuremberg International Military Tribunal (1945–1946), the IMTFE put 28 defendants on trial, including former Ministers of War and those who had been high-ranking field commanders in Manchuria and China. Since Nazi Germany had refrained from using its chemical weapons during the war and had not developed biological weapons, the Protocol was fairly irrelevant at Nuremberg. The Chinese calculated that the Tokyo trial was the right forum to settle its old CBW scores.

3 Retrieving CBW Evidence from China, 1946

Eleven nations were represented at the IMTFE: the United States, the United Kingdom, the Soviet Union, France, China, Australia, New Zealand, the Netherlands, India, and the Philippines. Each was entitled to send an Associate Prosecutor to head its division at the International Prosecution Section (IPS), the IMTFE's cooperative organization for selecting defendants, composing the Indictment, and arguing the Allied cases. The IPS was also in charge of organizing war crimes evidence against those accused. The Chinese Section had virtually no staff, just its lead prosecutor, Hsiang Che-chun (Xiang Zhejun), and his assistant, Henry Chiu. In an unusual gesture of support, Joseph B. Keenan, the IPS Chief of Counsel, sent an investigative team to China on a month-long investigation to track evidence and identify witnesses who could testify at the trial on behalf of the Chinese Division (Keenan 1946).

On March 12, 1946, two American IPS lawyers, Colonel Thomas Morrow and civilian David Nelson Sutton, along with Henry Chiu, began their investigation, which took them to Shanghai, Beijing, Chongqing, and Nanjing (Guillemin 2017, 105–157).

Sutton was responsible for investigating the BW allegations, which proved difficult. He interviewed knowledgeable public health physicians who had reported on the plague outbreaks, but none had actually seen the Japanese air attacks, which had targeted small war-torn cities in Central and South China. No testimony at the time had been taken from victims and Sutton had no mandate to conduct an intensive field inquiry. General Ishii Shiro, the leader of the Japanese germ weapons program, centered at Unit 731 in Manchuria, had cleverly intended to mask the plague attacks so that they would seem to be naturally caused—and he had largely succeeded. Sutton was able to gather substantial evidence for Japan's 1937 conquest of Nanjing and the ensuing seven weeks of massacres, rapes, and looting, and for the Japanese opium enterprise, which had enslaved many Chinese communities, but not for the plague attacks.

After returning to Tokyo, Sutton submitted his report on allegations of Japanese bacteriological warfare to Chief of Counsel Keenan. He also sent a copy to General Douglas MacArthur, Supreme Commander for the Allied Powers (SCAP) and head of the Allied Occupation, the ultimate authority in Tokyo, even though he professed to keep a distance from the IMTFE. Sutton's conclusion was succinct: "As the case now stands, in my opinion the evidence is not sufficient to justify the charge of bacteria warfare" (Sutton 1946, 1).

Morrow and Sutton were both aware that a CWS officer was in the process of interviewing General Ishii and a dozen of his former Unit 731 staff; the news had been leaked to the military newspaper *Stars and Stripes* the previous February (Kalisher 1946). That inquiry, Morrow had been advised, was for CWS and G-2 "war booty" purposes alone. Although Sutton later made other inquiries about Japan's BW program, IPS was left with his original assessment: the evidence for Japanese BW, which killed hundreds and terrorized thousands, was insufficient for prosecution. Another question troubling the Chinese charge concerned the problem of linking any such attacks to particular defendants in the dock. How could germ weapons be developed and used without the knowledge of the high command in Tokyo?

In contrast to Sutton's BW inquiry, Colonel Morrow's March–April 1946 investigation of Japanese chemical weapons use in China yielded documents and eye-witness accounts highly suitable for trial. On returning to Tokyo, Morrow wrote up a detailed summary of what he had acquired and, on May 13, he and Kenneth N. Parkinson, another IPS attorney, included it in their "Form of Brief" on "All China Military Aggression, 1937–1945," which they submitted to Chief of Counsel Keenan and to General MacArthur (Parkinson and Morrow 1946a). Following a detailed IPS outline, the draft described each of the relevant counts of the Indictment as they related to China's charges against Japan and it listed the names of witnesses ready to testify. The summary covered the Sino-Japanese War from July 7, 1937 until Japan's surrender in September 1945.

Colonel Morrow's special contribution to the draft was his evidence for Japanese CW against China, which he included with Japan's other war crimes:

> This waging of war by Japan in China was characterized by gross violations of international law and treaties, by massacre of civilians and Chinese soldiers, prisoners of war, and by the outlawed use of poison gas (Parkinson and Morrow 1946a, 7–8).

Starting with potential trial witnesses, Morrow drew on an affidavit from a Major General Chang, deputy director of China's Army Medical Corps, who stated that the Japanese used poison gas at Ichang (outside Shanghai), where Chang "personally saw men who were burned about the eyes, arm pits, and the crotch whose cases were diagnosed by himself. He saw 30 or 40 soldiers affected this way." (Parkinson and Morrow 1946a, 8). A photographer at the scene, a gas defense officer of the 34th Army group, could testify to having observed the soldiers burnt by vesicant gas. In addition, Brigadier General Wang Chang Ling, director of the gas defensive administration, could testify that in 1943 he found on the battlefield fifteen-centimeter howitzer shells containing hydrocyanic acid and that he saw a dozen soldiers suffering from gas poisoning by the Japanese, three of whom died. The general still had his notebook describing the analyses of contents of the shells. Finally, the director of a museum of chemical munitions near Chongqing was willing to describe the spent Japanese vesicant bombs and shells that he and others had brought back from battlefields.

As for official data, Major Woo Chia Shing of the Chinese Army, a custodian of records obtained from the Japan's Ministry of War in Tokyo, stated that 26,968 persons were injured by poison gas in the Sino-Japanese War, of whom 2086 died. These records, he said, verified that gas was used by the Japanese 1312 times in ten battles (Parkinson and Morrow 1946a, 8). Morrow then referred to Japan's secret chemical plant at Okunoshima.

> Weapons from there, including mustard and lewisite, were shipped to China; Documents and testimony from Japanese officers connected with chemical warfare indicated that gas warfare was used in emergencies with the permission of Tokyo up to and including 1942 (Parkinson and Morrow 1946a, 8).

Meanwhile, the US Chemical Warfare Service (CWS) had an interest in not having Japanese chemical weapons use prosecuted as a war crime; that notoriety could only limit its own agenda to develop offensive weapons. While Morrow was still in China, Lieutenant Colonel John Beebe from the CWS office in Tokyo began conducting his own inquiry—which was part of larger G-2 project to assess Japan's chemical weapons. First Beebe interviewed former Premier Tojo Hideki and former War Minister Hata Shunroko, then being held at Sugamo Prison, and then, in addition, four other Japanese officers who might know about Japan's chemical arsenal and use.

In response to Beebe's questioning, Tojo insisted that although chemical stocks had been produced, none had been used and could not have been used without his permission. He had heard President Roosevelt's 1942 warning about retaliation in kind:

> I thought, as I had from the beginning, that the use of gas would be disadvantageous for Japan because of tremendous America's industrial capacity and this statement of the President strengthened my own ideas (Interrogation of Tojo 1946, 4).

Tojo admitted that the Japanese did use "harassing agents" such as sneezing agents, tear gas, and smoke (as opposed to "casualty agents"). When asked if these non-lethal weapons were banned by international treaties, he replied defensively that police forces all over the world used them with impunity, including those in the United States who wanted to quell riots. "And what about the atom bomb?" he added. Changing the subject, Beebe confronted him with Japanese field reports that Chinese soldiers had died of asphyxiation from gas attacks, with blood running out of their noses and mouths. Tojo attributed these deaths to the Chinese overreactions to harassing agents and to their sometimes fatal inexperience with using gas masks.

General Hata had been in command of the Central China Expeditionary Force in 1938 and in 1941–1944 he was the Commander-in Chief of all the China armies. He denied any Japanese use of chemical weapons, allowing only that sneezing and tear gas might have been used (Interrogation of Hata 1946), nor had he ever heard of China's complaint to the League of Nations. The four other officers Beebe interviewed similarly denied that any chemical warfare had been waged against the Chinese. One, a former commander at the Narachino Chemical Warfare School, admitted to the production of some munitions but insisted that the high command had forbidden chemical attacks. "If gas were used in China," he said, "it was just on the spur of the moment and not on the orders of high authorities" (Interrogation of Yokoyama 1946, 5).

If presented in court, would Colonel Morrow's CW evidence be enough to counter these denials? The opportunity never arose, due to the influence of CWS working with US military intelligence in Tokyo. In mid-May, Morrow was advised of the content of Beebe's interrogation and the argument for "the outlawed use of poison gas" was erased from the "All China Military Aggression" report. In its final version, dated May 24, Japan's use of "poison gas" was briefly mentioned in passing on its first page, and the Tokyo document citing Chinese chemical casualties of 26,968, including 2086 who died, was repeated from the earlier version, along with the admission that Japan was known to have manufactured "various types of poisonous gases." But all the trial-worthy evidence—the eyewitness testimony, diaries, photographs, and medical records of victims, the retrieved battlefield munitions—was excised (Parkinson and Morrow 1946b).

In addition, reasons were inserted, taken straight from Lieutenant Colonel Beebe's interviews of Tojo and Hata, why allegations against Japan should not be "over-emphasized." As the authors explained:

> 1) It does appear that gas was used only in emergencies and for the most part tear, sneezing, and vomiting gas was used and not the vesicants, 2) the amount of casualties inflicted on the Chinese as evidenced from their statistics was a very small proportion of the total casualties suffered by the Chinese during the war, which is well over 3,800,000 according to their own records, and 3) in their interrogations Generals Hata and Togo refer to the fact that in the United States we have used poisonous gases such as sneezing, vomiting and tear gasses in labor disputes and General Togo in his interrogation about gas, raises the question, "How about the atomic bomb?' which he claims is a much more outrageous weapon of warfare than poisonous gas (Parkinson and Morrow 1946b, 31–32).

The intimation of this last point was that if chemical warfare charges were made against Japan, the defense would have leeway to introduce the Hiroshima and Nagasaki bombings as a way of undermining the legitimacy of American prosecutors.

The revised report then dismissed China's first report to the League, and with it any reference to existing treaties:

> The reference to the use of poison gas in this warfare appears to have been made in the form of a complaint by the Chinese to the effect that the Japanese army used gas in Shanghai three and four October 1937, but which the Japanese emphatically deny. We do not intend to offer this in evidence but merely invite the committee's attention (Parkinson and Morrow 1946b, 32).

From then on, as IPS was informed, all future evidence about Japan's alleged chemical weapons use would be distributed to staff by the head of the CWS office in Tokyo, Colonel Geoffrey Marshall.

With its criminal charges of chemical warfare reduced to complaints and its germ weapons charges without solid proof, the Chinese Division, heavily reliant on American support, focused on the prosecution of the more flagrant Japanese crimes resulting from "aggressive war" and on their "total casualties." These charges were successfully argued, with convictions of top Japanese leaders from the time of the conquest of Manchuria to the 1937 instigation of the Sino-Japanese War and through to August 1945 and Japan's surrender to the Allies. In November 1948, seven defendants, including Tojo, were sentenced to death by hanging, while the rest were given jail sentences ranging from life to seven years, with the exception of two of the accused who died during the trial and another who spent the trial in a mental institution (Brackman 1987, 454–462).

4 Conclusion

Over the years, the IMTFE has been criticized as an example of hypocritical "victor's justice" for its prosecution of war crimes—like the killing of civilians or abuses of POWs—of which the Allies and particularly the United States were themselves guilty (Totani 2008, 218–245).

The obstacles to justice that G-2 and the Chemical Warfare Service posed at IPS were too covert to attract public notice. By the summer of 1946, General Willoughby controlled all witness interrogations at IPS—in the name of national security—and he later engaged leading lawyers at IPS and General MacArthur in his project to protect General Ishii and some two dozen of his scientists from war crimes prosecution for inhumane medical experimentation and their role in multiple mass germ attacks on Central China in 1942 (Guillemin 2017, 244–284).

In the 70 years since the IMTFE, the history of chemical and biological weapons proceeded from state-level proliferation on an extravagant scale, at immeasurable cost, to eventual international restraints. The proliferation began with the United States and its Western allies and with the Soviet Union, which legitimized the Cold War excesses. In combination with nuclear arms, chemical and biological weapons became integral to the "weapons of mass destruction" model for small and developing nations in troubled regions of the world. The legal restraints began with

President Richard Nixon's unexpected 1969 renunciation of offensive biological weapons on behalf of the United States and the 1972 Biological Weapons Convention and, after the end of the Cold War, the culmination of the 1993 Chemical Weapons Convention and its organized destruction of state CW munitions—a process that both increases political stability and public safety.

In retrospect, chemical weapons were early on recognized as deplorable and, had the world known of Japan's germ attacks on the Chinese, those weapons, too, would have been reviled and perhaps criminalized at the IMTFE. On a purely practical level, major states were late in understanding that, compared to conventional arsenals, chemical and biological weapons were cumbersome, liable to endanger friendly troops and civilians, and inefficient, given their unpredictable impact on targets. Nor had chemical and biological arms the annihilating power of atomic bombs, no matter how much their advocates over the years were able to persuade those who controlled state military budgets that strategic capability was always one step away.

A post-facto evaluation of chemical weapons as archaic is cold comfort for those populations who suffered from the post-war proliferation. These victims, as might have been predicted, were always defenseless populations. Egypt's attacks on Yemen in the 1960s and, worse, Saddam Hussein's use of poison gas during the 1980s Iran-Iraq War reached a crescendo with his nerve gas attacks on Kurdish villages.

The latest spill-over from the long years of proliferation is the 2013 sarin attacks on Syrian civilians in Ghouta, outside Damascus, a war crime for which proof of accountability has been lost, for the time being, in the fog of war, along with the 2017 sarin attacks in Kahn Sheykun in northern Syria. As with the Chinese in 1945, one hopes for justice in the name of victims of chemical or biological attacks. But what might that forum be? In 1949, Harvey Northcroft, the New Zealand judge at the IMTFE, wrote passionately of the need for a disinterested international war crimes court:

> No nation or nations which established their own tribunals, when such a Permanent Court existed, could escape the imputation that their action was dictated by the desire for vengeance, or by other improper motives. A Permanent Criminal Court would, therefore, provide the greatest possible measure of insurance against the unscrupulous use of power by victorious nations in the future (Northcroft 1982, 136–137).

More than fifty years later, in 2003, the International Criminal Court (ICC), a United Nations initiative, opened as an independent resource for victims and states with no other avenues to justice. While the ICC represents a step forward in international criminal law, its few cases have been selected with care (prosecution on behalf of child soldiers in the Congo wars was its first), adjudicated slowly, and have addressed mostly sub-Saharan African conflicts. Echoing the Geneva Protocol, Article 8 of its statute refers to criminal sanctions against "employing poison or poisoned weapons" or "prohibited gases, liquids, materials or devices" such that any would cause "serious damage to health in the ordinary course of events, through its asphyxiating or toxic properties" (Schabas 2001, 305).

The statute is clear, yet the court's mission focuses on international armed conflict rather than civil wars or internal conflicts. It is questionable whether weapons-specific atrocities would by themselves make a compelling case, even when civilians were murdered. In addition, in its brief history, the ICC has had to navigate between pressures from the United Nations Security Council and the uncertain cooperation of states in conflict zones, while also being burdened by the complexities of its own organizational evolution (Minow et al. 2015). The future will see whether those responsible for the sarin attacks on Ghouta will someday sit accused in the dock at the ICC or another court and whether the charges are framed as war crimes or, in broader acknowledgement of the harm done to civilians, as crimes against humanity.

References

Ames, C.B. 1922. Article Eight of the League of Nations. *The Southwestern Political Science Quarterly* 2: 302–310.

Boserup, Anders. 1973. *The problem of chemical and biological warfare: CBW and the Law of War*, vol III. New York: Humanities Press.

Brackman, Arnold C. 1987. *The other Nuremberg: The untold story of the Tokyo war crimes trials*. New York: William Morrow.

Brown, Frederic J. 2006. *Chemical warfare: A study in restraints*. New Brunswick: Transaction Books.

Buckley, John. 1999. *Air power in the age of total war*. Bloomington: Indiana University Press.

Buruma, Ian. 2013. *Year zero: A history of 1945*. New York: Penguin Press.

Cirincione, Joseph. 2007. *Bomb scare: The history and future of nuclear weapons*. New York: Columbia University Press.

Douhet, Giulio. 1942. *Command of the air*. Trans. Dino Ferrari. London: Faber & Faber.

Gordon, Andrew. 2014. *A modern history of Japan: From Tokugawa to the present*. New York: Oxford University Press.

Grunden, Walter E. 2005. *Secret weapons and World War II: Japan in the shadow of Big Science*. Lawrence: University Press of Kansas.

Guillemin, Jeanne. 2005. *Biological weapons: From state-sponsored programs to contemporary bioterrorism*. New York: Columbia University Press.

Guillemin, Jeanne. 2017. *Hidden Atrocities: Japanese germ warfare and American obstruction of justice at the Tokyo trial*. New York: Columbia University Press.

Home, R.W., and Morris F. Low. 1993. Post-war scientific intelligence missions to Japan. *ISIS* 84: 527–537.

Hudson, Manley O. 1931. Present status of the Hague Conventions of 1899 and 1907. *American Journal of International Law* 25: 114–117.

International Criminal Court (ICC). 2012. *History of the PCIJ*. http://www.icj-cij.org/en/history. Accessed 10 Mar 2016.

Interrogation of General Shunroku Hata. 1946. US National Archives, Japanese War Crimes (JWC), RG496, Entry UD UP53, Box 342, Folder C-306, April 11.

Interrogation of Hideki Tojo. 1946. US National Archives, JWC, RG496, Entry UD UP53, Box 342, Folder C-306, April 2.

Interrogation of Takeshige Yokoyama. 1946. US National Archives, JWC, RG496, Entry UD UP53, Box 342, Folder C-305, April 17.

Jacobsen, Annie. 2014. *Operation Paperclip: The secret intelligence program that brought Nazi scientists to America*. New York: Little, Brown.
Jansen, Marius B. 2002. *The making of modern Japan*. Cambridge: Harvard University Press.
Kalisher, Peter. 1946. SCAP locates and questions Gen. Ishii. *Pacific stars and stripes*, February 27.
Keenan, Joseph B. 1946. Directions to Sutton from Keenan. Tokyo war crimes trial collection, University of Virginia Law School, Section 1, Folder 1, MSSI Su 863a, March 3.
Kennedy, David. 1987. The move to institutions. *Cardozo Law Review* 8: 841–988.
Lepick, Olivier. 1998. *La Grande Guerre chimique*. Paris: Presses Universitaires de France.
MacMillan, Margaret. 2013. *The War That Ended Peace: The Road to 1914*. New York: Random House.
Minow, Martha, True-Frost, C. Cora, and Whiting, Alex, eds. 2015. *The First Global Prosecutor: Promises and Constraints*. Ann Arbor: University of Michigan Press.
Northcroft, Harvey. 1982. The Former New Zealand Member, International Military Tribunal for the Far East, to the Prime Minister, 17 March 1949. In *Documents on New Zealand external relations*: *The surrender and occupation of Japan*, vol. 2, ed. Robin Kay, 1736–1737. Wellington, New Zealand: P.D. Hasselberg.
Parkinson, Kenneth N., and Colonel Thomas H. Morrow. 1946a. Trial Brief All China Military Aggression, 1937–1945. Tokyo War Crimes Trial Collection, University of Virginia Law School, Box 2, Folder 2, MSS 93–4, May 13.
Parkinson, Kenneth N. and Colonel Thomas H. Morrow. 1946b. Trial Brief All China Military Aggression, 1937–1945. Tokyo War Crimes Trial Collection, University of Virginia Law Library, Box 3, Folder 4. MSS93–4, May 24.
Sartori, Mario. 1943. *The war gases*. New York: D. Van Nostrand.
Schabas, William A. 2001. *An introduction to the International Criminal Court*. Cambridge: Cambridge University Press.
SIPRI (Stockholm International Peace Research Institute). 1971. *The Problem of Chemical and Biological Warfare*: *CB Disarmament Negotiations, 1920–1970*, vol. 4. New York: Humanities Press.
Smart, Jeffrey K. 2004. History notes: Chemical and biological research and development during the civil war. *CBIAC Newsletter* 5: 3–11.
Sutton, David Nelson. 1946. Subject: Bacteria warfare. David Nelson Sutton Collection, University of Richmond Law Library, Folder 13, April 25.
Taylor, Telford. 1992. *The anatomy of the Nuremberg trials*. New York: Knopf.
Totani, Yuma. 2008. *The Tokyo war crimes trial: The pursuit of justice in the wake of World War II*. Cambridge: Harvard University Asia Center.

Part III
Dual Use, Storage and Disposal of Chemical Weapons Today

The Reconstruction of Production and Storage Sites for Chemical Warfare Agents and Weapons from Both World Wars in the Context of Assessing Former Munitions Sites

Johannes Preuss

Abstract This chapter begins by listing the quantities and sites of chemical agent production during both world wars and outlining the relative importance of these new weapons. Using the example of the production sites of World War II, the setting in which the construction and operation of these factories took place will be described, as well as the structure of the facilities. It will be shown that it was not only Fritz Haber's former colleagues who made important contributions to the research of chemical warfare agents and their production, but that an important role was also played by students of his successor at the Kaiser Wilhelm Institute for Physical Chemistry and Electrochemistry. In order to be employed militarily, chemical warfare agents must be put into grenades, bombs, mines, warheads, and spray tanks. This took place at seven filling plants, five of the army and two in air munitions facilities. Gaseous and particularly dangerous modern warfare agents were filled in the chemical factories where they were produced. This article is based on extensive research in the context of the investigation, ongoing since 1979, into former armaments sites, the methodology of which will be briefly outlined. It will be shown that the effects of chemical warfare agents—their production and deployment at the frontline—continue to pose a risk 100 years later. In consideration of general public health, the disposal of these agents of must be prioritized. Also in Germany, these agents have been exploded, burned, and buried, and the residues pose environmental risks. Some of the demolition sites of these agents are still unknown today.

J. Preuss (✉)
Geographisches Institut, Johannes Gutenberg-Universität Mainz, Mainz, Germany
e-mail: jpreuss@uni-mainz.de

B. Friedrich et al. (eds.), *One Hundred Years of Chemical Warfare: Research, Deployment, Consequences*, DOI 10.1007/978-3-319-51664-6_16

1 Introduction

The importance of chemistry and the chemical industry to human well being and prosperity has grown constantly since the mid-nineteenth century. The discoveries of this science are, however, also suited to improving weapons and other armaments employed by states against one another in conflicts.

Whereas the war of 1870–71 was still fought on the German side with around 2,000 metric tons of gunpowder for propellant and explosives (ca. 26 million rounds of infantry munition and ca. 845,000 rounds of artillery munition) (*Der deutsch-französische Krieg* 1881, 816–19), the situation changed crucially with the development of Granatfüllung 88 (Grf. 88 = picric acid) in 1888 and of Füllpulver 02 (Fp 02 = trinitrotoluol) in 1902. It was no longer gunpowder factories that were providing propellant and explosives but rather the private chemical industry. Over the course of the First World War, 33 predominantly private factories in Germany produced 510,000 metric tons of explosives and 36 factories 285,000 metric tons of propellant. Seven chemical factories manufactured 47,400 metric tons of chemical warfare agents. In addition, around 450,000 metric tons of intermediate products were required, some of which were taken from limited food supplies for the population (see Table 1).

In Germany during the Second World War, war-related chemical production increased to 977,500 metric tons of explosives (28 factories), 974,000 metric tons of propellant (43 factories), and 69,500 metric tons of chemical warfare agents (10 factories) as well as around 805,000 metric tons of intermediate products (see Table 1 and Table 2).

This increase in production was made possible by, among other things, the method Fritz Haber and Carl Bosch developed in the early twentieth century to synthesize ammonia, which was honored with a Nobel Prize and made it unnecessary to import saltpeter from Chile for fertilizer and as a raw material to produce munitions. Over a period of just under seventy-five years the production of explosives and propellants in Germany increased by a factor of 975.

Even today, the legacies of both world wars have considerable negative significance for the livelihoods of the population, considering that military chemical substances contaminate the soil and ground water in many locations and unexploded munitions pose a threat to public safety.

Table 1 Quantities of propellant and explosives produced for the wars of 1870–71, 1914–18, and 1939–45 (in Metric Tons) (*Der deutsch-französische Krieg* 1881, 816–19; Preuss and Schneider 2005, 31–47, supplemented)

	Explosives	Gunpowder	Chemical agents	Total
War of 1870–71	–	ca. 2000	–	2000
First World War	508,198	284,808	47,395	840,401
	33 factories	36 factories	7 factories	76 factories
Second World War	977,492	974,188	69,500	2,021,180
	28 factories	43 factories	10 factories	81 factories

Table 2 Production sites for chemical warfare agents and quantities produced in the First and Second World Wars (Preuss and Schneider 2005, 36–37, supplemented)

Site (WW I)	Company	Quantity	Site (WW II)	Company	Quantity	Products
Berlin-Adlershof	Kahlbaum	935	Ammendorf	Orgacid	25,976	Mustard gas
Frankfurt	Farbwerke	12,877	Berlin-Haselhorst	Lonal	1455	Clark, Dora
Leverkusen	Bayer	19,033	Dyhernfurth	Anorgana	12,753	Tabun
Ludwigshafen	BASF	288	Gendorf	Anorgana	4110	Mustard gas
Mainkur	Casella	1042	Hüls	Chemische Werke Hüls	62	Mustard gas
Offenbach	Griesheim-Elektron	950	Ludwigshafen	IG Farben	1180	Phosgene
Wolfen	Agfa	2770	Seelze	Riedel de Haen	7139	Chloroacetophenone
	Total	47,395	Stassfurt	Ergethan	8224	Arsine
			Uerdingen	IG Farben	3881	Adamsite
			Wolfen	IG Farben	4720	Phosgene
				Total	69,500	

As the cleanup of the former munitions site of Stadtallendorf, near Marburg in the region of Hesse demonstrates, the cleanup of a single munitions waste site can cost taxpayers 160 million euros (Reile 2005, 424).

Since the mid-1980s, former munitions sites in Germany have been tested for waste. The tipping point was September 6, 1979, when an explosion in the basement of a residence in Hamburg caused the death of a child. The explosion was caused by chemicals from the abandoned Chemische Fabrik Dr. Hugo Stoltzenberg. The subsequent investigations revealed that munitions, incendiary materials, and chemical warfare agents were present in large quantities there (Scholz 2004; Brauch and Müller 1985, 331–59). At the time, the Bundesarchiv (Federal Archives) was tasked with researching whether similar events—but especially the occurrence of chemical warfare agents—was also possible in other locations.[1]

The weekly magazine *Stern* published the results of its own research, which roused the authorities and the citizens. Politicians discovered the topic and funds were made available to study waste sites. In the 1980s, the munitions waste sites of Hessisch-Lichtenau and Stadtallendorf in Hesse were studied first and a survey was conducted (König and Schneider 1985; Preuss 1990; Preuss et al. 1992a). Even earlier Lower Saxony had begun conducting a survey of munitions waste sites and organized the first conference of experts in the field in 1989.

As part of the environmental research plan of the Bundesminister für Umwelt, Naturschutz und Reaktorsicherheit (Federal Minister for the Environment, Nature Conservation, and Nuclear Safety), research projects were launched to take stock of suspected locations of munitions wastes in Germany. They provided comprehensive lists of sites, studied materials on typical combinations of explosives and chemical warfare agents, and compiled finding aids for federal and state archives (see Thieme 1996). This work provided a practical entry point to the complex theme of "munitions wastes." The Umweltbundesamt (UBA, German Environmental Agency) also supported and assisted exemplary research and study projects in which the guidelines that had been developed could be tested for feasibility. We were able to carry out one such exemplary study (Preuss et al. 2002).

In addition to these UBA programs, handbooks and waste studies were published by the German states (Preuss and Eitelberg 1996) and information gathered on international experience with collecting data, studying, evaluating, and cleaning up the waste of former sites of the armaments industry and the military (Schaefer et al. 1997, 139–44). For federal sites, the process is divided into three phases:

Phase I: Collecting data and initial analysis. This consists of studying the historical development and use of a site and reviewing as applicable the resulting suspicion of waste and describing the site and suspect areas. To that end, the relevant authorities and eyewitnesses are interviewed and promising archives in Germany and abroad visited. In addition to federal archives, state archives,

[1]Bundesarchive Koblenz und Freiburg, "Fertigung, Lagerung und Beseitigung chemischer Kampfstoffe unter besonderer Berücksichtigung des Territoriums der Bundesrepublik Deutschland," 1979, Preuss private collection.

municipal archives, local history museums, and residents with an interest in history can be important sources. Not infrequently, the companies that operated former munitions factories have documentation. Maps, building plans, building descriptions, descriptions of production, production figures, and aerial photographs are also evaluated. Although the client does not always want this, it can make sense to visit the site and make an initial assessment of the potential risk to the protected property. Occasionally, this revealed that immediate action was required.

Phase II: Risk assessment. Risk assessment is based on two steps: orientation studies (Phase IIa) and detailed studies (Phase IIb). The goal of Phase IIa is either to disprove the suspicion of the presence of waste through limited site studies or to confirm that suspicion through studies of the presence of hazardous materials, their release and spread, and their effects on the protected property. On former munitions and military sites, the presence of chemical agents is often suspected, so that the sites to be studied first have to be cleared before disturbing the soil. Chemical weapons and hazardous materials also necessitate that a work safety plan be prepared. In addition to suspect areas typical of the site, such as facilities for smelting, mixing, and casting in munitions filling factories or sewage facilities, in many cases a suspicion arises immediately upon entering owing to the presence of munition parts, residues of explosives, and traces of decontaminants or other chemicals. This suspicion can be checked with quick tests or laboratory analysis of soil and material samples. Phase IIa also calls for conducting groundwater studies downstream from potential waste concentrations and, regarding the type of hazardous materials, studies of soil gases. If the initial suspicion is disproven, the area can be removed from the study program. If the initial suspicion is confirmed or the studies are not yet sufficient, in Phase IIb (detailed studies) a program appropriate to the type and extent of the measures based on earlier findings must be worked out. The objective is to contain and assess the contamination in media that can spread it such as soil, groundwater, and soil gases and to assess the risk to the protected property and to make recommendations for further action. If Phase IIb should reveal the necessity for an immediate measure, it must be implemented without delay. If it should reveal the necessity for a safety or cleanup measure (Phase III), additional site studies for these areas may be necessary until it is possible to develop a comprehensive project proposal that addresses all of the questions relevant to a local measure. In practice, moreover, studies that could be conducted in the context of dissertations have also been necessary (see Szöcs 1999; Bausinger 2007; Mense-Stefan 2005).

Figure 1 shows the relevant sites for the production, processing, and storage of chemical warfare agents (see also Tables 2 and 3). In preparation for war, production plants were built and operated in a secret collaboration between the Oberkommando des Heeres (OKH; Supreme High Command of the Army) and companies from the chemical industry. With few exceptions munitions with chemical warfare agents were filled by five munitions plants of the army and two of the air force. Phosgene, which is a gas at room temperature, was filled directly into bombs at the production plants in Wolfen and Ludwigshafen. For safety reasons, the modern nerve gas tabun was filled in bombs and shells in the factory in Dyhernfurth (now Brzeg Dolny), near Breslau (now Wrocław) where it was produced.

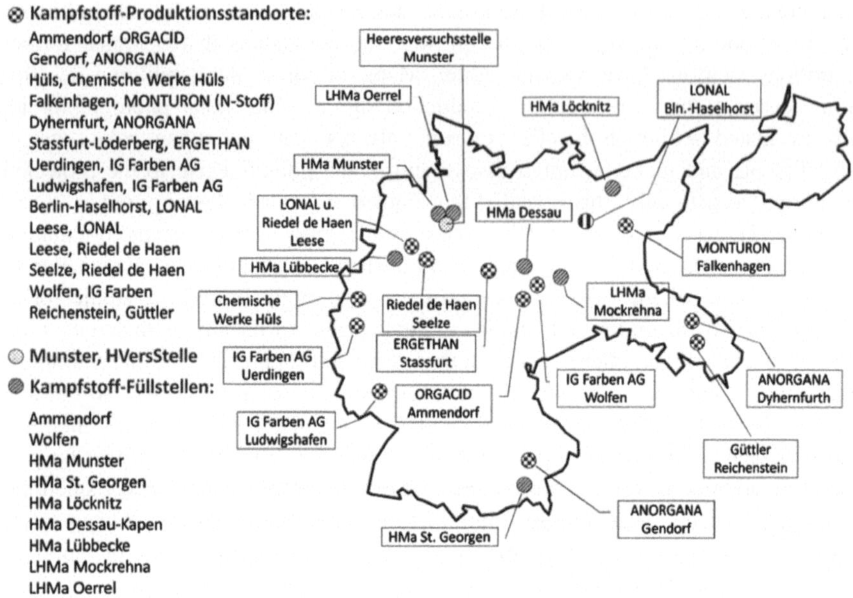

Fig. 1 Working production and filling sites in the Second World War (Kampfstoff-Produktionsstandort (Chemical agents production sites) Munster, HVersStelle [Heeresversuchsstelle] (Munster, army testing site) Kampfstoff-Füllstellen (Chemical munition filling sites))

Table 3 lists 24 production sites for chemical warfare agents, of which just 13 produced a total of 69,500 tons of chemical warfare agents. Five sites were operated directly by private companies; two of them in Ludwigshafen and two in Uerdingen which belonged to IG Farbenindustrie A.G. The necessary intermediate products were already available there. The fifth factory was Riedel de Haen in Seelze, near Hanover.

As part of the collaboration between the OKH and various companies from the chemical industry, 19 plants were built. The army-owned holding company Montan mediated between the contract partners.

Montan was founded in 1916 as Verwertungsgesellschaft für Montanindustrie GmbH:

> §1 Messrs. geh[eimer] Kommerzienrat [Privy Commercial Councilor] Ernst Fromm and Dr. Otto Kahn hereby establish a company with limited liability under the name *Verwertungsgesellschaft für Montanindustrie, Gesellschaft mit beschränkter Haftung*. §2 Company headquarters are in Munich. §3 The objective of the company is to acquire mining rights, shares in unions, mining shares, the commercialization of such rights and assets, and participation in similar companies.[2]

[2]Justizrat (Legal Councilor) Franz Pündtner, notary in Munich, Articles of association, February 3, 1916, notarized copy, reg. no. 180, Preuss private collection.

Table 3 Factories for chemical warfare agents in the former German Reich (1935–1945). (Preuss and Schneider 2005, 36–37, supplemented = PS); and (Ehmann 1948, 720–40 = E)

Type, material	Company, location	Completed	Remark	Owner	Capacity (metric tons monthly)	Total production (metric tons)	Source
White ring (omega salt)							
Phenacyl chloride	Riedel de Haen, Seelze	1935–42	Closed in 1942	Riedel de Haen	140	4006	PS
Phenacyl chloride	IG Farben AG, Ludwigshafen	1935–44	Intermediate product in Ludwigshafen, destroyed in 1944	IG Farben AG	70	3133	E
	Chem. Fabrik Hahnenberg, Leese	1942	Subsidiary of Riedel de Haen	Montan	500	Not in operation	E
Green ring							
Phosgene (F-Oel)	IG Farben AG, Wolfen	1940–1942	Filling of bombs at production site	Montan	400 (filling)	4720	PS
	IG Farben AG, Ludwigshafen	1942	Filling of bombs at production site	IG Farben AG	120 (filling)	1180	PS
	IG Farben AG, Uerdingen	1942	Filling of bombs at production site	IG Farben AG	180 (filling)	Not in operation	E
	IG Farben AG, Auschwitz	–	Planning	Montan	700	Not built	E
Nitrogen Mustard Gas/T 9/N-Lost	Orgacid, Ammendorf, near Halle/Saale	1937	Subsidiary of Auergesellschaft AG and Th. Goldschmitt AG	Montan	50	1928	PS
Tabun (T 83)	Anorgana GmbH, Dyhernfurth	1942	Subsidiary of IG Farben AG, filling of bombs and shells	Montan	1000	12,753	PS

(continued)

Table 3 (continued)

Type, material	Company, location	Completed	Remark	Owner	Capacity (metric tons monthly)	Total production (metric tons)	Source
Sarin (T46)	Anorgana GmbH, Dyhernfurth	Under construction	Subsidiary of IG Farben AG	Montan	100	Not finished	E
	Monturon GmbH, Falkenhagen	Under construction	Joint operation of Montan and IG Farben AG	Montan	500	Not finished	E
Cyanogen chloride (T150)	Anorgana GmbH, Dyhernfurth	1943–44	Subsidiary of IG Farben AG	Montan	20	?	E
Hydrogen cyanide (T155)	Anorgana GmbH, Dyhernfurth	1943–44	Subsidiary of IG Farben AG	Montan	20	?	E
Bi IV 99 (T 300)	Anorgana GmbH, Dyhernfurth	1944	Subsidiary of IG Farben AG	Montan	100	Construction halted	E
Arsine							
Blue ring							
Adamsite (Azin)	IG Farben AG, Uerdingen	1940–44	Intermediate product in Uerdingen, demolished in 1944	IG Farben AG	200	3881	PS
Clark I (C I)	Lonal-Werke, Berlin-Haselhorst	1940	Prof. Dr. Engelhard	Montan	90	1396	PS
Clark II (CII, Dora)	Lonal-Werke, Berlin-Haselhorst	1940	Prof. Dr. Engelhard, Production for experiments	Montan	45	59	PS
Blue-yellow ring							
Arsine oil (As Oel)	Ergethan GmbH, Stassfurth	1938	Subsidiary of Auergesellschaft and Kalichemie AG	Montan	270	8224	PS
	Lonal-Werke, Leese	1940–41	Prof. Dr. Engelhard	Montan	400	Not in operation	E

(continued)

Table 3 (continued)

Type, material	Company, location	Completed	Remark	Owner	Capacity (metric tons monthly)	Total production (metric tons)	Source
Yellow ring							
Oxol-Lost (O)	Orgacid GmbH, Ammendorf, Halle/Saale	1936	Subsidiary of Auergesellschaft AG and Th. Goldschmitt AG	Montan	700	24,048 18,045 OL, 5,739 OKM, 264 OB.	PS
	Anorgana GmbH, Gendorf	1941	Subsidiary of IG Farben AG	Montan	800	Not in operation	E
	Chemische Werke Hüls GmbH, Marl	1941/42	Chemische Werke Hüls GmbH	Montan	600	62 test operation in 1942	PS
Direkt-Lost (D)	Anorgana GmbH, Gendorf	1943	Subsidiary of IG Farben AG	Montan	1500	4,110	PS
	Chemische Werke Hüls GmbH, Marl	1943	Reich-owned experimental plant	Montan	200	Test operation in 1943	E
					Total:	69,500	PS

Very soon thereafter it was in the hands of the Eisenwerksgesellschaft Maximilianshütte, Sulzbach-Rosenberg. In 1934 the Geräte- und Apparate Handelsgesellschaft (GERAP) in Berlin acquired ninety-five percent of the shares. That same year the Reich Ministry of War purchased the shelf corporation Montan from GERAP and Maxhütte (Preuss and Schneider 2005, 25–26). The sale had probably been arranged by the former internal corporate auditor of Maxhütte: the businessman Dr. Johann Martin, known as Dr. Max Zeidelhack,[3] who in early 1934 had transferred to the Heereswaffenamt (HWa; Army Weapons Agency) of the OKH with the rank of Regierungsrat (Senior Civil Servant). Zeidelhack rose to the rank of Ministerialdirigent (Ministerial Director) by 1940. He headed Montan as its first chief executive officer from 1935 to 1942.[4] During this period, 108 army-owned businesses with 180,000 employees were assembled under Montan's roof, with invested capital of 4.5 billion reichsmarks and a turnover of 1.8 billion reichsmarks in 1942. According to a curriculum vitae in tabular form of 1948, Zeidelhack was born in Bayreuth in 1891. He completed secondary school in Ingolstadt and studied six semesters in Munich (German, History of economics, English, French) before the First World War.

Zeidelhack then served as a lieutenant in the reserve of the Bayerische Schwere Artillerie (Bavarian Heavy Artillery); later he served as first lieutenant, pilot, and fighter pilot. After the war he continued his studies in Munich in the subjects of law and political science, graduating in 1922 and receiving his doctorate. After an apprenticeship in a bank, he worked as an auditor. On January 1, 1934, he began work as a business and contract consultant at the Heereswaffenamt of the Reichswehrministerium (Reich Defense Ministry). From November 1, 1934, he was Oberregierungsrat and head of the department of business administration with responsibility for all the financial contracts with German and foreign industry, in particular for army-owned companies as well as for trust companies and Montan. By his own account, he was dismissed from his Ministerialdirigent post on January 14, 1943, by Reichsminister für Bewaffnung und Munition (Reich Minister for Weapons and Munition) Albert Speer, because he was not prepared to sell the army-owned companies for "a fifth of their share value to friends of the minister in the party." After the war, as part of de-Nazification, Zeidelhack was ranked as a "fellow traveler" (*Mitläufer*) by Spruchkammer [Sentencing Chamber] VII in Munich and on January 30, 1948, benefited from the Christmas amnesty.

Montan's task was to acquire properties and plants in trust for the OKH/HWa and to supervise the construction and operation of factories. In order to camouflage them, the factories appeared on Montan's balance sheets and were thus not immediately recognizable as the property of the Reich. The relationships between

[3]Lebenslauf Max Zeidelhack, March 27, 1948, Bayerisches Hauptstaatsarchiv, Munich, Office of Military Government, Bavaria, 13/83-2/5, as well as National Archives, Washington, DC: Microfilm Publication 12065 Sect. 3-402/NNDG no. 775037.

[4]Wehrmacht-Fernsprechverzeichnis [Army Telephone Records] for Greater Berlin, part I, p. 147, February 1, 1943, Dr. Zeidelhack, department head, business administration (Wa Z 3), National Archives, Washington DC., Microfilm T-77, reel 342.

the OKH, Montan, the companies, and the subsidiaries to be founded were regulated by the master or framework agreement and the lease agreement. Under these agreements, the company could produce without capital or risk and profit from the agreements. Shortly before the end of the war, the Reichsminister für Rüstungs- und Kriegsproduktion transferred nearly all of the shares to Montan as free property (Preuss and Schneider 2005, 23–24).

During the postwar period, the factories were demolished and many buildings and plants suitable for armaments were blown up. Refugees were housed in several former factories in West Germany, which led to the founding of the so-called New Cities (e.g., Stadtallendorf, Waldkraiburg, Geretsried, Espelkamp and Traunreut) (Preuss 1990, Preuss et al. 1992a, 1994; Preuss and Eitelberg 2009, 91–115).

In some cases, the occupying troops used the facilities for their own ends. Other munitions plants are even now collections of ruins and rubble in forests that have become shrouded in myths (e.g., Clausthal-Zellerfeld, Reinsdorf, Forst) (Preuss et al. 1993; Bausinger et al. 2005).

Against this backdrop, in what follows we will discuss three factories for the production of chemical warfare agents: the Orgacid GmbH factory in Ammendorf, near Halle an der Saale; the Monturon GmbH factory in Falkenhagen, near Fürstenwalde; the Anorgana GmbH factory in Dyhernfurth (now Brzeg Dolny) on the Oder River. We will also look at the storage and processing of chemical warfare agents in munitions dumps.

2 The Ammendorf Factory of Orgacid GmbH

Just four kilometers from the center of the city of Halle an der Saale, Germany's largest mustard gas factory lies northeast of Ammendorf. The Ammendorf chemical weapons factory of Orgacid GmbH had a floor area of around twelve hectares. In its final form, it had plants to produce various types of the chemical weapon mustard gas as well as the necessary intermediate products; it also had a plant for filling bombs with chemical warfare agents, which could also be used for shells, and a large storage bunker for O mustard (oxol mustard) and a smaller one for N mustard (nitrogen mustard).

Within a radius of a kilometer around the mustard gas bunker, there were railroad tracks and the buildings for the Buckau chemical factory, which produced chlorine gas, among other things, as well as a sports field, the Rosengarten inn, and residential buildings with gardens, which provided the "ideal" camouflage for this important munitions factory. The toxicologically risky sewage of the factory was routed through Ammendorf parallel to an existing trench in a piped section with fourteen shafts, underneath Hallesche Strasse, Hindenburgstrasse, and Hauptstrasse and along Badstrasse to the Elster River. There was also a detour line with an iron pipe sixty centimeters in diameter, which led into the lower trench and also led to the Elster, which reached the Saale River about two and a half kilometers downstream.

On November 23, 1934, the Degea Aktiengesellschaft (Auergesellschaft), represented by board members Dr. Adolf Gerdes and Dr. Ing. Karl Quasebart, both of Berlin, and the Chemische Fabrik Buckau, represented by its authorized director, Hermann Cordes of Essen, formed a limited liability company called Orgacid GmbH, with headquarters in Berlin. The purpose of the company was the "production and distribution of chemical products of all kinds, especially Orgacid."[5] The chemist Dr. Hermann Engelhard,[6] also of Berlin, was appointed its chief executive office. The company had an supervisory board of three to five members, on which the OKH/HWa was also represented, by Messrs. Zeidelhack and Zahn.[7]

By a resolution of a special general meeting of December 22, 1937, the assets of the Chemische Fabrik Buckau were transferred to Th. Goldschmitt A.G. in Essen. After recording this change in the commercial register, Chemische Fabrik Buckau was dissolved. The party to the contract was now Th. Goldschmitt AG in Essen.[8] On December 10, 1934, within three weeks of signing the articles of association, Orgacid's chief executive officer Engelhard received three preliminary notices from the Waffenamt for contracts to build plants to produce polyglycol M, a conversion plant, and a polyglycol M storage facility in Ammendorf. That meant that the groundwork for the constructing the factory, the ordering of machines, and the setting up of the equipment could begin immediately.[9]

At this time, there was not yet a sales agreement for the property on which the factory was to be constructed. It was not concluded until June 3, 1935. The commercialization company for Montanindustrie GmbH, in Munich, represented by its authorized agent, the businessman Dr. Johann Martin, known as Dr. Max Zeidelhack, from Berlin, then purchased a property in Ammendorf of 75,500 square

---

[5]Gesellschaftsvertrag: Urkunde Nr. 331/1934, Notar Dr. F. Jacke, November 23, 1934, Preuss private collection.

[6]Engelhard was born in Trier, on November 1, 1896. In August 1914, after taking emergency school leaving exams, he became a soldier at seventeen. After the war he began studying chemistry at the Technische Hochschule in Karlsruhe and completed his degree in Munich. Then from 1921 to 1923 he was working at the Kaiser-Wilhelm-Institut für Physikalische Chemie und Elektro-Chemie in Berlin-Dahlem under Fritz Haber. Under the direction of Haber, he received his doctorate at the TH Karlsruhe with a paper on locating gold in seawater. Subsequently, Engelhard worked at the Auer-Gesellschaft AG. By taking over a teaching position on colloid chemistry at the TH Berlin from 1934, he qualified for an honorary professorship. From 1946, Prof. Engelhard was active at the Physiologisch-Chemisches Institut of the Universität Göttingen. See Lebenslauf von Prof. Dr. H. Engelhard, February 15, 1960, Universitätsarchiv Göttingen, UAG-Kur. 10221. Karl Quasebart has also been employed by Fritz Haber. He explored the Atlantic on a research journey of the Meteor as part of the Gold aus Meerwasser (Gold from Seawater) project (Stoltzenberg 1994, 497).

[7]Wehrmacht-Fernsprechverzeichnis [Army Telephone Records] for Greater Berlin, part I, p. 144, June 6, 1943, Dr. Zahn, department head, business administration, head chemist Wa A. National Archives, Washington, DC., Microfilm T-77, reel 342.

[8]Transcription of the copy from the commercial register, Amtsgericht Halle an der Saale, Department B, December 23, 1937, Preuss private collection.

[9]Entwürfe der Schreiben von Wa B 4 VII an die Orgacid GmbH, December 10, 1934, The National Archives, London, FO1031/204.

meters located between the Chemische Fabrik Buckau grounds and the Dessau–Merseburg train line. The directors of the Buckau factory, Hermann Cordes of Essen and Dr. Camillo Irmscher of Chemische Fabrik Buckau, which still existed at this time, appeared at the signing of the agreement.

On September 20, 1935, the Chemische Fabrik Buckau, Ammendorf; the Degea-Aktiengesellschaft (Auergesellschaft), Berlin; and the German Reich (treasury of the Wehrmacht's department of the army), represented by the Reichskriegsminister (Rkm, Minister of War), signed a legal contract to build a polyglycol factory and a conversion facility.[10] For purposes of secrecy, the agreement had a passage stipulating that the companies were to be active on behalf of and for the account of the Rkm, but in their own names vis-à-vis outsiders. The plant was to take into account the requirements of air defense, with roof constructions to withstand fire bombing and capable of being blacked out. According to Sect. 2, the land and the existing buildings of Orgacid GmbH were to be

> made available on the basis of a lease agreement still to be concluded by the authorized agent of the Rkm and the commercialization company for Montanindustrie GmbH, Munich, into whose possession all new plants that the companies will operate in trust under section 1 will also be transferred after completion and before beginning operation.

Degea/Auergesellschaft and Chemische Fabrik Buckau were permitted to contract their subsidiary Orgacid GmbH to build the factory, with which a lease agreement for operating the factory would be signed as well.

The factory grounds were divided into the following sections of production:

- A-Plant for the production of ethylene oxide, ethylene chloride, polyglycol M, and polyglycol MI.
- B-Plant (old) for the production of Oxol mustard gas (O mustard); winter mustard gas (OKM, OB).
- B-Plant (new) for the production of Oxol mustard gas (O mustard); winter mustard gas (OKM, OB).
- T9 plant for the production of nitrogen mustard gas (N mustard).
- Plant for filling KC 250 Gb bombs.
- Sewerage.

The annual report for the fiscal year 1937–38 reveals that production of polyglycol M, ethylene oxide, and ethylene chloride began in May 1937.[11]

[10]Vertrag zwischen der Chemischen Fabrik Buckau (Ammendorf), der Degea-Aktiengesellschaft (Auergesellschaft), Berlin, und dem Deutschen Reich (Fiskus des Wehrmachtsteils Heer), vertreten durch den Reichskriegsminister (Rkm) über die Errichtung einer Polyglykolfabrik und einer Umsetzungsanlage, September 20, 1935, Preuss private collection.

[11]Orgacid GmbH, Bericht über das Geschäftsjahr 1937–38, December 21, 1938, Preuss private collection.

Table 4 Construction projects for chemical weapons plants managed by the building department of Orgacid GmbH, later Lonal GmbH (Preuss and Eitelberg 2003b, 65)

1937–38	1939	1940–41	1943–44 (Lonal from 1941–42)
Ammendorf	Ammendorf	Ammendorf	Ammendorf
Trostberg	Gendorf	Gendorf	Gendorf
Hahnenberg	Hahnenberg	Hahnenberg	Leese
–	Hörpolding (St. Georgen)	Hörpolding (St. Georgen)	Hörpolding (St. Georgen)
Munster	Munster	Munster	Munster
–	–	–	Oerrel
–	Löcknitz	Löcknitz	Löcknitz
–	–	Dessau (Kapen)	Dessau (Kapen)
–	–	Lübbecke	Lübbecke
–	–	–	Mockrehna

In addition, Orgacid GmbH was granted several construction contracts late in the fiscal year 1937–38 (see Table 4).[12]

The annual report contains no production figures for the fiscal year 1938–39. The report on the events of the fiscal year reads:

> In the new fiscal year the company was contracted to begin producing final products as well. Manufacture began in May [1938] and is slowly increasing. During the critical days of September, the leadership did everything in its power to reach maximum production. Unfortunately, a series of accidents could not be avoided. Now all measures have been taken to reduce such accidents to an absolute minimum.[13]

For the fiscal year 1939–40, the production of various types of mustard gas in B-Plant is documented.[14] A-Plant continued to produce polyglycol M, polyglycol M I, ethylene oxide, and ethylene chloride.

In A-Plant, which was brought into operation in 1939, production was increased after the war began. In B-Plant, 572 metric tons of Oxol mustard gas were produced per month. In the winter of 1939–40, parts of the plants were closed to make repairs. The experimental plant for nitrogen mustard (T 9) was opened without a hitch (Table 5).

Of the 30,148 metric tons of mustard gas produced in the German Reich by the end of the war, 25,976 metric tons (86.2%) were from Ammendorf. In the filling plant in Ammendorf, 61,108 bombs (KC 250 Gb) were filled with ca. 90.2 kg for a total of 5512 metric tons of mustard gas.

[12]Orgacid GmbH, Bericht über das Geschäftsjahr 1938–39, December 21, 1939, Preuss private collection.

[13]Orgacid GmbH, Bericht über das Geschäftsjahr 1938–39, December 21, 1939, Preuss private collection.

[14]Orgacid GmbH, Bericht über das Geschäftsjahr 1939–40, Preuss private collection.

Table 5 Production of Orgacid GmbH Ammendorf, May 1937–March 1945 (tons, piece)

Time/Production	Eth. ox.	PM	PMI	O Mustard	OB Mustard	OKM Mustard	T 9 Mustard	KC 250 Gb
–	–	–	–	–	–	–	–	–
5/1937–3/1938 (G)	756	2085	–	–	–	–	–	–
4/1938–3/1939 (F)	15	3507	–	2183	–	–	19	–
4/1939–3/1940 (E)		1872	1473	3742	–	1215	118	–
4/1940–3/1941 (E)	669	1828	1481	3476	–	1908	293	6867
4/1941–3/1942 (D)	2,206	827	565	3868		789	361	53,859
4/1942–3/1943 (C)	2,115	1608	396	862	264	399	432	382
4/1943–3/1944 (B)	982	2361	848	2652	–	1030	516	–
4/1944–12/1944 (A)	2,314	248	128	972	–	398	189	–
1/1945–3/1945 (A1)	?	?	?	290	–	–	–	–
5/1937–3/1945	9,057	14,336	4891	18,045	264	5739	1928	61,108
				25,976 tons of mustard gas				

Key: *Eth. ox.* ethylene oxide; *P M* Polyglycol M; *PMI* Polyglycol M I; *OL* Oxol mustard gas; *OB* Winter mustard gas based on S mustard gas/O mustard gas; *OKM* mustard gas and Dichlordipropylsulfid (mixed/winter mustard gas); *T 9* nitrogen mustard gas; *KC 250 Gb* chemical bombs filled with 90.2 kg of mustard gas, of which 47% were sent to L.H.Ma. Mockrehna, 26% to L.Ma. Krappitz, 22% to L.Ma. Domnau, 5% to L.Ma. Regny, and 7 bombs to Munster. Unfilled chemical warfare agents (O mustard gas, N mustard gas) were first delivered to Munster and to H.Ma. Löcknitz. The quantities indicated in Table 5 correspond to deliveries to the OKH. For all products, the exception of KC 250 Gb, there was additonal production and deliveries to private parties, such as IG Farbenindustrie AG, Frankfurt am Main, and the Lonal-Werk GmbH Berlin. Of the 28,800 metric tons of mustard gas produced in the German Reich by the end of the war, 25,976 metric tons (= 86.2%) were from Ammendorf. In the filling plant, 61,108 KC 250 Gb bombs were filled with 5512 metric tons of mustard gas. Table 5 is based on the sources A1–G (Preuss private collection and BArch R 8135/7003 and 7798)

(*A1*) Boyne, J.G., Lanfear, W.E., Calcott, W.S. and P.J. Leaper (1945): Production of Vesicant Agents at Ammendorf. CIOS,Item No. 8, File No. XXXII-7, p. 8. The British Library, Boston Spa, Wetherby, West Yorkshire

(*A*) Report der Orgacid GmbH Ammendorf/Saale District, on the fiscal year 1943–44 and on the current fiscal year (additions through December 1944), 4th version, p. 3, January 1945

(*B*) Report (no. VI/12270) by the Deutsche Revisions- und Treuhand-AG, Berlin on the audit of the annual statement of Orgacid GmbH Ammendorf, near Halle, on March 31, 1944, copy no. 6, p.23

(*C*) Report (no. VI/11517) by the Deutsche Revisions- und Treuhand-AG, Berlin on the audit of the annual statement of Orgacid GmbH Ammendorf/Saale District, on March 31, 1943, copy no. 6, p. 22

(*D*) Report (no. VI/10240) by the Deutsche Revisions- und Treuhand-AG, Berlin on the audit of the annual statement of Orgacid GmbH Berlin, on March 31, 1942, copy no. 6, p.21

(*E*) Report (no. VI/7840) by the Deutsche Revisions- und Treuhand-AG, Berlin on the audit of the annual statement of Orgacid GmbH Berlin, on March 31, 1941, copy no. 11, appendix, p. 28

(*F*) Report (no. 12795) by the Deutsche Revisions- und Treuhand-AG, Berlin on the audit of the annual statement of Orgacid GmbH Berlin, on March 31, 1939, copy no. 6, p. 6 and appendix, p. 21

(*G*) Report (unnumbered) by the Deutsche Revisions- und Treuhand-AG, Berlin on the audit of the annual statement of Orgacid GmbH Berlin, on March 31, 1938, copy no. 7, appendix, p.21

In the fiscal year 1940–41, production of chemical warfare agents continued. Production of Oxol mustard gas and winter mustard gas could not be continued at full capacity because of a shortage of storage space, so it was halted in the months of August and November.[15] Production of T 9 (nitrogen mustard gas) was slowed by inadequate supply of intermediate products. The expansion of the factory was continued.

During the fiscal year 1941–42, Orgacid was split into two divisions. From June 1941, there was Orgacid GmbH, building department, Berlin, and Orgacid GmbH, operations, Ammendorf. The building department was transferred to Lonal-Werke GmbH, Berlin, in October 1941.[16]

Production of mustard gas (OL and OKM) in B-Plant averaged 476 metric tons per month. A shortage of workers caused considerable difficulties. B-Plant had to be closed for three weeks beginning in mid-December 1941 because of health problems among the workers. During the fiscal year 1942–43, A-Plant achieved its highest ever production numbers for ethylene oxide. But production of polygycol M or polygycol M I continued for only five months, so that only 1,608 metric tons could be produced. This led to a considerable decline in production in B-Plant. For that reason, almost no Oxol mustard gas at all was produced in the fiscal year 1942–43. There were 399 metric tons of OKM produced. In addition, preparations for the production of OB were affected, and 230 metric tons were produced as part of an experimental production.

In T 9-Plant, its production capacity of 50 metric tons monthly could not be exploited fully because of reduced supplies of intermediate products.

The filling plant was closed for the entire year, apart from April 1942, because of a lack of orders. In April, 382 bombs were filled.

The expansion of the factory continued in 1943–44. In order to increase production of OB, B-Plant was further expanded. The storeroom of T-9-Plant was enlarged by installing a fourth vat of 120 metric tons.[17]

On April 18 and 19, 1945, Ammendorf was occupied by American troops and guarded by about thirty American soldiers until it could be cleared out later. At this time, stores consisted of 600 metric tons of arsenic powder from Leese and 625 metric tons of mustard gas. When American troops arrived, numerous files and documents were confiscated.[18] That may have been connected with the visit by a CIOS team, which presented an extensive report including plans and process diagrams.[19]

---

[15]Orgacid GmbH, Vorläufiger Geschäftsbericht über das Geschäftsjahr 1940–41, from April 1, 1941 to December 31, 1940, p. 1, Preuss private collection.

[16]Orgacid GmbH, Vorläufiger Geschäftsbericht über das Geschäftsjahr 1941–42, from April 1, 1941 to December 31 1941, p. 1, Preuss private collection.

[17]Orgacid GmbH, Bericht über das Geschäftsjahr 1943–44, p. 3, Preuss private collection.

[18]Sekretariat K, Essen, Bericht: Betrifft: Orgacid GmbH Essen, May 28, 1945, Preuss private collection.

[19]Boyne et al. 1945. Production of Vesicant Agents at Ammendorf. M52.D92., CIOS Target No. 8/30 Chemical Warfare, Item No. 8, File No XXXII-7, Combined Intelligence Objectives

It is not known what happened on the grounds during the course of dismantling by the Soviet Military Administration in Germany (Sovyetskaya Voyennaya Administratsya v Germanii, SVAG). The first studies and simple cleanups were carried out in the 1950s.[20] A study by Chemiewerk Kapen, near Dessau, in 1953 produced the following results: Some of several bunkers and underground cisterns were still well preserved, while others had been blown up. A lead-lined iron container with a capacity of 2–3 cubic meters in one part of the facility was found to be contaminated with chemical warfare agents. The former grounds of the filling plant smelled of chemical agents; empty bombs and artillery shells were lying around. The remains of the structure of the filling plant were recognizable. The storage bunker for mustard gas had eight cells. They contained

> ca. 1,400 cubic meters of water with small amounts of mustard gas, including ca. 150 cubic meters of concentrated nitrogen mustard, which forms a layer of insulation from the water above by means of hydrolysis.[21]

Later "ca. 110 metric tons of sulfur mustard were found in the reinforced-steel bunkers," of which 52.5 cubic meters mustard gas were destroyed in Kapen, and 855 cubic meters of neutralized liquids were directed into the Elster. Nearly two years later, tests of the mustard sludge in the cells of the mustard gas storage bunker still had high levels of mustard gas. Substances active in mustard gas represented as much as 50%; thiodigycol, sulfone, and sulfoxide were also found. The water above the sludge had a pH of 1.0.

The final report on measures and determinations from 1956 reads as follows:

> Work was begun on April 16, 1956, and conducted according to the instructions of May 11, 1956. All of the containers and trenches were examined for chemical warfare agents, then decontaminated and filled with gravel. The mustard sludge located in the bunker cells was destroyed by us at the site. […] After emptying the cells, the floors and walls were decontaminated and cleaned. The cell openings were then walled up. It was therefore unnecessary to fill the cells, since the cells are empty and decontaminated. Around 4,000 cubic meters of gravel were moved to fill communication trenches, funnels and pits. Work was completed on December 21, 1956.[22]

It also reported that 75 cubic meters of mustard sludge were neutralized and destroyed between April 1956 and April 1957; 15 cubic meters of mustard sludge were said to have been found in each of the seven bunker cells. Around 50% of the

---

(Footnote 19 continued)

Sub-Committee G-2 Division, SHAEF (Rear) APO 413, National Archives, Washington, RG 338 and British Library, Document Supply Center, Boston Spa.

[20]The discussions referred to in what follows are based on Arbeitsberichte Chemie Werk Kapen, p. 1247 (April 17, 1957), p. 1249 (February 11, 1957), p. 1722 (February 2, 1957) sowie Zusammenfassungen von Berichten zwischen dem January 29, 1953, to July 23, 1956, Preuss private collection.

[21]Ibid.

[22]Arbeitsberichte Chemie Werk Kapen, p. 1247 (April 17, 1957), p. 1249 (February 11, 1957), p. 1722 (February 2, 1957), sowie Zusammenfassungen von Berichten zwischen dem January 29, 1953, to July 23, 1956, Preuss private collection.

factory grounds (southern part) were decontaminated; that area was said to be usable under certain conditions. The plan was to close the northern part for fifty years. The situation did not change in the 1960s. According to notes on proposals concerning the problems of the Orgacid buildings made to an advisory committee of the District of Halle on June 6, 1978,[23] the division of the grounds was still recommended, with the northern part to be closed for fifty years (with entry prohibited), but with the involvement of the chemical weapons expert Professor Karlheinz Lohs it was to be further divided into the "bunker area" and the "remainder of the northern part." Permission was granted to add soil to a height of at least three meters, on which it was said to be possible to build without a foundation. The conditions in the southern part were unchanged. The area around the factory up to a distance of 50 meters were included in these measures.[23] In February 1990, members of the officers' college of the ground forces of the Nationale Volksarmee (NVA; National People's Army) in Löbau-Zittau visited the city of Halle.[24] The subject of the visit was "guaranteeing order and safety on the former site of Orgacid, Halle-Ammendorf." One of them was the chemical weapons expert Colonel Professor Siegfried Franke, who had already been involved in the decontamination efforts of the Chemie-Werk Kapen. He explained that chemical warfare agents had still been present only in the mustard gas bunker, but they had been completely removed. At the time, sulfur mustard had been found but not nitrogen mustard or arsenic. Based on this assessment of the situation, it seemed necessary to the participants of the event to "rethink earlier arrangements."[25]

Additional site studies have been conducted since 1990 (Fig. 2, Table 6).

3 The History of Falkenhagen Factory of Monturon GmbH, Development, the Structure of the Buildings, and Production

The following text is based on an assessment of the Falkenhagen factory of Monturon GmbH as part of an investigation of this former munition site (Preuss and Eitelberg 1994).

The area of around nine square kilometers where the former factory grounds of the "Seewerk" Falkenhagen (Falkenhagen "Lake Factory") of Monturon GmbH are located is around 40 km east of Berlin and 10 km west of Frankfurt an der Oder, in the forested area of the Falkenhagener Heide (Falkenhagen Heath) between Falkenhagen to the west, Döbberin to the northeast, and Petershagen to the south.

[23]Niederschrift zur Beratung über die Problematik Orgacid-Gebäude, June 6, 1978, District Council of Halle, Preuss private collection.

[24]Aktennotiz zum Besuch der Offiziershochschule Löbau-Zittau der Landstreitkräfte der NVA, February 21, 1990, Preuss private collection.

[25]Ibid.

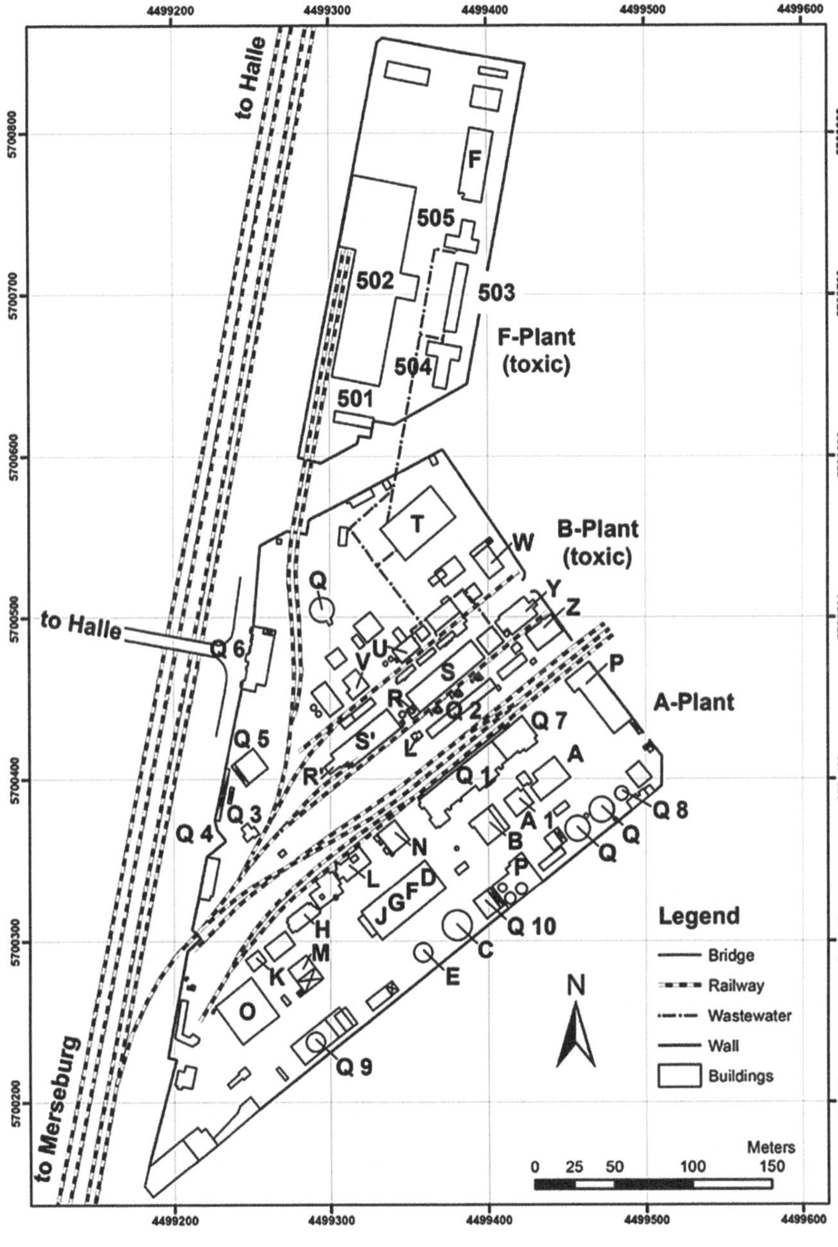

Fig. 2 Layout of Orgacid GmbH Ammendorf (Werkspläne von 1922–1928; Old building fabric, factory plan, A-Plant, 1935, Preuss private collection; Factory plan of 1940, scale 1:500: F-Plant (filling plant); Factory plan of May 1940, revised by VEB Chem. Werke Buna, Schkoppau, 1977; Factory plan of autumn 1942, scale 1:500: Reichsbahnanschluss Orgacid GmbH Ammendorf, Preuss private collection. Boyne, J.G., E.W. Lanfear, W.S. Calcott and P.J. Leaper. 1945. Production of Vesicant Agents at Ammendorf. M52.D92., CIOS Target No. 8/30 Chemical Warfare, Item No.8, File No. XXXII-7, Combined Intelligence Objektives Sub-Committee G-2 Division, SHAEF (Rear) APO 413, National Archives, Washington DC., RG 338 and British Library, Document Supply Center, Boston Spa, Weatherby, West Yorkshire)

Table 6 Inventory of buildings: Orgacid GmbH Ammendorf

Plant	Building No.	Function
A-plant	A/II	Alcohol rectification and storage
	A1/II	Alcohol storage
	B/I	Ethylene furnaces
	C/XIV	Ethylene storage
	D/XIX	Ethylene purification
	E/XV	Ethylene storage (pure), 1000 m^3
	F/XIX	Chlorhydrine manufacture
	G/XIX	Ethylene oxide manufacture
	H/XVIII	Ethylene oxide storage
	J/XIX	Thiodiglycol manufacture
	K/-	Sulfur storage
	L/XVII	Chlorine storage
	M/-	Hydrogen sulfide manufacture
	N/XVI	Lime slaking and solution
	O/X	Neutralization, waste treatment
	P/II, VII, VIII	Thiodiglycol storage
	P/-	Thiodiglycol storage
	Q/XI	Hydrogen storage
	Q/XII	Storage, 3000 m^3
	Q1/-	Washrooms and changing rooms, decontamination
	Q2/-	First aid
	Q3/-	Guard room
	Q4/-	Main gate
	Q5/-	Stores
	Q6/-	Administration offices
	Q7/IX	Workshop, warehouse
	Q8/XIII	Storage, 200 m^3
	Q9/XXI	Cooling tower
	Q10/XXIII	Warehouse
B-plant	R/-	Hydrogen chloride burners
	R′/-	Hydrogen chloride burners
	S/-	Mustard gas reactor
	S′/-	Mustard gas reactor
	T/-	Mustard gas storage
	U	Decontamination of toxic wastes
	V	Nontoxic wastes
	W/-	Triethanolamin-thionylchloride reactors(T-9)
	X/-	Soda storage and solution
	Y/-	T-9 neutralization, etc.
	Z/-	T-9 storage

(continued)

Table 6 (continued)

Plant	Building No.	Function
F-plant	501	Administration office
	502	Large aboveground workshop with charging machines
	503	Common rooms with kitchen
	504	Washrooms and changing rooms, nontoxic
	505	Washrooms and changing rooms, toxic
	F	Static water tank

Werkspläne von 1922-1928, Gebäudealtbestand, Werksplan A-Anlage, 1935; Werksplan von 1940, scale 1:500: F-Plant (filling plant); Werksplan vom May 1940, revised by VEB Chem. Werke Buna, Schkoppau, 1977; Werksplan vom Herbst 1942, scale 1:500: Reichsbahnanschluss Orgacid GmbH Ammendorf, Preuss private collection; Boyne, J.G., E.W. Lanfear, W.S. Calcott and P.J. Leaper. 1945. Production of Vesicant Agents at Ammendorf. M52.D92., CIOS Target No. 8/30 Chemical Warfare, Item No. 8, File No. XXXII-7, Combined Intelligence Objektives Sub-Committee G-2 Division, SHAEF (Rear) APO 413, National Archives, Washington, RG 338 and British Library, Document Supply Center, Boston Spa, Wetherby, West Yorkshire

In the Falkenhagener Heide in November 1938, groundwork was begun for the building of an industrial site to produce the incendiary material chlorine trifluoride (ClF_3) (code name N-Stoff [Substance N]). According to a decision in May 1942, another large facility for sarin production was to be built in the same location.[26] This construction project was set to begin in September 1943. In contrast to the cramped site of the Orgacid factory in Ammendorf, the grounds on the Falkenhagener Heide were oversized. The history of the factory's development and its background were presented in detail by Schmaltz (2005) under the motto "Wunderwaffe N-Stoff" (N Substance: The Miracle Weapon). It was crucial that Hitler was fascinated by this particular incendiary material and had great hopes for it. On the other hand, there must have been a group of interested parties behind the project that made it possible for more than 60 million reichsmarks to have been spent already by the end of March 1944 without any N-Stoff being produced. At the same time, the competition between the research department of the Heereswaffenamt and the laboratories of IG Farbenindustrie AG also played a role, as did competition with the increasingly powerful SS later.

The Forschungsabteilung des Waffenamtes (WaF; Research Department of the Weapons Agency) was engaged in November 1938 to immediately begin groundwork for a new facility in Falkenhagen (Muna-Ost [Muna East] or M.O.) (Schmaltz 2005, 152). Owing to the project's urgency, the acquisition of the site and the construction were to be handled by Montan or carried out under its control.[27]

[26]Klenck, Jürgen E. von "History of the 'Seewerk' (Falkenhagen)" 26. March 1945, p. 1/69. The National Archives, London, WO 208/2186.

[27]Schreiben der Waffenamt-Forschungsabteilung (WaF) an den Chef Wa J Rü betreffend Muna-Ost, December 1938, Preuss private collection.

From mid-May 1943, there were plans to produce the new chemical weapon sarin in a large facility (500 metric tons monthly) in Falkenhagen (Schmaltz 2005, 159). That decision was preceded by an order to transfer Falkenhagen to IG Farben by February 12, 1943. The head of the agency, General Leeb, wanted Professor Thiessen of the Kaiser Wilhelm Institute to participate in a preliminary discussion on this.[28] The construction and management of this large facility exceeded the capabilities of the Waffenamt's research department, so the Falkenhagen construction site was transferred to IG Farbenindustrie AG or more precisely to its Turon GmbH subsidiary on September 1, 1943, after the transfer of the Falkenhagen construction site had been discussed at the site itself on July 23, 1943.[29] Turon was still in the process of being founded at this time. The legal form was the lease agreement of September 1, 1943, which concerned the factory and all its associated facilities. Only the forest and the management thereof remained in the hands of Montan.[30] On September 3, 1943, Turon GmbH was founded, with headquarters in Ludwigshafen. The share capital of the company was split equally between Montan Industriewerke GmbH and IG Farbenindustrie AG. Turon was later renamed Monturon to avoid a confusion of names.[31] The chief executive officer of Monturon was Director Dr. Otto Ambros; his representative was Dr. Jürgen E. von Klenck; von Klenck, chief engineer Bilfinger, and Röhr (MBA) were named as internal auditors.[32]

The significance of the new product sarin is clear from a report on the production of chemical warfare agents of February 1, 1944.[33] In addition to the experimental facility in Heidkrug, near Munster (capacity 40 metric tons monthly), two expansion stages were constructed at this time following different procedures. These were the experimental facility Sarin I in Dyhernfurth with 100 metric tons monthly and

[28]Chef des Stabes (Scholz) an WaF, Betr.: Sondervorhaben Falkenhagen, February 2, 1943, Preuss private collection.

[29]Aktenotiz des Wa J Rü (Mun) - Az. 70 o 40-19 Wa J Rü/Mun 3 zbV/IX; Nr. 1320/43 g.Kdos. betreffend die Ausweichstelle des Reichsamtes für Wirtschaftsausbau in Falkenhagen, July 26, 1943, Preuss private collection.

[30]Schreiben von Oberst a.D. Hammer, Falkenhagen an die Verwertungsgesellschaft für Montanindustrie GmbH, München, September 18, 1943, Preuss private collection.

[31]Monturon GmbH. Abschlussbericht, 1943–44, Fabrikhauptbuch Seewerk, Preuss private collection, and Schmaltz 2005, 160 n. 520.

[32]Dr. von Klenck was born in Bromberg on June 8, 1909. He attended school in Frankfurt am Main, Frankfurt an der Oder, Berlin, and Cologne. After graduating secondary school, he studied chemistry in Cologne and Göttingen. In December 1933 he passed his doctoral exams under Professor Thiessen. His first position was at IG Farbenindustrie A.G. in Höchst. From 1935 to 1940 he was a member of the Allgemeine SS in Frankfurt. He claimed to have been a lieutenant with the 29th flak regiment (Frankfurt) from the time the war broke out. In February 1942 he was conscripted to work at IG Farbenindustrie in Ludwigshafen and was appointed vice deputy director of Monturon in Falkenhagen (Seewerk). He experienced the end of the war in the Anorgana factory in Gendorf, where he was arrested on May 20, 1945, because of his previous membership in the SS and transferred to Dustbin, Enemy Personnel Exploitation Section, FIAT, Control Commission for Germany (BE), B.A.O.R. In FIAT EP 254–84 (von Klenck), December 14, 1945, Interrogation Report, p. 5. The National Archives, London, FO 1031/97.

[33]"Deutsche Kampfstoff-Produktion, Ludwigshafen am Rhein," February 1, 1944, BArch R3/1894, p. 6.

the large facility Sarin II in Falkenhagen with 500 metric tons monthly. The large facility in Falkenhagen had a planned budget of ca. 44 million reichsmarks,[34] of which 380,000 reichsmarks had been spent by January 1944. The planned start date was "mid-1945." These anticipated costs for the sarin plant contrast starkly with the cost value of the Falkenhagen N-Stoff factory reported on March 31,1944.[35] At that time it was more than 61 million reichsmarks. This recalls the mismanagement in the production of infantry munitions, whose production in 1939–40 was reduced in the middle of the war at one third (Preuss and Eitelberg 2010, 185–89).

The involvement of IG Farbenindustrie AG is also evident in the draft agreement of June 13, 1944, between the German Reich (office of the army), the Verwertungs gesellschaft für Montanindustrie GmbH, IG Farbenindustrie AG, and Monturon GmbH concerning the expansion of the N-Stoff plant of the OKH by adding a fabrication facility to produce 500 metric tons monthly of sarin.[36] According to that contract, IG Farben was also supposed to determine how the N-Stoff plant with provisional capacity of 10 to 15 metric tons monthly could be put into operation.

Precisely a year after the deadline for preparations to transfer the Falkenhagen factory to Monturon, matters took an astonishing turn. On July 23, 1944, the building inspector Glupe (Wa F), who was responsible for the opening of the N-Stoff plant, arrived in the company of SS officers and handed Monturon's management a letter from the brigade leader Dr. Otto Schwab (head of the technical office in the main office of the SS leadership) (SS-Führungs-Hauptamt) that read as follows:

> With reference to the order, known to you, from the Führer and head of OKH 1731/44 of July 7, 1944, you are hereby informed that the Führer also ordered that the N-Stoff factory be taken over immediately by the Waffen SS. The agreements between the Heeres-Waffenamt and IG Farben are to be annulled effective immediately and measures already undertaken reversed. […] Its seamless transfer […] is the responsibility of the current head of the factory, Dr. Glupe.[37]

The expansion was to be continued; the production of N-Stoff was to begin in August 1944.[38] The new owners, the Waffen-SS, wanted the N-Stoff facility to be under the direction of Dr. Siegfried Glupe, a former employee of the

---

[34]Entwurf eines Bau- und Errichtungsauftrages für die Errichtung der Sarin II Anlage in Seewerk (Auftrags-Nr. 3/IX-4888-9026/43). The National Archives, London, FO 1031/179.

[35]MONTURON GmbH, Aufstellung des Buchwertes zum 31.03.1944 (= Anschaffungswert 1944), Werk Falkenhagen, Preuss private collection.

[36]Draft agreement: Vertrag zwischen dem Deutschen Reich (Wehrmachtsfiskus), vertreten durch das Oberkommando des Heeres (OKH), der Verwertungsgesellschaft für Montanindustrie GmbH, der IG Farbenindustrie AG und der Monturon GmbH über die Erweiterung der OKH-eigenen N-Stoff-Anlage durch eine Fabrikationsstätte zur Herstellung von 500 moto Sarin, June 3, 1944, Preuss private collection.

[37]Monturon to the Verwertungsgesellschaft für Montanindustrie, August 2, 1944, concerning the demand by the SS that the factory in Falkenhagen be transferred, July 23, 1944, Preuss private collection. The letter is from Director Ambros and signed by his representative, v. Klenck. On this, see also Schmaltz (2005, 171).

[38]Aktenvermerk der Montan betreffend Seewerk, July 25, 1944, Preuss private collection.

Kaiser-Wilhelm-Institut für physikalische Chemie und Elektrochemie, whose doctoral advisor had been Prof. Thiessen (Schmaltz 2005, 146). At this time Glupe was section head of the research department of the Heeres-Waffenamt (Wa F, Gruppe IIc) His group leader was Prof. Eschenbach.[39] Glupe guaranteed that ten metric tons would be produced in September. Experimental production seems to have begun in October 1944; a production report from December 1944 mentions a quantity of 5 metric tons produced[40]; according to von Klenck, a total of 22 metric tons were produced.[41]

The end of the Falkenhagen factory is documented by a letter from Monturon GmbH, informing Montan in February 1945 that the N-Plant of the factory had been closed when the Oderbruch became part of the battle zone.[42] It can be assumed that the factory was occupied by Soviet troops by April 19, 1945, at the latest (Griess 1985, 82).

At the instructions of OKH and in agreement with the technical office in the headquarters of the SS, already on February 10, 1945, sixty freight cars with special equipment and machines and five empty tank cars departed for Stulln, Bavaria. After the war Glupe reported that in February 1945 a total of five trains were prepared with goods to be relocated to Stulln, but only four arrived there, and one remained behind in Prague. Other material, he claimed, had been transported to Leese including things from the Anorgana factory in Dyhernfurth.[43] According to a letter from the Luranil-Baugesellschaft mbH, material was also transported from the Falkenhagen construction to the Gendorf factory of GmbH.[44] The company assets belonging to the "Seewerk" Falkenhagen were first seized and later expropriated by the Soviet Military Administration in Germany.[45]

4 Production at the Falkenhagen Factory

Originally, the factory in Falkenhagen was only supposed to produce the incendiary material chlorine trifluoride (N-Stoff). The plan was to have an experimental facility with a monthly capacity of 10 metric tons, which would be increased to 50 metric tons.

---

[39]Abschrift/Bu. des Fernsprechverzeichnisses des H Wa, Forschungsabteilung (mit Hochschulzentralstelle), z. Zt. Kummersdorf-Schiessplatz, Preuss private collection.

[40]Fertigungsbericht C-Stoff, Fertigung in December 1944, p. 87, BArch R 3/1894.

[41]Klenck, Jürgen E. von, "History of the 'Seewerk' (Falkenhagen)," March 26, 1945, p. 3/71, The National Archives, London, WO 208/2186. The author, von Klenck, was the vice deputy director of Monturon GmbH.

[42]Rundschreiben der Monturon GmbH, Nr. 914/45/IVa1/Pr. an die Montan-Industriewerke GmbH betreffend Betriebsstillegungen und Verlagerungen, February 1945, Preuss private collection.

[43]Aktenvermerk (Nr. 8) der IVG betreffend nach Stulln verlagertes Gut aus Falkenhagen, May 23, 1952, Preuss private collection.

[44]Schreiben der Luranil-Baugesellschaft mbH (in Auflösung) an das IG Farben Control Office (Liquidation section), Frankfurt am Main betreffend Silbermaterial, Frankfurt am Main, January 18, 1951, Preuss private collection.

[45]Verfügung der Landesregierung Brandenburg, Enteignungsurkunde für das beschlagnahmte Betriebsvermögen der Firma M.O. Falkenberg (!), Falkenberg, Kreis Lebus, July 7, 1848, Preuss private collection.

The deadline to begin operations was October 1944 and the total quantity produced 22 metric tons was stated, probably accurately, by the vice deputy director of Monturon, von Klenck.[46]

In May 1943, it was also decided to move the planned large production 500 metric tons monthly of the nerve gas sarin to Falkenhagen as well. Construction for that facility began in September 1943[47]; completion was planned for April or Mai 1945.[48] In early February 1945, the Falkenhagen factory was cleared out, because the front was moving closer, so the sarin facility was neither completed nor put into operation.

4.1 *Brief Description of the Facilities at the Factory*

The area surveyed for the "Seewerk" Falkenhagen was about nine square kilometers in size, of which just 8.24 km^2 had been transferred to Montan in the land register on April 31, 1944.[49] Originally, the area of the "Gut Falkenhagen" property had been 7.35 km^2.[50] That area included a site intended for a research institute on the former grounds of the Falkenhagen Castle. Montan had acquired, or intended to acquire, another 23.51 km^2 in the Falkenhagen area, so that in the end it would have more than 31.75 km^2.

In July 1943, the "Gesamtplanung Seewerk" (Overall Plan for the Lake Factory) included 5 groups of buildings (A–E), of which only Building Group D, the N-Stoff plant, was built at that time. These groups were to serve the following purposes[51]:

- (A) General operations: Briesen train station, energy distribution, storage for raw materials, administration, main workshops (including transportation workshops), housing for factory security guards, cafeteria for the entire staff, and central kitchen to distribute food to the satellite kitchens.
- (B) Five scientific institutes with semitechnical testing facilities (by the lake).
- (C) Experimental field with "Sprenggarten" (detonation area) and shooting range.
- (D) N-Stoff plant.
- (E) Unknown, presumably later (S) for sarin.

[46]Klenck, Jürgen E. von, "History of the 'Seewerk' (Falkenhagen)," March 26, 1945, p. 3/71, The National Archives, London, WO 208/2186.

[47]Ibid., p. 1/69.

[48]"Die Deutsche Kampfstoff-Produktion," Ludwigshafen am Rhein, February 1, 1944, p.7, BArch R3/1894.

[49]Entwurf eines Schreibens der Montan über Grunderwerb Falkenhagen, June 23. 1944, Preuss private collection.

[50]Anhang zum Bericht der Deutschen Revisions- und Treuhand-Aktiengesellschaft Berlin über die bei der Verwertungsgesellschaft für Montanindustrie GmbH, Berlin-Charlottenburg, vorgenommene Sonderprüfung betr. Forstabteilung, p. 17, BArch Berlin, R 8135/4782.

[51]Baustelle Seewerk: Sachdarstellung über die Gesamtplanung, das Abrechnungswesen, das Sozialwesen, den Werkschutz, Werkfeuerwehr und Werkluftschutz, July 15, 1943, The National Archives, London, FO 1031/179.

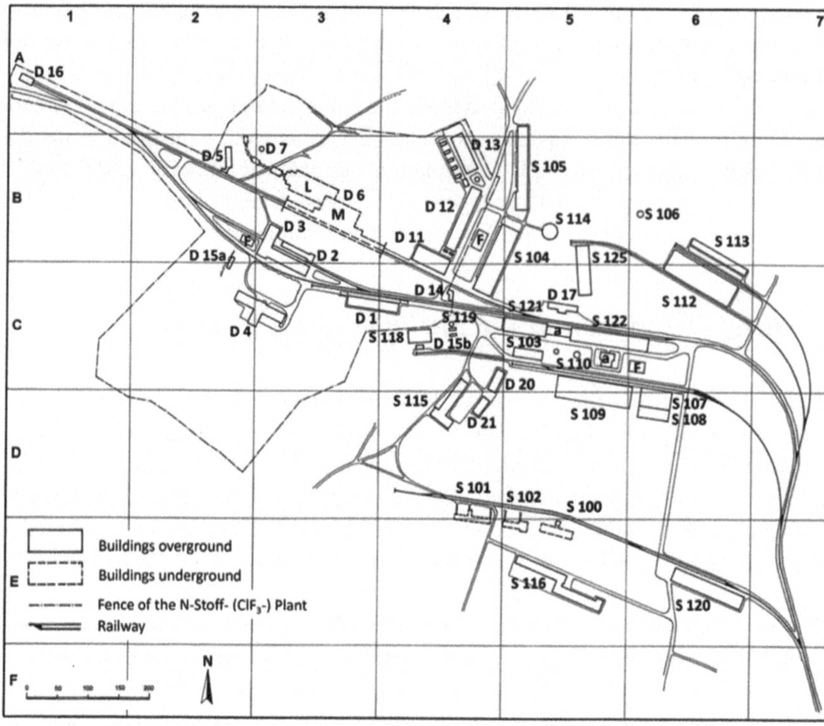

Fig. 3 Layout of Monturon GmbH Falkenhagen ("Seewerk") (Gesamtlageplan - Seewerk S/o (M 1 : 2.500), Dipl. Ing. Max Haaf, Stuttgart, December 21, 1943, Preuss private collection)

In an overall site plan for the Seewerk (Falkenhagen) project of 1943,[52] the buildings of the factory are marked only with the letters D for the N-Stoff plant and S for the sarin plant.

The five scientific institutes (Building Group B) were located on the eastern shore of the Schwarzer See (Black Lake). There were laboratories and a vacuum tunnel to test the ballistic properties of a new type of weapon; these were also referred to as "Institut Ost" (East Institute).[53]

The function of the various buildings could be determined from the lists of buildings on the site plan. The lists of buildings were supplemented by evaluating other written sources (see Fig. 3, Table 7).[54]

[52] Gesamtlageplan–Seewerk S/o (M 1 : 2.500), Dipl. Ing. Max Haaf, Stuttgart, December 21, 1943, Preuss private collection.

[53] Schreiben der Waffenamt-Forschungsabteilung (WaF), Nr. 595/39 gKdos Wa F/A2, an Wa J Rü 10 und Wa J Rü (Mun3) betreffend MO, Büro Prof. Loos. January 30, 1940, Preuss private collection.

[54] Bauvorhaben Seewerk, Zusammenstellung der bis zum 30.06.1943 angefallenen Kosten, The National Archives, London, FO 1031/179.

Table 7 Inventory of buildings: Monturon GmbH Falkenhagen ("Seewerk")

Building-No.	Function	Plant	Coordinates
D 1	Preparation	N-Stoff	C 3/4
D 2	Water purification plant	N-Stoff	B 3
D 3	Workshop, central heating	N-Stoff	B/C 3
D 4	Common room, administration	N-Stoff	C 2/3
D 5	Neutralization	N-Stoff	B 2
D 6	Manufacturing plant L (Tanks)	N-Stoff	B 2/3
D 6	Manufacturing plant M	N-Stoff	B 2/3
D 7	High-level water tank	N-Stoff	B 3
D 8	Workshop	N-Stoff	
D 8a	Workshop	N-Stoff	
D 9	Storehouse	N-Stoff	
D 9a	Storehouse	N-Stoff	
D 10	Defroster	N-Stoff	
D 11	Diesel powerhouse	N-Stoff	B/C 4
D 12	Generator building	N-Stoff	B 4
D 13	Apparatus building	N-Stoff	B 4
D 14	Keeper's lodge	N-Stoff	C 4
D 15a	Purification plant I	N-Stoff	B/C 2
D 15b	Purification plant II	N-Stoff	C 4
D 16	Locomotive shed	N-Stoff	A 1
D 17	Garage	N-Stoff	C 5
D 18	Pump room (lake)		
D 18a	Pump room		
D 20	Laboratory	N-Stoff	C 4
D 21	Laboratory school	N-Stoff	D 4
F	Static water tank		
S 100	Storehouse (underground)	Sarin	E 5
S 101	Storehouse with tanks	Sarin	D/E 4
S 102	Storehouse with tanks	Sarin	E 5
S 103	Storehouse with tanks	Sarin	C 5
S 104	Manufacturing plant	Sarin	B/C 4/5
S 105	Product purification	Sarin	A/B 5
S 106	Chimney	Sarin	B 6
S 107	Boiler house	Sarin	C/D 6
S 108	Turbine house	Sarin	D 6
S 109	Coal yard	Sarin	C/D 5
S 110	Chloromethane gasometer	Sarin	C 5
S 111	Storehouse	Sarin	
S 112	Filling plant	Sarin	B/C 6
S 113	Product purification	Sarin	B/C 6

(continued)

Table 7 (continued)

Building-No.	Function	Plant	Coordinates
S 114	Cooling tower	Sarin	B 5
S 115	Special bath	Sarin	C/D 4
S 116	Casualty ward	Sarin	E 5
S 117	Chimney for acids	Sarin	C 5
S 118	Sewage purification plant	Sarin	C 4
S 119	Sewage purification plant	Sarin	C 4
S 120	Warehouse	Sarin	E 6
S 121	Manufacturing plant	Sarin	C 5
S 121a	Manufacturing plant	Sarin	C 5
S 122	Manufacturing plant	Sarin	C 5/6
S 122a	Manufacturing plant	Sarin	C 5/6
S 123	Mesh net station	Sarin	
S 125	HCl basin	Sarin	B/C 5
S 126	Gas decomposition	Sarin	
S 127	Subway	Sarin	C 4

5 The Development of N-Stoff Production

In 1937 representatives of the Waffenamt visited the laboratories of IG Farbenindustrie AG in Leverkusen. The skin-damaging properties of chlorine trifluoride, and its reactivity when organic and inorganic materials combine with high heat or flame, attracted their interest. Between 1937 and 1944, between three and five metric tons of chlorine trifluoride, which had been produced in a small laboratory-sized plan in Leverkusen, were delivered to the OKH. The Waffenamt saw possible military applications as an incendiary material and later as filling for artillery shells, as propellant for underwater shells and ramjet torpedoes, and as a rocket fuel additive (Schmaltz 2005, 148). Interesting results from experiments with chlorine trifluoride led the research department of the Waffenamt to grant permission as early as 1938 to build a factory for its mass production in Falkenhagen. In addition to the large plant in Falkenhagen, a smaller experimental plant was built under the direction of the OKH on the Kummersdorf testing grounds in Gottow. It was said to have been in operation in 1943.[55]

[55]Enemy Personnel Exploitation Section FIAT, Control Commission for Germany (BE), B.A.O.R. In FIAT EP 254-84 (von Klenck) 14. December 1945, Interrogation Report, pp. 16–19. The National Archives, London, FO 1031/97. Anonymus: Chlortrifluorid (ClF3), Anlage 9(5/-), p. 1p1–19, after 1945, Preuss private collection.

5.1 *The Buildings of the N-Stoff Plant at the "Seewerk" Falkenhagen*

The buildings of the N-Stoff plant are identifiable in Fig. 3 by the initial letter D. They are located in the northwest of the factory grounds inside a fence around the manufacturing plant (D 6). The fenced area is entered via the gate (D 14). Purification plant II (D 15b), a garage (D 17), a laboratory building (D 20), and the laboratory school (D21) were located outside the fence. Located within the fence were, from northeast to southwest, the apparatus building (D 13), the generator building (D 12), the electrical center (D 11), and preparation (D 1). To the west of this last, followed the water purification plant (D 2), the workshop with the central heating, purification plant I (D 15a), and neutralization (D 5). The common building held, among other things, the cafeteria and the rooms for the administration (D 4). Buildings D 8, D 9, D 10, D 18, and D 19 were presumably located outside the area depicted on the factory plan.

6 The Development of the Nerve Gases Tabun and Sarin

In the pest control laboratory of the Elberfeld factory of IG Farbenindustrie AG in 1934 Gerhard Schrader began working on the development of a means to combat aphids. The Heereswaffenamt, to whom this invention was reported in 1935, showed interest. Laboratory experiments on its military usefulness followed in 1936. From 1939, the Wehrmacht conducted experiments with the new chemical weapon on its experimental grounds in Raubkammer, near Munster, and in the army's gas protection laboratory in Berlin-Spandau. The new chemical weapon was assigned the code names tabun, T-83, and G (for Gelan). A plant was built in Dyhernfurth to mass produce it, and from June 1942 to January 1945 a total of 12,753 metric tons of tabun were produced.[56]

In 1938, again in the Elberfeld factory of IG Farbenindustrie AG, Schrader developed another chemical weapon, which after several tests by the Heereswaffenamt was to begin mass production under the name sarin. The code name for sarin was T 46. However, by 1945 just 0.5 metric tons of this chemical warfare agent could be produced in a pilot plant in Raubkammer.[57]

[56]"Die Deutsche Kampfstoff-Produktion, Ludwigshafen am Rhein," February 1, 1944, p. 4, BArch R3/1894. Sonderausschuss C beim Reichsminister für Rüstung und Kriegsproduktion, Bericht über die Lage auf dem Kampfstoffsektor, Auszug aus dem Vortrag im Führerhauptquartier, May 15, 1943, mit einer Gegenüberstellung der Situation vom March 1, 1944, p. 6/32, BArch, R 3/1894. Aufstellung über K-Stoffe, Gesamtfertigung Stand December 1, 1944, p. 67, BArch R3/1894. Sonderausschuß C, Arbeitsbüro Dr. v. Klenck an Dr. Pfaundler, I.G. Frabenindustrie A.G. Produktionszahlen, December 1944, p. 93, BArch R3/1894.

[57]This is an occasion to point out that Eibl (1999, 157 n. 217) wrote that, according to a statement given on the telephone by J. Preuss, some 50 metric tons of N-Stoff and some 500 kg of sarin had been produced in Falkenhagen. This is probably the result of a misunderstanding on the telephone, since the 500 kg of sarin should be credited to Plant R VIII in Munster. Hahn's unsourced

7 The Manufacturing Processes for Sarin

The manufacturing process intended for Falkenhagen was based on the four-step salt method.[58] Mass production of sarin was supposed to reach 500 metric tons monthly in Falkenhagen. A simplified four-step method was planned. The first step was to follow the procedure of IG; the second step could be avoided after a redesign of the chemical processes. The third step was supposed to be produced according to a process developed by the OKH. The production of the final stage—that is, sarin itself—was based on a continuous process that had been tested by the OKH in Building R VIII in Munster. The technical design of the processes and plants was to be in the hands of IG.

On the map of the buildings of the Falkenhagen factory (Fig. 3), the buildings of the sarin plant are identifiable by the initial letter S. In addition to three manufacturing buildings (S 104, S 121, S 122), there are two buildings identified as purification buildings (S 105, S 113); located between the two were a cooling tower (S 114), a chimney (S 106), and a hydrochloric acid basin (S 125). The buildings numbered S 101, S 102, and S 103 had aboveground tanks; there was also an underground storehouse (S 100). There was additional storage in a warehouse (S 120), a chloromethane gasometer (S 110), and a coal yard (S 109). The last of these belonged functionally to the boiler house (S 107) and turbine house (S 108). A separate power plant was planned for the sarin plant. The sewage purification plants (S 118, S 119) would presumably have caused problems when operating the plant, because of the minimal slope. Because a filling plant (S 112) was planned, the dangerous chemical warfare agent sarin was to be filled in bombs and shells at the production facility itself, just like tabun in Dyhernfurth.

In the aerial photograph, another built area with connection to the railroad is identifiable south of the casualty ward (S 116) as well as one between the tank area and the coal yard (S 109). The broad curve of the railroad line is striking and suggests that additional plants were planned on the Falkenhagener Heide.

8 Vereinigte Flussspatgruben GmbH in Stulln

In November 1939 Montan was contracted by the Waffenamt to acquire fluorite mines in the Bavarian districts of Stulln, Lissenthan, and Brudersdorf (near Nabburg). There was particular interest in those owned by the Vereinigte

(Footnote 57 continued)

assertion that sarin production in Falkenhagen had "partially begun," Hahn (1986, 229) is likewise unprovable.

[58] Aktennotiz der I.G. Farbenindustrie Aktiengesellschaft, Ludwigshafen am Rhein, Zwischenprodukten-Gruppe, betreffend Sarin, July 21, 1943, The National Archives, London, FO 1031/179; Aktennotiz der I.G. Farbenindustrie Aktiengesellschaft, betreffend Sarin, Leverkusen, July 26, 1943, The National Archives, London, FO 1031/179; "Die Lage auf dem K-Stoffgebiet," December 1, 1942, BArch Berlin, R 3112/191.

Flussspatgruben Lissenthan GmbH.[59] Fluorite is the raw material for hydrogen fluoride, which was an intermediate product in N-Stoff production. In the summer of 1942, following a meeting with the head of the central division of the Waffenamt, there was a plan to found a company with headquarters in Falkenhagen for M.O. Falkenhagen that would be merged with Vereinigte Flussspatgruben GmbH, Nabburg, in order to ensure the supply of raw materials for the Falkenhagen factory. The firm Riedel de Haen AG, Berlin was contracted to manage in trust the building of a factory at Flussspatgruben GmbH in Stulln. Hence Riedel de Haen was also asked whether it was prepared to take over as leaseholder the management of the army-owned factory in Falkenhagen.[60]

After the war a team from the British Intelligence Objectives Sub-Committee (BIOS) visited the plant in Stulln. Its report indicates that it was not just a large plant to manufacture hydrofluoric acid and a replacement plant that had been built. The factory in Stulln was so large that it would also have been possible, using the equipment moved from Falkenhagen, to produce chlorine trifluoride (N-Stoff) in a quantity similar to that produced there. Glupe gave the BIOS team a tour of the factory in Stulln; in the text he is referred to as an employee of Riedel de Haen. According to his statements, he had built both the plant in Stulln and the one in Falkenhagen. After he transferred to the Waffen-SS, in October 1944 he had taken on the task of starting N-Stoff production in Falkenhagen for the SS. Glupe had developed the production process for chlorine trifluoride used in Falkenhagen in a laboratory at the Kaiser Wilhelm Institut für physikalische Chemie und Elektrochemie in Berlin. The next step was to build a somewhat larger plant in Gottow. Then Glupe could translate his findings and developments in Falkenhagen and in Stulln on a large technical scale (Schmaltz 2005, 145).[61]

9 The Dyhernfurth Factory of Anorgana GmbH

The site of the factory in Dyhernfurth (now Brzeg Dolny) was between the Oder River and the town of Seifersdorf (now Radecz). The properties were acquired by Montan beginning in 1940. The tabun plant began production in 1942; it was the

---

[59]Schreiben des Oberkommandos des Heeres (Ch H Rüst u BdE), Wa J Rü Stab IV d an das Oberkommando der Wehrmacht (OKW), W Stab W Rü, betreffend M.-O., Falkenhagen, Erwerb von Flussspatgruben, Antrag auf Bestimmung einer Bedarfsstelle, January 1940, Preuss private collection.

[60]Schreiben des Oberkommandos des Heeres (Ch H Rüst u BdE), 70 o 30 18 Wa J Rü (Mun 3 zbV/VIII), Nr. 10006/43 g.Kdos., an I.D. Riedel - E. de Haen AG, Berlin-Britz betreffend Sonderbauvorhaben OKH, February 18, 1943, Preuss private collection.

[61]BIOS. Final report no. 1595, item no. 22, "German Fluorine and Fluoride Industry," London, p. 78, Preuss private collection; W. Archer, W. J. V Ward, and O. S. Whitson, "Hydrofluoric Acid, Vereinigte Flusspatgruben GmbH Stulln," 1946, BIOS target no. C22/2012, C.I.O.S. Black List Item 22, Miscellaneous Chemicals, British Intelligence Objectives Sub-Committee, Preuss private collection.

second-largest producer of a single chemical warfare agent. The operating company of the state-owned Montan plant was Anorgana GmbH, a subsidiary of IG Farbenindustrie AG. The factory's capacity was 1,000 metric tons of tabun monthly in 1944. Altogether, from June 1942 to January 1945, 12,753 metric tons of tabun were produced and filled in bombs and shells.[62] The capacity of the tabun filling plant was 770,000 shells for the light field howitzer (lFH 10.5 cm), or 250,000 shells for the heavy field howitzer (sFH 15 cm), or 12,500 bombs monthly.[63] The map of the factory makes it clear that it was possible to store tabun and the intermediate products for it at the factory. The shells filled with tabun (Green Ring 3) were picked up by the army munitions facilities, which were responsible for chemical weapons, with munitions trains. Consequently, the filled Green Ring 3 bombs were sent to the air munitions institutions for which they were intended. Other chemical warfare agents produced in Dyhernfurth included cyanogen chloride (T 150) beginning in 1943 or 1944, with a capacity of 20 metric tons monthly, and, in addition from 1943 or 1944 hydrogen cyanide (T 155) as well, with a capacity of 20 metric tons monthly, and Bi IV 99 (T 300), an alloy of arsenic, magnesium, and aluminum produced by wetting arsine. The planned capacity of T 300 production was 100 metric tons monthly.[64] An experimental station for filling bombs with chemical warfare agents (e.g., Aeroform) in powder form was built at the end of the war in the HMa St. Georgen (Powder Filling Plant, Building W4 or 1003) (Preuss and Eitelberg 2001,162–65).

10 The Filling and Storage of Chemical Weapons in the Munitions Facilities of the Army and Air Force

With the exceptions of phosgene and tabun, chemical weapons were filled in five army and two air force munitions facilities. The bomb-filling plant in the Ammendorf factory was another exception, but it was active only until the Lufthauptmunitionsanstalt (LHMa; Main Airforce Munitions Facility) in Mockrehna, east of Leipzig, could be put into operation.

---

[62]List of K-Stoffe, total production as of December 1, 1944, p. 67. R3/1894; Report on the production of C-Stoffe, December 1944, p. 86, R3/1894.

[63]Oberkommando des Heeres (Ch.H.Rüst u. BdE.) Firma IG Farbenindustrie A.G. to the attention of the director, Dr. Ambros o.V. Ludwigshafen am Rhein, concerning the construction contract to build a factory to manufacture Product G, contract no. 9/IXa-240-9018/39, p. 1, The National Archives, London, FO 1031/223.

[64]E. Ehmann, U.S. Army Chemical Warfare Project, A.4, "Produktionsstätten und Produktionshöhen, a) Kampfstoffe," 1948, pp. 720–40, Preuss private collection; report (no. VI/ 11302) by the Deutschen Revisions- und Treuhand AG, Berlin, on the audit by Anorgana GmbH, Ludwigshafen a/Rh., Dyhernfurth factory, of the annual report on March 31, 1943, copy no. 1, p. 4, The National Archives, London, FO 1031/165.

There were only seven filling plants for chemical weapons in the former German Reich. The first to be built was the filling plant of the HMa Munster, which can be considered the model for the filling plants of the HMa Löcknitz and the HMa St. Georgen (Traunreut). The third generation of buildings were the filling plants of the HMa Dessau and the HMa Lübbecke, which were considerably smaller. The filling plants planned for the Luftwaffe were the LHMa Mockrehna and the LHMa Oerrel (Dethlingen) (Preuss 2002; Preuss and Eitelberg 2003a, b). These plants were planned and built by Orgacid/Lonal.

11 The Chemical Weapons Complex in Munster

The chemical weapons complex in Munster was located within the Truppenübungsplatz Munster (Munster military training area), the history of which will be sketched below based on a Festschrift produced 1983 by the Kampftruppenschule 2.[65]

The moor, forest, and heath areas to the south and southwest of Munster, a small village on the around 40 km north of Celle on the Lüneburger Heide, were expanded from 1892 onward into a military training area of circa 49 km^2. The first troops arrived on June 7, 1893. There they found around fifty newly constructed buildings, and that number grew over the next two decades until the beginning of the First World War to more than 144 buildings. In January 1916, construction began on another training area north of Munster, around Breloh and along the railroad line to Uelzen, in the forest area of Raubkammer. It was referred to as Munster-Nord (Munster-North) to distinguish it. Within this area, between the Heidkrug outwork to the east, Breloh to the west, and north of the railroad line to Uelzen, the Gasplatz Breloh (Breloh gas area), of around 4.4 km^2, was built in 1917. It served as a filling plant for gas shells. In addition, field experiments with the use of chemical warfare agents were carried out in Munster-Nord. The Clark factory served to fill Blaukreuz munition. Grünkreuz (Green Cross) shells were produced in the Klopper factory, and Gelbkreuz (Yellow Cross) shells in the Lostwerk. The filling plants were headed by Dr. Hugo Stoltzenberg, who had also played a role in Munster after the First World War and in the secret rearmament between the wars (Stoltzenberg 1994, 333–34).

At the end of the First World War, the Gasplatz Breloh had around 48,000 metric tons of gas munitions and several thousand tons of seized munitions as well as 100 tank cars and containers with liquid chemical warfare agents.[66]

Of these, around a million chemical warfare shells and 230,000 chemical warfare mines as well as 40 tank cars and containers with chemical warfare agents were

[65]Spezialstab ATV, Gruppe Ausbildungsmaterial, Kampftruppenschule 2, *90 Jahre Truppenübungsplatz Munster, 1893–1983* (Munster 1983).

[66]Ibid., p. 24.

exploded on October 24, 1919. In addition, 42 buildings of the Grünkreuzwerk and the Gelbkreuzwerk were destroyed. Around 950 metric tons of chemical warfare agents in liquid and solid form, which had been stored in approximately 60 tank cars, survived. The glass bottles filled with 500–1000 metric tons of Blaukreuz that had been uncovered when the munitions were dismantled were stored in Munster-Nord at this time. Because the remaining 60 tank cars were needed for other purposes, they were supposed to be emptied as quickly as possible. Hugo Stoltzenberg received instructions from the Reichstreuhandgesellschaft (Reich Trust Company) to destroy the chemical weapons. Previously, in a meeting between the Ministry of Finance, the Kaiser-Wilhelm-Institut für physikalische Chemie und Elektrochemie (Fritz Haber), the Ministries of Trade, Commerce, Transportation, and Labor, as well as the Reichstreuhandgesellschaft on October 28, 1920, there had been negotiations about how Stoltzenberg should destroy the chemical weapons. In the meeting, Privy Councilor Mente had described Dr. Stoltzenberg as very trustworthy. Haber explained that Stoltzenberg had worked under his direction and that he could give him the best recommendation. In a meeting on December 17, 1920, it was revealed that Stoltzenberg had been director of the filling plant in Breloh during the war. In the meeting on October 28, 1920, the representative of the Reichstreuhandgesellschaft explained that it was estimated that 1,000 metric tons of gas artillery shells and cylinders (filled with Blaukreuz) were scattered about the German Reich. "In total, there ca. 2,000,000 such shells in around 35 locations. Of those 500–600,000 were stored in Unterlüss."[67] It is largely unknown what happened with those shells. Hence the discussions that follow are still significant today. There are three types of sites for burying gas shells and cylinders filled with Blaukreuz, in which they have also demonstrably been found.

First, factories in which, during the First World War, chemical warfare agents had been produced and filled in glass bottles, metal containers or shells, and the filling plants in which glass bottles and metal containers filled with Blaukreuz were inserted into empty shells and covered with explosives (15 sites). Examples: At the former chemical factory of E.-Schering AG in Berlin, several thousand 10-centimeter Gelbkreuz shells from the First World War were buried in three pits after the Second World War. In Wahn, near Cologne, in 1976, 1064 drums of 200 L each were filled with Clark bottles from the First World War and brought to the mine in Herfa-Neurode. Secondly, it is to be expected that agents and munitions were buried at former dismantling sites as well, where, according to existing files, more than 25,000 gas shells were present (31 sites).[68] Example: In Hallschlag, after

[67]Besprechung im Reichsschatzministerium III betr. Gasplatz Breloh, unter Vorsitz des Herrn Ministerialdirektors Kautz, October 28, 1920, BArch R 2201/3305; HQ-ETO-US-Army CWS, War Office, "Intelligence Division Report, no. 3961" Report on German CW Dump at Münster-Ost" (= Munster), June 23, 1945, National Archives, Washington, DC., RG 338; Bericht über die Vernichtung von Gas-Kampfstoffen, Berlin, December 17, 1920, BArch, R 2201/3305.

[68]Reichsschatzministerium, Listen der Orte, an denen Zerlegungsarbeiten von Munition vorgenommen wurden oder wo dies beabsichtigt ist, 1. Zerlegestelle für Eisenmunition und Metallkartuschen (1919–22), BArch 2201-Nr. 3225.

the First World War, some 950,000 shells were dismantled, of which around 1.3% were filled with chemical warfare agents. Most of them were destroyed in Hallschlag as well. Intense research in the 1990s was able to identify another circa 500 shells from the First World War.

Thirdly, chemical warfare agents have also been found at sites where they were supposed to be used for commercial purposes after the First World War. For example, Monzingen, near Bad Kreuznach, where 8,000 glass bottles filled with Blaukreuz and 30,000 7.5-cm shells filled with phosgene had been buried and were uncovered after the Second World War. The Stolzenberg factory in Hamburg is another example.

Even after the explosion in Munster, additional chemical munitions were brought to Munster from dismantling sites in the 1920s to be dismantled or exploded as part of contracts between the Reichsschatzamt and Evaporator AG. It is not known whether all 35 dismantling sites delivered shells with chemical munitions and filled glass bottles to Evaporator AG. It is, however, known that shells and/or glass bottles with chemical warfare agents were buried at some of these sites.

From 1934–35, the Munster-Nord exercise area was reconverted into a testing ground for chemical warfare agents and from 1939 onward expanded to 108 km^2.[69]

On April 17, 1945, Munster was occupied by British troops, who found several large facilities for chemical warfare agents in its forests. These included the Raubkammer military exercise area with the Raubkammer army experimental site and the Munster-Nord filling plant as well as the air force testing site of Munster-Nord, the HMa Munster-Ost, the LHMa Örrel, and various facilities of the Nebeltruppen (smoke-mortar troops). To the surprise of the British, however, there were also employees and documents from the Heereswaffenamt and the army's gas-protection laboratory in Berlin-Spandau found in Munster, where they had been moved to protect them from bombing and the approaching Soviet troops.[70]

Directly north of Munster, the British discovered the Munster-Raubkammer (Munster-Nord) military exercise area, which was around 16 km long and 8 km wide and covered with pines and heather. Roughly in its center was a large testing ground for chemical artillery and to the east of that a place to drop chemical bombs. The Heeresversuchsstelle Raubkammer (Raubkammer Army Testing Grounds), including several buildings, was located in the southwest part of Munster-Nord. It was subdivided into *Bereiche* (areas), which were numbered from R I to R IX, not including the administration. Bereich R I was responsible for field testing and photographic documentation. Bereich R II had the chemistry laboratory. Bereich R III consisted of several buildings for decontamination, gas detection, and the maintenance and repair of

[69]Spezialstab ATV, Gruppe Ausbildungsmaterial, Kampftruppenschule 2, *90 Jahre* Truppenübungsplatz *Munster, 1893–1983* (Munster 1983), pp. 33, 54.

[70]"Report on the C. W. Experimental Station at Raubkammer bei Munster and related Establishments," 1945, C.D.R-5. Enemy C. W. and Smoke Intelligence Summary, no. 83, The National Archives, London, WO 208/3576; A. K. Mills, "Investigation of Chemical Warfare Installations in the Munsterlager Area, including Raubkammer," 1945, report no. XXXI-86, CIOS Item 8, Chemical Warfare, Combined Intelligence Objectives Sub-Committee (CIOS), G-2 Division, SHAEF (Rear) APO 413, Preuss private collection.

protective clothing. Bereich R IV had parking garages and an auto repair shop. In Bereich R V, medical aspects of the field tests and toxicology were studied, and R VII was the medical area. East of this complex of related buildings was Bereich R VI, which measured around 600 by 400 meters and was an experimental plant for the production of chemical munitions, with equipment to melt and mold chemical agents and to manufacture munitions. In addition, experiments with hot and cold storage were conducted in this area, and munitions and chemical agents could be stored in six munitions buildings and six warehouses respectively. East of that, between R VI and R VIII, was the Nebelfüllstelle (Smoke Filling Plant) Munster-Nord. The *Nebel* (smoke) was replacing the term *Kampfstoff* (chemical warfare agent). Initially, it was used for testing filling methods for chemical warfare agents. Increasingly, however, the plant was used for ordinary filling work. In addition to the Nebelfüllstelle in Munster, there were chemical filling plants for the army in St. Georgen, Löcknitz, Dessau, and Lübbecke. The plant at Lübbecke, however, was partially functional, but was never in operation (Preuss und Eitelberg 2003a, 164). The air force also had a filling plant in Mockrehna and one in Örrel. The Nebelfüllstelle in Munster had around 60 buildings. The central plant was the bunker for chemical agents, with eight tanks holding 450 cubic meters of chemical agents each (a maximum of ca. 4700 metric tons) and two filling buildings. Füllhaus I had two automatic filling machines for shells, below which were twelve tanks for intermediate storage of mustard gas and arsine oil. In Füllhaus II stood two semiautomatic filling machines for 15-cm rockets and four semiautomatic machines for shells. Three buildings were available for preparing and finishing munitions.

The testing site for the air force was immediately adjacent. This post consisted of five small and one medium-sized two-story buildings. Their task was to develop additional ways for the air force to employ chemical warfare agents. In addition to a laboratory and rooms for filling and testing bombs, spraying and molding equipment for liquid chemical agents was studied and tested. Outside the buildings, the British found 250-kg bombs marked with three green rings. They claimed they had contained sarin that had been destroyed shortly before the Allies arrived. In fact, however, the three green rings indicated they were filled with tabun (Preuss and Eitelberg 2003a, 414). Sarin would have been marked with five green rings. A little farther to the east followed Bereiche R VIII (Vorwerk Heidkrug), which was camouflaged to look like a farm. For a time, tabun, excelsior (10 metric tons), and sarin (0.5 metric tons) were produced there in small quantities in succession. Bereich R IX housed animals and facilities for animal experiments. On the grounds of Munster-Nord, there were more than ten areas for testing and experimenting the use of chemical munitions and decontaminants.

Southeast of Munster was the LHMa Örrel. The British found 131,000 bombs of various sizes there, stored in bunkers. There was also a mustard gas filling plant for 250 and 500 kg bombs with two semiautomatic filling machines and one destroyed mobile filling plant for mustard gas, mounted on a railroad car; additional mobile filling plants were said to have been stationed in St. Georgen and Löcknitz.[71]

---

[71]A.K. Mills, "Investigation of Chemical Warfare Installations in the Munsterlager Area, including Raubkammer," report no. XXXI-86, CIOS Item 8, Chemical Warfare, Combined Intelligence

The Heeresmunitionsanstalt Munster-Ost was located about two kilometers southwest of Munster. It had around 17,000 metric tons of chemical munitions in the form of 763,580 filled gas shells and rockets and 7000 bombs.[72]

In November 1945 quantities of unfilled mustard gas in Munster were burned in a continuous process in an open trough. It was located 20–25 m from the mustard gas bunker of the Nebelfüllstelle. It was determined that arsenic compounds could be burned together with it. In addition, the contents of a large number of Italian one- and two-ton mustard gas containers were burned. They were stored outdoors and in warehouses. The incineration of chemical warfare agents was proposed by the British as a way of destroying chemical agents in the American zone as well. They were so convinced of this method that they also used it for British chemical warfare agents (mustard gas, lewisite) in the Bowes Moor munitions depot of the R.A.F.[73]

After the war, the heavy contamination of many areas prevented the use of the Truppenübungsplatz Munster-Nord for military training purposes. For that reason, efforts to study and decontaminate the grounds began in 1950. At first, the areas cleared for use were used only by British troops for exercises. In 1954, the Bundeswehr took over the grounds, and so from May 1956 German soldiers were trained there as well. At the same time, the Entgiftungskommando Raubkammer (Raubkammer Decontamination Command) was working on removing chemical traces from two world wars. The pressure on the Entgiftungskommando was great, because there was a lack of training areas at the time. That led to chemical munitions that had already been removed being transported to another site to be dismantled. In Munster, by April 1960, after thirty-eight months, 38 km^2 of contaminated grounds had been scoured by 60–70 employees, removing circa 48,000 shells, bombs, and containers for chemical warfare agents.[74] Even after that, however, parts of the site were considered unsafe, so that a small group continued to remove munitions from the First and Second World War from the military training area.

On September 6, 1979, these specialists had an opportunity to be active in the civilian realm as well, when a child playing near the former site of the Chemische Fabrik Dr. Hugo Stoltzenberg in Hamburg was fatally injured. The Bundeswehr bomb disposal group from Munster was brought into investigate the grounds. It found circa 100 metric tons of munitions and chemicals on the abandoned industrial

(Footnote 71 continued)

Objectives Sub-Committee (CIOS), G-2 Division, SHAEF (Rear) APO 413, Preuss private collection.

[72]HQ-ETO-US-Army CWS, The War Office: Intelligence Division Report No. 3961, Report on German CW Dump at Münster-Ost (= Munster), June 23, 1945. National Archives, Washington, DC, RG 338.

[73]Williams, C., Burning of Mustard Gas and Lewisite at N0. 81 M.U., R.A.F., Bowes Moor, Estimation of Risk. Summary of Porton Report No. 2744. 1947. National Archives, Washington, DC., RG 338.

[74]Spezialstab ATV, Gruppe Ausbildungsmaterial, Kampftruppenschule 2, *90 Jahre Truppenübungsplatz Munster, 1893–1983* (Munster 1983), pp. 49–50.

site (Scholz 2004). Dr. Hugo Stoltzenberg had attracted attention previously, because in 1928 a phosgene cloud was emitted from a tank car on his factory grounds that injured and even killed people.[75]

The Stoltzenberg factory drew attention again when the British visited it on June 21, 1945, and found dangerous chemicals, hand grenades filled with Blaukreuz and Weisskreuz, as well as smoke candles, incendiary materials, chemical warfare agents for training purposes, and 100 one-liter bottles with chloropicrin (PS). After his work in Breloh during and after the First World War, Stoltzenberg, probably on the recommendation of Fritz Haber, worked as a specialist in chemical warfare agents and munitions in a chemical factory in Hamburg and built chemical plants in Russia, Yugoslavia, Brazil, and Spain and hence participated in secret projects of the Reichswehr.[76]

The removal of munitions from the First and Second World War in Breloh continued in the 1980s. A first incineration plant was built for the destruction of chemical warfare agents. The second plant, which met all the requirements of the authorities, was operated by a federal organization: the Gesellschaft zur Entsorgung von chemischen Kampfstoffen und Rüstungsaltlasten mbH (GEKA mbH); (limited company for the removal of chemical warfare agents and armaments waste). With its 150 employees, it is capable of incinerating shells up to a caliber of 15 cm without dismantling them. It has facilities to clean contaminated soil, a plasma plant, a detonation chamber, and facilities to dismantle munitions. In terms of its construction, the facility is pioneering, and it also deals with munitions and excavated soil from other former munitions locations in Germany and abroad.

12 Other Munitions Sites with Filling Plants for Chemical Weapons

After the war and during the initial phase of the Allied occupation, large quantities of munitions were located in sites for chemical weapons that had filling plants. In the HMa Lübbecke, from April to August 1945, a total of 530,000 shells and mines with chemical warfare agents were stored, 117,000 of which were from Italy (Preuss and Eitelberg 2003b, 351). In the HMa St. Georgen, as a result of the United States Army's collection actions, 1,655,000 chemical shells and mines were

[75]"Note de la Delegation Française au sujet de l'explosion de gaz phosgens à Hambourg," June 14, 1928, "Notes d'un Secretair prises au cours d'un réunion tenne au Quai d'Orsay le lundi, December 3, 1928, Service historique de la Défense, Vincennes, 4N91 Dossier 1 Conférence des Ambassadeurs.

[76]Chemische Fabrik Dr. Hugo Stoltzenberg, Hamburg, Germany, Intelligence Division Report No. 3953, CIOS, Headquarters European Theater of Operations United States Army Chemical Warfare Service. National Archives, Washington, DC, RG 338.

stored, along with 5870 metric tons of unfilled German chemical agents and 1600 m^3 (ca. 2100 metric tons) Italian and Hungarian chemical agents in drums (Preuss and Eitelberg 2001, 204, 216).

The enormous quantity of Germany munitions but also the battlefield munitions of the Allies were dismantled or exploded in Germany. Large quantities were sunk at sea by order of the Allies (Frondorf 1993).[77]

13 Dismantling Work After the First World War

All of the dismantling work after the First World War was done under contracts between either states or the occupying armies and private companies. One example of this is the former factory of Espagit AG in Hallschlag, where 992,000 shells were dismantled after the First World War, 23,800 of which were gas shells. This factory was also ultimately destroyed by a large explosion (Preuss and Eitelberg 1999) It took years to collection the munitions that had been scattered by explosions. The project of studying the site in Hallschlag and making it safe cost around 50 million euros in the 1990s.

14 Summary and Prospects

The production of the chemical weapons industry and the processing and storage of its products cost many human lives and hundreds of millions of reichsmarks by the end of the Second World War. Even today, considerable funds have to be spent removing old munitions and waste left over from the production of chemical armaments in the First and Second World Wars from the soils and groundwater. The front lines—of the First World War, for example—were never thoroughly cleaned up because the project seemed hopeless; instead, in large areas only superficial cleanup was done. However, initial steps in this direction can be observed (Hubé 2016). The areas affected in that period are in Belgium, France, Italy, and Poland. The "Zone Rouge" identified in France after the First World War measured around 10,000 km^2. Even today, it is largely reserved for forestry.

After a century, isn't it time to set ourselves the task of removing the remains of the two world wars in European solidarity using modern concepts and methods?

[77]See also: The History of Captured Enemy Toxic Munitions in the American Zone, European Theater, May 1945 to June 1947, Section VI Operation Davey Jones Locker, Chemical Corps 1946. Office of the Chief of Chemical Corps, Headquarters European Command, Preuss private collection.

References and Source Materials

Aktennotiz der IG Farbenindustrie Aktiengesellschaft, betreffend Sarin, Leverkusen, July 26, 1943. The National Archives, London, FO 1031/179.

Aktennotiz der IG Farbenindustrie Aktiengesellschaft, Ludwigshafen am Rhein, Zwischenprodukten-Gruppe, betreffend Sarin. July 21, 1943. The National Archives, London, FO 1031/179.

Aktennotiz des Wa J Rü (Mun), Az. 70 o 40-19 Wa J Rü/Mun 3 zbV/IX, Nr. 1320/43 g.Kdos. betreffend die Ausweichstelle des Reichsamtes für Wirtschaftsausbau in Falkenhagen. July 26, 1943. Private Collection Preuss.

Aktenvermerk (Nr. 8) der IVG betreffend nach Stulln verlagertes Gut aus Falkenhagen. May 23, 1952. Private Collection Preuss.

Aktenvermerk der Montan betreffend Seewerk. July 25, 1944. Private Collection Preuss.

Anhang zum Bericht der Deutschen Revisions- und Treuhand-Aktiengesellschaft Berlin über die bei der Verwertungsgesellschaft für Montanindustrie GmbH, Berlin-Charlottenburg, vorgenommene Sonderprüfung betr. Forstabteilung, BArch Berlin, R 8125/1691.

Anonymus, Chlortrifluorid (ClF3), Anlage 9(5/-), p. 9–11, post 1945. Private Collection Preuss.

Arbeitsberichte Chemie Werk Kapen, Seite 1247 (April 17, 1957), 1249 (February 11, 1957), 1722 (February 2, 1957) sowie Zusammenfassungen von Berichten zwischen dem January 29, 1953 und dem July 23, 1956, Private Collection Preuss.

Archer, W., W.J.V. Ward and O.S. Whitson, Hydrofluric Acid, Vereinigte Flusspatgruben G.m.b. H. Stulln. 1946. BIOS Target No. C22/2012, C.I.O.S. Black List Item 22, Miscellaneous Chemicals, British Intelligence Objectives Sub-Committee. Private Collection Preuss.

BIOS. Final Report No. 1595, Item No. 22, Carter, W. J., M. A. Howick, J. R. Burston, R. Y. Eagers and A. B. Crooks, German Fluorine and Fluoride Industry, London 1945, Private Collection Preuss.

Bausinger, Tobias. 2007. *Ausbreitung von Nitroaromaten über das Sickerwasser der ehemaligen Espagit AG in Hallschlag. Ein Beitrag zur geoökologischen Bewertung eines kontaminierten Standortes*. PhD dissertation, University of Mainz.

Bausinger, Tobias, Ulrich Dehner, and Johannes Preuss. 2005. Nitrodiphenylaminen und verwandten Verbindungen im Sickerwasser einer Rüstungsaltlast. *Umweltwissenschaften und Schadstoff-Forschung* 17 (1): 7–12.

Baustelle Seewerk: Sachdarstellung über die Gesamtplanung, das Abrechnungswesen, das Sozialwesen, den Werkschutz, Werkfeuerwehr und Werkluftschutz. July 15, 1943. The National Archives, London, FO 1031/179.

Bauvorhaben Seewerk, Zusammenstellung der bis zum 30.06,1943 angefallenen Kosten. The National Archives, London, FO 1031/179.

Bericht über die Deutsche Kampfstoff-Produktion. February 1, 1944. BArch Berlin, R3/1894.

Bericht über die Lage auf dem K-Stoffgebiet. December 1, 1942. BArch R 3112/191.

Bericht über die Vernichtung von Gas-Kampfstoffen, Berlin, December 17, 1920. Bundesarchiv, BArch Berlin, R 2201/3305.

Besprechung im Reichsschatzministerium III betr. Gasplatz Breloh, unter Vorsitz des Herrn Ministerialdirektors Kautz. October 28, 1920. Bundesarchiv, BArch R 2201/3305.

Boyne, J.G., E.W. Lanfear, W.S. Calcott, and P.J. Leaper. 1945. Production of Vesicant Agents at Ammendorf. M52.D92., CIOS Target No. 8/30 Chemical Warfare, Item No. 8, File No XXXII-7, Combined Intelligence Objektives Sub-Committee G-2 Division, SHAEF (Rear) APO 413, National Archives, Washington, RG 338 and British Library, Document Supply Center, Boston Spa.

Brauch, Hans Günter, and Rolf-Dieter Müller (eds.). 1985. *Chemische Kriegführung. Chemische Abrüstung. Dokumente und Kommentare*. Berlin: Berlin Verlag Arno Spitz.

Bundesarchive Koblenz und Freiburg. 1979. Fertigung, Lagerung und Beseitigung chemischer Kampfstoffe unter besonderer Berücksichtigung des Territoriums der Bundesrepublik Deutschland. BArch Koblenz, 35 pages. Private Collection Preuss.

Burning of Bulk Mustard, Raubkammer, Germany, HQ-ETO (REAR) CWS-APO No. 887 to Chief CW Officer. November 6, 1945. National Archives, Washington, DC., RG 338.

Chemische Fabrik –Dr. Hugo Stoltzenberg, Hamburg, Germany, Intelligence Division Report No. 3953, CIOS, Headquarters European Theater of Operations United States Army Chemical Warfare Service. June 21, 1945. National Archives, Washington, DC., RG 338.

Der deutsch-französische Krieg 1870–71. 1881. Redigiert von der kriegsgeschichtlichen Abteilung des großen Generalstabes. Zweiter Teil. *Geschichte des Krieges gegen die Republik*. Dritter Band, des Gesamtwerkes V. Band. Berlin: Mittler.

Ehmann, Emil. 1948. U.S. Army Chemical Warfare Project, A.4. Produktionsstätten und Produktionshöhen, a) Kampfstoffe, 720–740. Private Collection Preuss.

Eibl, Christina. 1999. *Der Physikochemiker Peter Adolf Thiessen als Wissenschaftsorganisator (1899–1990). Eine biographische Studie*. Diss. phil. Dreieich: Historical Institute of the University of Stuttgart, Abteilung für Geschichte der Naturwissenschaften und Technik.

Entwürfe der Schreiben von Wa B 4 VII an die Orgacid GmbH, Vorbescheide zur Errichtung von Anlage, December 10, 1934. National Archive, London, FO 1031/204.

Frondorf, Mark J. 1993. Special Study on the Sea Disposal of Chemical Munitions. Science Applicatios International Corporation, Sponsored by U.S. Arms Control and Disarmament Agency, Contract Number: AC92RC1005, Task Order: 92-01, Mclean, Virginia.

Gesamtlageplan—Seewerk S/O (M 1: 2,500). December 21, 1943. Private Collection Preuss.

Gesellschaftsvertrag: Urkunde Nr. 331/1934, Notar Dr. F. Jacke. November 23, 1934. Private Collection Preuss.

Griess, T.E. (ed.). 1985. *Atlas for the Second World War, Europe and the Mediterranean*. Wayne: Avery Publishing Group.

Gutachten Nr 48. Entwurf eines Bau- und Errichtungsauftrages für die Errichtung der Sarin II Anlage in Seewerk (Auftrags-Nr. 3/IX-4888-9026/43), ohne Datum. The National Archives, London, FO 1031/179.

Hahn, Friedrich. 1986. *Waffen und Geheimwaffen des Deutschen Heeres 1933–1945*. Band 1: *Infanteriewaffen, Pionierwaffen, Artilleriewaffen, Pulver, Spreng- und Kampfstoffe*. Koblenz: Bernhard & Graefe Verlag.

HQ-ETO-US-Army CWS, The War Office: Intelligence Division Report No. 3961, Report on German CW Dump at Münster-Ost (= Munster-Ost), June 23, 1945. National Archives, Washington, DC. RG 338.

Hubé, Daniel. 2016. *Sur les traces d'un secret enfoui-Enquete sur l'heritage toxique de la Grande Guerre. Préface de Jean-Yves Le Naour*. Paris: Michalon Éditeur.

Klenck, Jürgen E. von. 1945. History of the 'Seewerk' (Falkenhagen), The National Archives, London, WO 208/2186.

König, Wolfram and Ulrich Schneider. 1985. *Sprengstoff aus Hirschhagen. Vergangenheit und Gegenwart einer Munitionsfabrik. Nationalsozialismus in Nordhessen*, Bd. 8. Kassel: Verlag Gesamthochschulbibliothek Kassel.

Lebenslauf Max Zeidelhack. March 27, 1948. Bayerisches Hauptstaatsarchiv München, Office of Military Government, Bavaria, 13/83-2/5, National Archives, Microfilm Publication 12065 Section 3-402/NNDG No. 775037).

Lebenslauf von Prof. Dr. H. Engelhard, February 15, 1960, Universitätsarchiv Göttingen, UAG-Kur. 10221.

Mense-Stefan, Anne. 2005. *Standortdifferenzierte Abschätzung von Sickerwasserraten in Hessen. Ein Beitrag zur Ermittlung von Stofffrachten aus dem Boden*. PhD diss. University of Mainz, Mainzer Geographische Studien, Bd. 53.

Mills, A.K., Investigation of Chemical Warfare Installations in the Munsterlager Area, including Raubkammer. 1945. Report No. XXXI-86, CIOS Item 8, Chemical Warfare, Combined Intelligence Objektives Sub-Committee (CIOS), G-2 Division, SHAEF (Rear) APO 413, Private Collection Preuss.

Monturon GmbH, Aufstellung des Buchwertes zum 31.03,1944 (= Anschaffungswert 1944), Werk Falkenhagen. Private Collection Preuss.

Monturon GmbH. Abschlussbericht 1943/1944, Fabrikhauptbuch Seewerk. Private Collection Preuss.

Note de la Delegation Francaise au sujet de l´explosion de gaz phosgens à Hambourg. June 14, 1928. Notes d´un Secretair prises au cours d´un réunion tenne au Quai d´Orsay le lundi 3. Dez. 1928. Service historique de la Défense, Vincennes, 4N91 Dossier 1 Conférence des Ambassadeurs.

Note de la Delegation Francaise au sujet de l´explosion de gaz phosgens à Hambourg. June 14, 1928. Notes d´un Secretair prises au cours d´un réunion tenne au Quai d´Orsay le lundi 3. Dez. 1928. Service historique de la Défense, Vincennes, 4N91 Dossier 1 Conférence des Ambassadeurs.Orgacid GmbH. Bericht über das Geschäftsjahr 1937/38. December 21, 1938. Private Collection Preuss.

Oberkommando des Heeres (Ch.H.Rüst u. BdE.) Firma IG Farbenindustrie A.G. to the attention of the director, Dr. Ambros o.V. Ludwigshafen am Rhein, concerning the construction contract to build a factory to manufacture Product G, contract no. 9/IXa-240-9018/39, p. 1, The National Archives, London, FO 1031/223.

Orgacid GmbH. Bericht über das Geschäftsjahr 1938/39. December 21, 1939. Private Collection Preuss.

Orgacid GmbH. Bericht über das Geschäftsjahr 1939/40. 1939. Private Collection Preuss.

Orgacid GmbH. Vorläufiger Geschäftsbericht über das Geschäftsjahr 1940/41. From April 1, 1941 to December 31 1940. Private Collection Preuss.

Orgacid GmbH. Vorläufiger Geschäftsbericht über das Geschäftsjahr 1941/42. From April 1, 1941 to December 31 1941. Private Collection Preuss.

Orgacid GmbH, Bericht über das Geschäftsjahr 1943–44. Private Collection Preuss.

Preuss, Johannes. 1990. *Vorerkundungen und Bewertungen des Gefährdungspotenzials altlastenverdächtiger Betriebsflächen der ehemaligen Rüstungsindustrie – Stadtallendorf, Landkreis Marburg, Biedenkopf.* Postdoctoral thesis, University of Marburg, unpublished.

Preuss, Johannes. 2002. *Zahlencode-System des Heeres von 1925 bis 1940.* Schwäbisch-Hall: Journal Verlag Schwend.

Preuss, Johannes, Ulrich Dehner and Frank Eitelberg, et al. 1994. Erkundung und Rekonstruktion des ehemaligen Werkes der Deutschen Sprengchemie (DSC) Werk Kraiburg, (unpublished for Stadtverwaltung Waldkraiburg).

Preuss, Johannes und Frank Eitelberg. 1994. *Rekonstruktion der ehemaligen Fabrik zur Herstellung von Brand- und Kampfstoff: N-Stoff und Sarin der Monturon GmbH in Falkenhagen.* Mainz: Johannes-Gutenberg-Universität Mainz, Geographisches Institut, (unpublished for Landesbauamt Frankfut/Oder).

Preuss, Johannes und Frank Eitelberg. 1996. *Ehemalige Anlagen und Produktionsverfahren auf Rüstungsaltstandorten und ihre Bewertung. Handbuch Altlasten*, Bd. 4 Wiesbaden: HLUG.

Preuss, Johannes und Frank Eitelberg. 1999. *Hallschlag. Historisch-genetische Studie zur ehemaligen Fabrik für die Herstellung von Trinitrotoluol, Dinitrobenzol und Presskörpern aus Sprengstoffen sowie zur Verfüllung und Entlaborierung von Munition der ESPAGIT AG. Mainzer Geographische Studien*, Bd. 45. Mainz: Johannes-Gutenberg-Universität Mainz, Geographisches Institut.

Preuss, Johannes und Frank Eitelberg. 2001. *Historisch-genetische Studie über die ehemalige Heeres-Munitionsanstalt St. Georgen*, (unpublished for Stadtverwaltung Traunreut).

Preuss, Johannes und Frank Eitelberg. 2003a. *Heeres-Munitionsanstalt Lübbecke. Vorgeschichte der Stadt Espelkamp.* Mainzer Geographische Studien, Sonderband 1. Mainz: Johannes-Gutenberg-Universität Mainz, Geographisches Institut.

Preuss, Johannes und Frank Eitelberg. 2003b. *Vertiefende Historisch-Genetische Studie und Geländebegehung über die ehemalige Kampfstoff-Füllstelle der HMa Dessau* (Kapen), (unpublished for Landesanstalt für Altlastenfreistellung, Magdeburg).

Preuss, Johannes and Frank Eitelberg. 2009. *Das Werk Kraiburg der Deutschen Sprengchemie GmbH 1938–1945. Waldkraiburg schaut zurück. Geschichte einer jungen Stadt.* 2nd rev. ed., Stadt Waldkraiburg.

Preuss, Johannes und Frank Eitelberg. 2010. *Die TEUTO-Metallwerke GmbH in Osnabrück. Rekonstruktion einer Fabrik für Infanteriemunition und Hülsen für Bordwaffenmunition der Jahre 1935–1945. Mainzer Geographische Studien*, Sonderband 6, Mainz: Johannes-Gutenberg-Universität Mainz, Geographisches Institut.

Preuss, Johannes, Frank Eitelberg, Gerd Albrecht, et al. 2002. Modellhafte Altlastenuntersuchung an ausgewählten Standorten der Munitionszerlegung und -vernichtung nach dem Ersten Weltkrieg. *UBA Text* 49(2): 141 pages.

Preuss, Johannes, Frank Eitelberg, and Tania Bader-Nia, et al. 1993. Erkundung und Rekonstruktion des ehemaligen Werkes der Westfälisch Anhaltinischen Sprengstoff AG (WASAG) in Reinsdorf, Lutherstadt Wittenberge, (unpublished for Stadtverwaltung Wittenberg), 2 vols.

Preuss, Johannes, Frank Eitelberg and Wolfgang Form. 1992. Rüstungsaltstandorte und Verdachtsflächen in Hessen (1. und 2. Weltkrieg), (unpublished for Hessisches Ministerium für Umwelt, Reaktorsicherheit und Bundesangelegenheiten).

Preuss, Johannes, Frank Eitelberg and Peter Wilmes. 1992. Rekonstruktion der ehemaligen Rüstungsbetriebe in Geretsried - Deutsche Sprengchemie (DSC) Geretsried und Dynamit Actien Gesellschaft - Gesellschaft zur Verwertung Chemischer Erzeugnisse (DAG/Verwert) Wolfratshausen, (unpublished for Stadtverwaltung Geretsried).

Preuss, Johannes and Ulrich Schneider. 2005. Rüstungsaltlasten in Deutschland. In *Boden gut gemacht. Die Sanierung des Rüstungsaltstandortes Stadtallendorf*, ed. Hessisches Ministerium für Umwelt. Wiesbaden: Hessisches Ministerium für Umwelt, 21–63.

Reichsschatzministerium, Listen der Orte, an denen Zerlegungsarbeiten von Munition vorgenommen wurden oder wo dies beabsichtigt ist, 1. Zerlegestellen für Eisenmunition und Metallkartuschen (1919–1922), BArch 2201/3225.

Reile, Heidrun. 2005. Kosten. In *Boden gut gemacht. Sie Sanierung des Rüstungsstandortes Stadtallendorf*, ed. Hessisches Ministerium für Umwelt. Meckenheim, 419–424.

Report on the C.W. Experimental Station at Raubkammer bei Munster and related Establishments. C.D.R-5. Enemy C.W. and Smoke Intelligence Summary, No. 83, 1945, The National Archives, WO 208/3576.

Rundschreiben der Monturon GmbH, Nr. 914/45/IVa1/Pr. an die Montan-Industriewerke GmbH betreffend Betriebsstillegungen und Verlagerungen. February 1945. Private Collection Preuss.

Schaefer, Karl Wolfram, Frank Boeren, Sergei Boutsev, et al. 1997. Internationale Erfahrungen der Herangehensweise an die Erfassung, Erkundung, Bewertung und Sanierung Militärischer Altlasten. *UBA Text* 4 (97): 139–144.

Schmaltz, Florian. 2005. *Kampfstoff-Forschung im Nationalsozialismus. Zur Kooperation von Kaiser-Wilhelm-Instituten, Militär und Industrie*. Göttingen: Wallstein.

Scholz, Ernst-Gerhardt. 2004. Stoltzenberg-Skandal. Zuerst starb ein Kind. *Hamburger Abendblatt,* September 7.

Schreiben der Luranil-Baugesellschaft mbH (in Auflösung) an das IG Farben Control Office (Liquidation section), Frankfurt am Main betreffend Silbermaterial. January 18, 1951. Private Collection Preuss.

Schreiben der Waffenamt-Forschungsabteilung (WaF) an den Chef Wa J Rü betreffend Muna-Ost. December 1938. Private Collection Preuss.

Schreiben der Waffenamt-Forschungsabteilung (WaF), Nr. 595/39 gKdos Wa F/A2, an Wa J Rü 10 und Wa J Rü (Mun3) betreffend MO, Büro Prof. Loos. January 30, 1940. Private Collection Preuss.

Schreiben des Oberkommandos des Heeres (Ch H Rüst u BdE), 70 o 30 18 Wa J Rü (Mun 3 zbV/VIII), Nr. 10006/43 g.Kdos., an I.D. Riedel - E. de Haen AG, Berlin-Britz betreffend Sonderbauvorhaben OKH. February 18, 1943. Private Collection Preuss.

Schreiben des Oberkommandos des Heeres (Ch H Rüst u BdE), Wa J Rü Stab IV d an das Oberkommando der Wehrmacht (OKW), W Stab W Rü, betreffend M.-O., Falkenhagen, Erwerb von Flussspatgruben, Antrag auf Bestimmung einer Bedarfsstelle. January 1940. Private Collection Preuss.

Schreiben von Oberst a.D. Hammer, Falkenhagen an die Verwertungsgesellschaft für Montanindustrie GmbH, München. September 18, 1943. Private Collection Preuss.

Schweer, Henning. 2008. *Die Geschichte der Chemischen Fabrik Stoltzenberg bis zum Ende des Zweiten Weltkrieges. Ein Überblick über die Zeit von 1923 bis 1945 unter Einbeziehung des historischen Umfeldes mit einem Ausblick auf die Entwicklung nach 1945*. Diepholz: GNT Verlag.

Sekretariat K, Bericht, Betrifft: Orgacid GmbH Der Betrieb liegt still und hat eine amerikanische Wache von etwa 30 Mann. Die Orgacis beschäftigt 42 Arbeiter..., Essen, May 28, 1945. Private Collection Preuss.

Sonderausschuss C beim Reichsminister für Rüstung und Kriegsproduktion, Bericht über die Lage auf dem Kampfstoffsektor, Auszug aus dem Vortrag im Führerhauptquartier am 15.03,1943 mit einer Gegenüberstellung der Situation vom 01.03,1944, March 3, 1944. BArch Berlin, R 3/1894.

Sonderausschuss C, Arbeitsbüro: Fertigungsbericht C-Stoffe für Dezember 1944. BArch Berlin, R 3/1894.

Spezialstab ATV, Gruppe Ausbildungsmaterial, Kampftruppenschule 2, 90 Jahre Truppenübungsplatz Munster 1883–1993, 6–21, Munster 1983.

Stoltzenberg, Dietrich. 1994. *Fritz Haber—Chemiker, Nobelpreisträger, Deutscher, Jude*. Weinheim: VHC Wiley.

Szöcs, Andreas. 1999. *Geoökologische Systemanalyse und Bestimmung der Nitroaromatenmobilität auf dem großflächigen Rüstungsaltstandort Stadtallendorf bei Marburg*. PhD dissertation, University of Mainz, Göttingen: Cuvillier Verlag.

Thieme, Jürgen. 1996. Bestandsaufnahme von Rüstungsaltlastverdachtsstandorten in der Bundesrepublik Deutschland. *UBA Text* 25/96-30/96.

Thieme, Jürgen. 1999. Umweltrelevante und technische Aspekte der Zerlegung von Munition und Waffen nach dem 1. Weltkrieg. *UBA Text* 3: 99.

Transcription of the copy from Handelsregister, Amtsgericht Halle an der Saale, Department B, December 23, 1937. Private Collection Preuss.

Verfügung der Landesregierung Brandenburg, Enteignungsurkunde für das beschlagnahmte Betriebsvermögen der Firma M.O. Falkenberg (!), Falkenberg, Kreis Lebus. July 7, 1848. Private Collection Preuss.

Vertrag zwischen dem Deutschen Reich (Wehrmachtsfiskus), vertreten durch das Oberkommando des Heeres (OKH), der Verwertungsgesellschaft für Montanindustrie GmbH, der IG Farbenindustrie AG und der Monturon GmbH über die Erweiterung der OKH-eigenen N-Stoff-Anlage durch eine Fabrikationsstätte zur Herstellung von 500 metric tons monthly Sarin. June 3, 1944. Private Collection Preuss.

Vertrag zwischen der Chemischen Fabrik Buckau (Ammendorf), der Degea-Aktiengesellschaft (Auergesellschaft), Berlin, und dem Deutschen Reich (Fiskus des Wehrmachtsteils Heer), vertreten durch den Reichskriegsminister (Rkm) über die Errichtung einer Polyglykolfabrik und einer Umsetzungsanlage. September 20, 1935. Private Collection Preuss.

Wehrmacht-Fernsprechverzeichnis Groß-Berlin Teil I, p. 147. February 1, 1943. National Archives, Washington DC., Microfilm T-77, reel 342.

Wehrmachts-Fernsprechverzeichnis Groß-Berlin, Ministerial Dirigent Dr. Zahn Abteilungs- Chef z. b. V, Abteilung Chef-Chemiker (Wa A). June 6, 1943, part I, p. 144. National Archives, Washington, DC., Microfilm T-77, reel 342.

Williams, C., Burning of Mustard Gas and Lewisite at No. 81 M.U., R.A.F., Bowes Moor, Estimation of Risk. Summary of Porton Report No. 2744. January 1, 1947. National Archives, Washington, DC., RG 338.

Zimmermann, K., and H. Martens. 1994. Kontamination von militärischen Gelände im Raum Munster durch Arsenverbindung. *Abfallwirtschaftsjournal* 6 (6/7): 510–513.

From Charles and Francis Darwin to Richard Nixon: The Origin and Termination of Anti-plant Chemical Warfare in Vietnam

Matthew Meselson

Abstract Anti-plant chemical warfare (CW), the use of chemicals to clear vegetation or destroy food crops as a method of warfare, was conducted on a large scale in the Vietnam War of the 1960s and 1970s. Unlike the anti-personnel CW of World War I, which continued until the Armistice, anti-plant CW in Vietnam was terminated while the war was still underway. Already subject to increasing controversy, its limitation and subsequent termination was precipitated by the appearance in late 1969 of a government-sponsored study suggesting that 2,4,5-T, a component of Agent Orange, the herbicide most heavily used for defoliation, might be teratogenic to humans. In consequence, its use in Vietnam was restricted and then prohibited altogether. Although another herbicide, Agent White, remained briefly in use, all large-area defoliation had ceased by May 1970, leaving crop destruction as the remaining form of large-area herbicide operations in Vietnam. After a review of the program requested by the U.S. Ambassador and the Commanding General in Saigon, the ambassador telegraphed Washington in early December 1970 their decision that chemical crop destruction should be phased out. Although secret, the content of the telegram became known to the press and was published a week later, followed shortly thereafter by President Richard Nixon's announcement that there would be "an orderly yet rapid phaseout of herbicide operations in Vietnam."

The development of anti-plant chemical warfare (CW) may be traced to discoveries made by Charles Darwin and his son Francis, described in their book "The Power of Movement in Plants," published in 1880 (Darwin and Darwin 1880; Holland et al. 2009). They found that the bending of oat and canary grass seedlings (specifically, the cotyledons) toward a light source does not occur if the tip of the seedling is shielded from light or excised. Observing that the bending occurs a short distance away from the tip, they concluded that "some influence is transmitted from the upper to the lower part, causing the latter to bend." They also repeated and

M. Meselson (✉)
Department of Molecular and Cellular Biology, Harvard University, Cambridge, MA, USA
e-mail: msm@wjh.harvard.edu

B. Friedrich et al. (eds.), *One Hundred Years of Chemical Warfare: Research, Deployment, Consequences*, DOI 10.1007/978-3-319-51664-6_17

confirmed disputed experiments by others, showing that the downward bending of roots in response to gravity is likewise "due to an influence transmitted from the apex to the adjoining part where the bending takes place."

Investigations early in the twentieth century showed the "influence" to be a material substance. Bending in response to light was found not to occur if the tip is separated by a thin sheet of mica from the region where bending would otherwise occur but does occur if the severed parts are separated instead by a layer of gelatin, through which a chemical substance might diffuse. The isolation and identification of the presumed substance was facilitated by the further discovery that a small block of agar that has been placed on the upper cut surface of a seedling cut through near the tip, when placed on one side of the lower surface, causes bending in the direction away from that side. This was interpreted to mean that a growth-promoting substance adsorbed in the agar block is transported downward on that side of the seedling, where it causes the observed bending. The angle of bending under defined conditions provided a quantitative assay for the growth-promoting substance that was then used to guide its isolation from two sources known to have such activity: human urine and the fungus *Rhizopus suinis*. The highly active substance isolated from both sources was found to be indole-3-acetic acid (IAA). Only much later was it established that IAA is the major naturally-occurring plant hormone involved in heliotropism and geotropism (Whippo and Hangarter 2006; Abel and Theologis 2010; Enders and Strader 2015).

Although IAA stimulates plant growth at low doses, higher doses were found to cause plant death. In 1941, Ezra J. Kraus, chair of the Botany Department at the University of Chicago, proposed that plant growth regulators might therefore find use as selective herbicides in agriculture and began a screen for compounds more stable than IAA that might be used for such purposes. Following Kraus' suggestion, a parallel screening program was undertaken at the U.S. Agricultural Research Center at Beltsville, Maryland under the direction of one of his former doctoral students. Late that year, in a memo written a few days after U.S. entry into WWII, Kraus proposed to a committee of the National Academy of Sciences formed to advise the War Department on biological warfare that a program be established to develop herbicides that might provide a "simple means of destruction of rice crops, the staple food supply of the Japanese" and which applied as "sprays or mists over enemy forests would, through the killing of trees, reveal concealed military depots" (Kraus 1942; Peterson 1967; Troyer 2001).

It had been found in 1942 that 2,4-dichlorophenoxyacetic acid (2,4-D) is a potent stimulator of plant growth (Zimmerman and Hitchcock 1942). But its powerful herbicidal activity and potential as a weed killer, discovered independently in Britain and in the U.S., remained secret until late in the war. Kraus, upon learning of the plant growth activity of 2,4-D, included it in the screens underway at Chicago and at Beltsville, thereby becoming one of the discoverers of its potential for use as an herbicide (Troyer 2001). Conducted under conditions of wartime secrecy, the work at Chicago was done in the University's botany department, just around the corner from the west stands of the track and football field where in the winter of 1942 Enrico Fermi and his colleagues were building the world's first nuclear reactor.

Starting in 1944, a large-scale project to screen chemicals for herbicidal activity and for plant species specificity and to develop methods for their military application was begun by the U.S. Army Chemical Warfare Service at the Army Biological Research Center at Camp (later Fort) Detrick, Maryland, established the year before. The main effort was on crop destruction with only limited work on defoliation. By late 1945, some one thousand substances had been tested for use against various food crops at Detrick or under its direction in field tests elsewhere in the U.S. Of the agents tested, 2,4-D and 2,4,5-trichlorophenoxyacetic acid (2,4,5-T) were considered to be the most effective, although later work showed them to be less so against grain crops. By the end of the war, substantial stocks of 2,4-D and other herbicides and equipment for their dissemination by aircraft had been procured but were not used in any theater. After the war, except for a severe cutback in 1957, reversed the following year, research and field testing continued on a substantial scale. Following aerial spray tests conducted in the 1950s at the U.S. Army chemical and biological proving grounds at Dugway, Utah, and in 1959 at Camp Drum, New York, formulations of 2,4-D and 2,4,5-T were chosen as defoliants and the plant metabolic disrupter and desiccant cacodylic acid (dimethylarsenic acid), more toxic for grain crops, was selected for use against rice and wheat (Cecil 1986; Young 2009).

Thus, unlike the anti-personnel CW of World War I, which began with little prior development and no established advocate organization within the military, anti-plant CW in Vietnam was preceded by many years of research, development, and testing by a dedicated organization within the U.S. Army. Further, a precedent had been set for anti-plant CW by the use of herbicides, mainly 2,4,5-T, for crop destruction and defoliation in British counterinsurgency operations in Malaya in the early 1950s (Connor and Thomas 1984).

Experimental testing of chemical crop destruction and defoliation in Vietnam began in August 1961 and continued until mid-January 1962 as part of Project Agile of the U.S. Defense Department's Advanced Research Projects Agency (ARPA), supported by the Crops Division of the Army Chemical Corps at Detrick (Brown 1962). The initial impetus for introducing herbicide warfare in Vietnam appears to have come from William H. Godel, an ARPA Deputy Director acting with the encouragement of Vietnam President Ngo Dinh Diem and assisted by the director of Detrick's Crops Division (FRUS 1961; Godel 1961; Brown 1962; Bundy 1972; *Washington Post* 1966). In the first tests, on August 10, 1961, a mixture of the n-butyl esters of 2,4-D and 2,4,5-T and the isopropyl ester of 2,4,5-T, known as Agent Purple, was sprayed by helicopter over manioc and rice fields and over roadside trees. While the tests were still underway, the Chief of the U.S. Military Assistance Advisory Group recommended to the Department of Defense in Saigon in October 1961 that defoliation and crop destruction be carried out against several designated targets (Olenchuk et al. 1963). In the following month, the Joint Chiefs endorsed a plan developed in Saigon to spray 334 square miles of manioc and rice with 2,4,5-T and cacodylic acid and defoliate 200 square miles of forest with 2,4-D and 2,4,5-T, warning, with respect to crop destruction, that "care must be taken to assure that the United States does not become the target for charges of employing chemical or biological warfare" (Buckingham 1982).

Late that same month, on November 30, 1961, responding to recommendations from the Deputy Secretary of Defense and the Secretary of State, President Kennedy agreed in principle to chemical defoliation and crop destruction in Vietnam, but on a far more limited and tightly controlled basis than had been envisaged by the Joint Chiefs, authorizing only a

> selective and carefully controlled joint (with the Republic of Vietnam) program of defoliant operations [...] proceeding thereafter to food denial only if the most careful basis of resettlement and alternative food supply has been created (Bundy 1961; Buckingham 1982).

Limited defoliation on an operational basis (intended to have a military effect rather than being only developmental) was begun early in January, after Kennedy reduced to only 16 miles the plan for defoliation of about 300 miles of roadside that had been submitted to him by the Departments of Defense and State. Authority to approve defoliation missions was retained in Washington until November 1962, when authority for defoliation of limited areas was delegated to the U.S. ambassador and the commander of U.S. forces in Vietnam, followed in May 1963 by the delegation to them of authority for all defoliation. Crop destruction was not authorized until October 1962. It required State and Defense Department authorization throughout the Kennedy administration and until July 1964, when it was delegated to Saigon (Collins 1967; Warren 1968; Clary 1971; Buckingham 1982; Cecil 1986).

Altogether, according to updated estimates, about 73 million liters of various herbicides were sprayed by fixed-wing aircraft and helicopters over an area of approximately 26,000 km^2, 15 percent of the land area of the Republic of Vietnam, most of it sprayed more than once. Most of the area sprayed was coastal or inland forest while about 10% was crop land. Of the total volume sprayed, about 63% was Agent Orange (a mixture of the n-butyl esters of 2,4-D and 2,4,5-T or, in smaller quantity, containing the iso-octyl rather than the n-butyl ester of 2,4,5-T), 28% Agent White (a mixture of 2,4-D and a chlorinated derivative of picolinic acid plus inert ingredients), and 7% Agent Blue (cacodylic acid and its sodium salt plus inert ingredients), and a few percent of other compounds or formulations that were employed mostly before the introduction of Agent Orange in 1965. After a gradual buildup in herbicide operations in earlier years, nearly 80% was dispensed in the four years 1966–1969, followed by a sharp drop in 1970, after the restriction and then cancellation of authorization for the use of Agent Orange (Buckingham 1982; Westing 1984; Stellman et al. 2003; Young 2009).

The limitation and subsequent cessation of the use of Agent Orange was precipitated when a study of possible carcinogenic, teratogenic, and mutagenic effects of a large number of pesticides and industrial chemicals was brought to the attention of the administration. The study, dated August 1968 and released to the public the following year, was commissioned by the U.S. National Cancer Institute and done by the Bionetics Research Laboratories of Kensington Maryland (Bionetics Research Laboratories 1968). It found that 2,4,5-T administered to pregnant mice and rats consistently caused a high frequency of malformations and death in fetuses and newborns and categorized 2,4,5-T as "probably dangerous."

Although the Bionetics report would have soon come to the attention of the White House one way or another, the swiftness of the response when it was brought to the attention of the President's Science Advisor is noteworthy. In the autumn of 1969, I was given a pre-release copy of the report. Aware of articles in Saigon newspapers claiming that herbicide exposure was causing birth defects and believing that the administration should be made aware of the Bionetics report, I and two colleagues called upon Lee DuBridge, the physicist and former president of CalTech who was then President Nixon's Science Advisor (Primack and von Hippel 1974; Hay 1982). After examining the report and while I was still with him in his office, DuBridge telephoned the Deputy Secretary of Defense, David Packard (co-founder of the Hewlett Packard Corporation), and they agreed on the spot to restrict the use of 2,4,5-T. In a White House press release late that same day, October 29, DuBridge announced that the Defense Department "will restrict the use of 2,4,5-T to areas remote from population"; that the Department of Agriculture "will cancel registrations of 2,4,5-T for food crops effective January 1"; and that the Departments of Agriculture and Interior "will stop using 2,4,5-T in their own programs" (Nelson 1969a, b). A few days later, DuBridge telephoned me at Harvard to say that representatives of the Dow Chemical Company had informed him that the likely teratogen was not 2,4,5-T itself but rather a highly toxic impurity, dioxin (2,3,7,8-tetrachlorodibenzodioxin). Dow had known of its toxicity following an outbreak of chloracne among workers at a Dow facility in 1964 (Baughman 1974; Crummett 2002). Dioxin is formed as an impurity in the Dow synthesis procedure for 2,4,5-trichlorophenol, a precursor of 2,4,5-T, particularly if carried out at too high a temperature (Young 2009). Knowing this, Dow had taken precautions to keep the concentration of dioxin in 2,4,5-T below 1 ppm. As found in research done much later, the extreme toxicity of dioxin is associated with its avid binding to a molecular receptor that regulates the expression of numerous genes (IOM 2014; Sorg 2014).

It was later found that the 2,4,5-T employed by Bionetics, not produced by Dow, contained 27 ppm of dioxin. Further tests were therefore undertaken to determine if purer 2,4,5-T also causes birth defects in rodents. Finding that 2,4,5-T containing only about 1 ppm of dioxin did so in mice, the Secretaries of Agriculture, Interior, and Health, Education and Welfare agreed in an announcement of April 15, 1970 to suspend registrations for uses of 2,4,5-T on agricultural land and in places likely to entail direct human exposure (US Department of the Interior 1970). Simultaneously, undersecretary Packard canceled authorization for all uses of Agent Orange in Vietnam. The cancellation of Agent Orange put a stop to nearly all large-area chemical defoliation, leaving only a few occasions on which Agent White, available in only limited supply, was used in this mode, bringing an end to all defoliation, except on the perimeters of fixed US installations, in May 1970 (Buckingham 1982).

While large-area defoliation had ceased, chemical crop destruction continued. Although strongly supported by the Joint Chiefs and the Secretary of Defense (Laird 1970), it was controversial ever since it had been conditionally authorized by President Kennedy in 1961. A 1968 interagency review of the herbicide program ordered by the U.S. ambassador in Saigon, Ellsworth Bunker, concluded that

> There is evidence that food shortages, for which crop destruction efforts were partly responsible, have at times created logistical problems for the enemy [...] The main impact of crop destruction, however, falls upon the civilian population [...] An estimated 90% of the crops destroyed in 1967 were grown, not by VC/NVA military personnel, but by civilians living there (American Embassy Saigon 1968; Clary 1971; Buckingham 1982).

The year before, a study of the military utility of the crop destruction program, based on some 2400 interviews conducted with Vietnamese familiar with the activities of the Viet Cong and the North Vietnamese army concluded that "the data consistently suggest that the crop destruction program has not in any major sense denied food to the VC" and that "the crop destruction effort may well be counterproductive. The VC continue to feed themselves while the peasant bears the brunt of the deprivation" (Betts and Denton 1967).

In 1972, after all aerial herbicide operations had ceased, Packard directed the Army Corps of Engineers to conduct a study of the military utility of herbicides in Vietnam. Based on a survey of U.S. military officers who had served in Vietnam, on the association of roadside defoliation missions with recorded friendly and enemy battle fatalities, and on earlier studies, the Engineers study concluded with respect to defoliation that "[h]erbicides were useful in supporting military operations in selected instances" and that "[m]any survey responses report that the use of herbicides around the perimeter of bases and installations is the most effective use of herbicides in Vietnam." Regarding crop destruction, the study concluded that "[h]erbicides destroyed enemy crops, but the enemy was able to compensate and overcome localized food supply shortages. At most, the crop destruction program harassed the enemy" (ESSG 1972).

An example of problems encountered in attempting to distinguish fields cultivated by military units from fields cultivated by civilians for their own consumption was encountered by a colleague and myself in August 1970. For five weeks in the summer of 1970, I was in Vietnam on behalf of the American Association for the Advancement of Science as part of a small team conducting a preliminary survey of the ecological and health effects of the military use of herbicides in Vietnam—interviewing farmers, photographing sprayed and not-sprayed forest, and collecting environmental and biomedical samples for mass spectrometric analysis for dioxin at Harvard (Constable and Meselson 1971; Meselson et al. 1972; Meselson and Baughman 1973, 1974). In order to inspect an area where crop destruction had recently taken place, the medical member of our team, Dr. John Constable, and I were flown by helicopter over a river valley in one of the northern provinces where Agent Blue had been sprayed along a 15 km path a few days before (Fig. 1). Flying along the length of the valley on two occasions, we saw rice fields browned by the herbicide but were too high to see much evidence of habitation. As indicators that the valley

Fig. 1 Aerial photograph of a portion of a valley sprayed with Agent Blue in August 1970. Many small dwellings may be seen throughout the valley. Terraced fields may be seen on the hillsides. In order to suppress groundfire against the slow and low-flying UC-123 spray aircraft, spray missions were generally preceded by fighter aircraft delivering 500 or 750 lb bombs, cluster bomb units, 20-mm ordnance, and/or napalm (Buckingham 1982). From the Meselson CBW Archive

AMERICAN ASSOCIATION FOR THE ADVANCEMENT OF SCIENCE

AAAS HERBICIDE ASSESSMENT COMMISSION
BOTANICAL MUSEUM, HARVARD UNIVERSITY
CAMBRIDGE, MASSACHUSETTS 02138, U.S.A.

MATTHEW S. MESELSON, CHAIRMAN
ARTHUR H. WESTING, DIRECTOR

12 November 1970

The Honorable Ellsworth Bunker
American Ambassador
Saigon, Republic of Vietnam
APO San Francisco, California 96243

Dear Ambassador Bunker,

On behalf of the Herbicide Assessment Commission of the American Association for the Advancement of Science, we wish to express our appreciation for the generous assistance provided by the Embassy during our recent tour of the Republic of Vietnam.

We are now preparing our report for presentation to the AAAS at the end of December and will send a copy to you as soon as it is available. Before that time, however, we wish to relay some observations that we believe merit your more immediate attention because they challenge the basis of the current chemical crop destruction program.

On August 21 and 28 we overflew an area in Quang Ngai province where chemical crop destruction operations had been conducted a few days before. The responsible chemical operations staff officer accompanied us on one of the overflights, and we spoke at length with other officers and civilian officials on the scene. It was explained to us that the targets were VC/NVA crop production areas and that most of the food destroyed would otherwise have been consumed by enemy forces. The reasons given for this assessment were that

1. The target area had only a very low population density.

2. The area under cultivation had expanded strikingly in recent years.

3. The cultivated area was much larger than that needed to support the small indigenous population.

4. The existence of numerous terraced rice fields indicated the influx of VC/NVA food production units, since the Montagnards who comprise the native population do not practice terracing.

Our observations are sharply at variance with all four of these points.

Fig. 2 Letter sent to Ambassador Ellsworth Bunker, General Creighton Abrams and Secretary of State William Rogers, November 12, 1970. From the Meselson CBW archive

- 2 -

Background information. One of the target areas is shown in photograph 1. Three C123 aircraft sprayed herbicide along the Song Re river valley from map coordinates BS 3455 to BS 3543. The photograph was taken near the latter coordinate, looking northward up the valley. The spray swaths are generally visible as brown streaks. Spraying was conducted continuously for a distance of about ten miles, resulting in coverage of approximately 1000 acres. Photographs 2 and 3 show the ground more closely. Numerous craters have been caused by the preparatory laying down of maximum suppressive fire, including the application of 300 per cent saturation with cluster bomb units, a measure required by the Seventh Airforce to protect the slow-flying C123's against hostile ground fire. Many dwellings may be seen in both close-ups, especially in photograph 3. The site of each of the photographs is indicated on the accompanying 1:50,000 scale U.S. Army map.

Population density. The map shows a high population density throughout the target area. In the twenty-seven 1 km^2 grid boxes through which the spraying passed, there are more than 900 dwellings, indicated as small black squares. Detailed comparison of the map with photographs 2 and 3 suggests that the number of dwellings in the target area is at least as great as it was in 1965, the year in which the map was last revised. Assuming, for example, an occupancy of six persons per dwelling, the population in the sprayed area would be approximately 5,000. This corresponds to 180 persons per square kilometer, hardly a low population density.

Expansion of cultivated area. The boundaries of cultivated fields on the enclosed map (and also on a matching 1:20,000 scale pictomap) agree well with the boundaries seen in our photographs, offering no evidence for any major expansion of crop production since 1965. Also, the fields seen in the photographs look well established and not of recent origin.

The question of surplus. The amount of land under cultivation in the target area may be estimated from the map to be approximately 800 hectares, about one hectare for each dwelling. Contrary to the view that a large surplus of food was being grown in the valley, one hectare of mountain land is just about enough to sustain a family unit.

Significance of terraced fields. Authoritative DoD publications on the Montagnard peoples as well as knowledgeable officials in Saigon state that the Montagnards of Quang Ngai, the Hre, have long grown rice on terraced fields.

We realize that this particular crop destruction mission may have been atypical. However, it was cited by the chemical staff officer and other officials as being particularly effective and well planned.

Moreover, our observations lend additional weight to several official studies done in Vietnam which have concluded that nearly all of the food destroyed by the chemical crop destruction program would normally be consumed by civilians, not by enemy soldiers.

Fig. 2 (continued)

- 3 -

As to the scale of the present crop destruction program, we were distressed by the implications of an analysis prepared by the GVN at the province level estimating the quantity of rice and other crops marked for chemical destruction under the 1970-71 herbicide program. Including missions requested by U.S. elements as well as those requested by the Vietnamese themselves, the total quantity of food scheduled for destruction is placed at 14,575 metric tons in Quang Ngai and Quang Tin provinces alone. This should be enough to sustain some 50-70,000 persons for a year. The targets are located mainly in upland regions where Montagnards are the traditional population. The Special Commission for Highland Affairs estimates the total Montagnard population of the two provinces as just under 70,000. Therefore if the areas we inspected are any indicator, the destruction of so much food or even of any substantial fraction of it would have devastating consequences for the Montagnard peoples of Quang Ngai and Quang Tin and for indigenous peoples in other provinces where similar conditions may prevail.

While we were in Quang Ngai province we had several occasions to interview Montagnard refugees whose lands had recently been sprayed with herbicide. We were impressed with the way in which they perceive the use of the chemicals. Apparently when the land is effected by herbicides, the Hre people consider it to be the manifestation of an evil spirit, and according to their tradition will abandon such land in the belief that it has fallen under a curse. Thus the folk beliefs of the Hre may intensify the effects of the crop destruction program on their lives.

Before leaving Vietnam we made a preliminary report of our observations to General Abrams and to Mr. Stephen Winship and Mr. Terence Grant of the Embassy Political Military section. Upon our return, one of us was requested to present a briefing at the State Department in Washington. The briefing took place on October 19 before a working level group consisting mainly of personnel from the Bureau of East Asian and Pacific Affairs and the Bureau of Intelligence and Research. We are also sending copies of this report to Secretary Rogers and to General Abrams.

We strongly hope that our observations can make a positive contribution toward bringing under review a program which seems to us very much in need of it.

Sincerely yours,

Matthew Meselson

John Constable

was an enemy food-production area, not home to a civilian population, the Chemical Corps officer who accompanied us in the helicopter and who had participated in planning the mission explained that the area under cultivation far exceeded the needs of the sparse population in the valley, that there had been a recent major expansion of rice fields, and that the presence of terraced rice fields on the hillsides, a form of rice culture practiced by ethnic Vietnamese but not by the indigenous Montagnard tribespeople, indicated that the area was an enemy crop production site.

Upon returning to the U.S., examining the high-resolution photographs we had taken from the air, comparing them with U.S. Air Force photographic coverage of the valley done in 1965 and consulting the Army's handbook on the Montagnard tribes of Vietnam and other sources, we found that all of the evidence for enemy crop production cited by the Chemical Corps officer was inaccurate or incorrectly interpreted. We therefore sent a letter describing our observations and a set of our photographs of the sprayed rice fields to Ambassador Bunker and General Creighton Abrams, Commander of U.S. forces in Vietnam and Secretary of State William Rogers (Meselson and Constable 1970) (Fig. 2). I had previously given a briefing on our observations at the State Department Bureau of Intelligence and Research and, in mid-December, had described our findings to President Nixon's National Security Advisor, Henry Kissinger (Guhin 1970; Hydle 1970; Buckingham 1982).

In November 1970, Bunker and Abrams initiated a review of the herbicide program with particular emphasis on crop destruction (Interagency 203 Committee 1970). After considering the resulting report they sent a telegram to Washington on December 9, saying they had decided that the crop destruction program should be phased out (Bunker 1970). Their recommendation leaked to the press and was published the following week (Jay 1970). On December 26, the day on which we reported our observations in Vietnam to the annual meeting of the American Association for the Advancement of Science in Chicago (Boffey 1971), President Nixon announced that "Ambassador Bunker and General Abrams are initiating a program for an orderly, yet rapid phase-out of the herbicide operations" and that during the phase-out, the use of herbicides in Vietnam would be restricted to perimeters of firebases and US installations or remote unpopulated areas (Office of the White House Press Secretary 1970). The last crop destruction mission was flown on January 7, 1971 (Cecil 1986; Young 2009). Thus, the large-area use of herbicides in Vietnam, already greatly reduced in 1970, came to an end at the start of the following year, two years before the Paris ceasefire agreement of January 1973 terminating direct US combat support for the Republic of Vietnam.

Two years later, in April 1975, President Gerald Ford proclaimed that

> The United States renounces, as a matter of national policy, first use of herbicides in war except use, under regulations applicable to their domestic use, for control of vegetation within U.S. bases and installations or around their immediate defensive perimeters (Executive Order 1975).

Postscript

While in Vietnam, I heard a spectrum of opinion from military officers, from pro to con, regarding the military utility of the herbicide program. A 1971 study conducted by the Department of Defense found that "[t]he military utility of herbicides has been conclusively established" (FRUS 1971). Particularly noteworthy, however, was the view expressed personally to me by General Creighton Abrams in his office in Saigon on September 3, 1970. "Do you want to know what I think? I think it's shit," he said, adding that his son John, then an Army captain who had served in Vietnam during 1967–69, was of the same view. When asked why the program continued even though he was Commander of U.S. forces in Vietnam, General Abrams replied that the decision to do so was made in Washington.

Acknowledgements I am grateful to General John N. Abrams, USA, Ret. and Captain Alvin L. Young, USAF, Ret. for helpful discussions.

References

Abel, Steffen, and Athanasios Theologis. 2010. Odyssey of auxin. *Cold Spring Harbor Perspectives in Biology* 2010 (2): a004572.

American Embassy Saigon. 1968. Report of the Herbicide Policy Review Committee, May 28. Meselson CBW Archive, Harvard University.

Baughman, Robert W. 1974. *Tetracholordibenzo-p-Dioxins in the Environment: High Resolution Mass Spectrometry at the Picogram Level*. Doctoral Dissertation, Harvard University.

Betts, Russell, and Frank Denton. 1967. *An Evaluation of Chemical Crop Destruction in Vietnam*. RAND Corporation, Santa Monica, CA, October 11. (RM-5446-ISA/ARPA). ADA032196. http://www.rand.org/pubs/research_memoranda/RM5446-1.html.

Bionetics Research Laboratories. 1968. *Evaluation of Carcinogenic, Teratogenic, and Mutagenic Activities of Selected Pesticides and Industrial Chemicals. Teratogenic Study in Mice and Rats*, vol II. Washington, D.C.: U.S. Department of Commerce. NTIS PB 223 160. http://www.nal.usda.gov/exhibits/speccoll/items/show/4100.

Boffey, Philip M. 1971. Herbicides in Vietnam: AAAS study finds widespread devastation. *Science* 171 (3966): 43–47.

Brown, James W. 1962. *Vegetational spray tests in South Vietnam*. Supplement. U.S. Army Chemical Corps Biological Laboratories, Fort Detrick, MD, April. AD368997. http://www.dtic.mil/cgi-bin/GetTRDoc?Location=U2&doc=GetTRDoc.pdf&AD=AD0368997.

Buckingham, William A. 1982. *Operation Ranch Hand: The Air Force and herbicides in Southeast Asia 1961–1971*. Washington, D.C.: Office of Air Force History, United States Air Force. ADA121709. https://media.defense.gov/2010/Sep/28/2001329797/-1/-1/0/AFD-100928-054.pdf.

Bundy, McGeorge. 1961. National Security Action Memorandum Number 115: Defoliant Operations in Vietnam, Nov 20. http://www.jfklibrary.org/Asset-Viewer/Z8-vBiaBdEW2Cz9g6NUB7g.aspx.

Bundy, William P. 1972. Recorded Interview by William W. Moss. John F. Kennedy Library Oral History Program, Mar 3. http://www.jfklibrary.org/Asset-Viewer/Archives/JFKOH-MGB-04.aspx.

Bunker, Ellsworth. 1970. Telegram to Secretary of State Washington. Subject: Herbicide Policy Review, Dec 9. Saigon 19374. Meselson CBW Archive.

Cecil, Paul F. 1986. *Herbicidal warfare: The Ranch Hand Project in Vietnam*. New York, London: Praeger.
Clary, James R. 1971. Project CHECO Report, Ranch Hand Herbicide Operations in SEA. Contemporary Historical Examination of Current Operations, Pacific Air Forces, July 13. ADA484753. http://www.dtic.mil/docs/citations/ADA121709.
Collins, Charles V. 1967. Project CHECO Report, Herbicide Operations in Southeast Asia, July 1961–June 1967, Oct 11. Contemporary Historical Examination of Current Operations, Pacific Air Forces. AD 779 796.
Connor, Steve, and Andy Thomas. 1984. How Britain sprayed Malaya with dioxin. *New Scientist* 101 (1393): 6–7.
Constable, John, and Matthew Meselson. 1971. The ecological impact of large scale defoliation in Vietnam. *Sierra Club Bulletin* 56: 4–9.
Crummett, Warren B. 2002. *Decades of dioxin: Limelight on a molecule*. Bloomington, IN: Xlibris.
Darwin, Charles, and Francis Darwin. 1880. *The power of movement in plants*. London: John Murray.
Enders, Tara A., and Lucia C. Strader. 2015. Auxin activity: Past, present, and future. *American Journal of Botany* 102 (2): 180–196.
Engineer Strategic Studies Group Office (ESSG). 1972. *Herbicides in military operations*, vol I, February. Washington, D.C.: Department of the Army. AD893214.
Executive Order. 1975. 11850. Renunciation of Certain Uses in War of Chemical Herbicides and Riot Control Agents, April 8. 40 Federal Register 16187.
Foreign Relations of the United States, 1961–1963, vol. 1, Vietnam. 1961. Document 96. https://history.state.gov/historicaldocuments/frus1961-63v01/d96.
Foreign Relations of the United States, 1969–1976, vol. E-2, Documents on Arms Control. 1971. Document 246. https://history.state.gov/historicaldocuments/frus1969-76ve02/d246.
Godel, W.H. 1961. Progress Report, Vietnam Combat Development and Test Center, September 13. Kennedy Papers, National Security File, Countries File, Vietnam (9/61) folder.
Guhin, Michael A. 1970. Memorandum to Henry Kissinger. Subject: Memcon of Meeting with Professor Meselson, Dec 17. Meselson CBW Archive.
Hay, Alastair. 1982. *The chemical scythe*. New York, London: Plenum.
Holland, Jennifer J., Diana Roberts, and Emmanuel Liscum. 2009. Understanding phototrophism: From Darwin to today. *Journal of Experimental Botany* 60 (7): 1969–1978.
Hydle, Lars H. 1970. U.S. State Department of Intelligence and Research. Memorandum to James D. Moffatt: Briefing by Matt Meselson on Herbicides, October 20. Meselson CBW Archive.
Institute of Medicine (IOM). 2014. *Veterans and Agent Orange: Update 2012*. Washington, D.C.: The National Academies Press. http://www.nap.edu/catalog/18395/veterans-and-agent-orange-update-2012.
Interagency 203 Committee. 1970. Report on the Herbicide Review, Nov 18. Hq. US MACV, Saigon.
Jay, Peter. 1970. US is urged to stop killing Vietnam crops. *Washington Post*, December 17.
Kraus, Ezra J. 1942. Plant Growth Regulators; Possible Uses, February 17. National Academy of Sciences Archives: Committees on Biological Warfare Records Group: WBC Committee: Report: Exhibit F,III. Plants 1.
Laird, Melvin. 1970. Memorandum for the Assistant to the President for National Security Affairs, Subject: United States Anticrop Warfare Program in Vietnam. With Enclosure: Assessment of the United States Anti-Crop Warfare Program in the Republic of Vietnam, July 18. Meselson CBW Archive.
Meselson, Matthew, and John Constable. 1970. Letter to Ambassador Ellsworth Bunker, November 12. Meselson CBW Archive.
Meselson, Matthew, and Robert Baughman. 1973. An analytical method for detecting TCDD (Dioxin): Levels of TCDD in samples from Vietnam. *Environmental Health Perspectives* 5: 27–35. http://www.ncbi.nlm.nih.gov/pmc/articles/PMC1474957/.

Meselson, Matthew S., Arthur H. Westing, John D. Constable, and Robert E. Cook. 1972. Preliminary Report and Background Material Relevant to Presentations at the 1970 Annual Meeting of the AAAS. Congressional Record, 92nd Congress, 2nd Session, vol. 118, part 6, Mar 3, 6806–6813.

Nelson, Bryce. 1969a. Studies find danger in defoliation herbicides. *Los Angeles Times*, October 30.

Nelson, Bryce. 1969b. Herbicides: Order on 2,4,5-T issued at unusually high level. *Science* 166 (3908): 977–979.

Office of the White House Press Secretary. 1970. President's Direction to Reduce the Use of Herbicides in Vietnam. Press Release, December 26.

Olenchuk, Peter G., Robert Burke, Oran Henderson, and Wayne Davis. 1963. *Task Force Saigon Herbicide Evaluation Team. Evaluation of Herbicide Operations in the Republic of Vietnam September 1962–September 1963,* October 10, 71–265, San Francisco: Military Assistance Commission, Task Force Saigon Herbicide Evaluation Team.

Peterson, Gale E. 1967. The discovery and development of 2,4-D. *Agricultural History* 41: 243–254.

Primack, Joel, and Peter von Hippel. 1974. *Advice and dissent: Scientists in the political arena.* New York: Basic Books.

Sorg, Oliver. 2014. AhR signalling and dioxin toxicity. *Toxicology Letters* 230: 225–233.

Stellman, Jeanne M., Steven D. Stellman, Richard Christian, et al. 2003. The extent and patterns of usage of Agent Orange and other herbicides in Vietnam. *Nature* 422: 681–687.

Troyer, James R. 2001. In the beginning: The multiple discovery of the first hormone herbicides. *Weed Science* 49: 290–297.

United States Department of the Interior. 1970. Press Release: Home Use of 2,4,5-T Suspended, April 15. http://www.nal.usda.gov/exhibits/speccoll/items/show/5799.

Warren, William F. 1968. A review of the herbicide program in South Vietnam. Alexandria, VA: Defense Technical Information Center, August. AD779797.

Washington Post. 1966. Embezzler Godel Sued to Repay Double, November 5, B4.

Westing, Arthur H. 1984. Herbicides in war: Past and present. In *Herbicides in war*, ed. Arthur H. Westing. London, Philadelphia: Taylor and Francis.

Whippo, Craig W., and Roger P. Hangarter. 2006. Phototropism: Bending towards enlightenment. *The Plant Cell* 18 (5): 1110–1119.

Young, Alvin. 2009. *The history, use, disposition and environmental fate of Agent Orange*. New York: Springer.

Zimmerman, P.W., and A.E. Hitchcock. 1942. Substituted phenoxy and benzoic acid growth substances and the relation of structure to physiological activity. *Contributions from the Boyce Thompson Institute* 12: 321–343.

The Indelible Smell of Apples: Poison Gas Survivors in Halabja, Kurdistan-Iraq, and Their Struggle for Recognition

Karin Mlodoch

Abstract On March 16, 1988 the Iraqi Army of Saddam Hussein's Baath regime attacked the Kurdish town of Halabja with poison gas, killing an estimated 5,000 people within a few minutes. In today's autonomous region of Kurdistan-Iraq, the "martyrs' town of Halabja" has become a symbol for the suffering of Iraqi Kurdish people under the Baath regime and a key element of Kurdish national identity. At the same time, the people of Halabja continue to suffer from the long-term psychological, health, and environmental consequences of the poison gas attack. The present account is based on the author's longstanding research and practical work among survivors of violence in Kurdistan-Iraq. It outlines the background and impact of the chemical attack on Halabja and provides an insight into the survivors' situation—from the immediate aftermath of the attack to this day; it details the constant struggle of the victims with the long-term psychological effects of the attack as well as their struggle for justice and recognition of their experience.

1 Introduction: The Indelible Smell of Apples

In 1987 and 1988, the Iraqi Baath regime under Saddam Hussein used chemical weapons against the Kurdish population in Iraq. On March 16, 1988 the Iraqi Army attacked the Kurdish town of Halabja by poison gas and killed an estimated number of 5,000 people in a few minutes.[1] Before and after the attack on Halabja, poison

[1]No accurate body count could be made at the time. The casualty figures are based on the testimonies of survivors, Kurdish *peshmerga* and Iranian soldiers, Iranian medical personnel, and journalists present in the immediate aftermath of the attack, (Hiltermann 2008). Human Rights Watch researcher Shorsh Resool collected 3,200 individual names of victims in interviews with survivors, Human Rights Watch (1993, 108). Kurdish and Iranian estimates ranged between 4,000 and 7,000 victims at the time (ibid.). Today, the figure of 5,000 victims is commonly used by Iraqi Kurdish sources and in the national Kurdish discourse and referred to in official memorial ceremonies and monuments such as the Central Halabja Monument.

K. Mlodoch (✉)
HAUKARI e.V, Berlin, Germany
e-mail: mlodoch@haukari.de

B. Friedrich et al. (eds.), *One Hundred Years of Chemical Warfare: Research, Deployment, Consequences*, DOI 10.1007/978-3-319-51664-6_18

gas was used in various other towns and villages during the so called Anfal Campaign of the Iraqi military against the Kurdish rural population in the border regions with Iran and Turkey, in the course of which thousands of villages were destroyed and some 100,000 people deported and killed. The Anfal Campaign is much less known to the outside world than the poison gas attack on Halabja.

The smell of apples is deeply imprinted into the memory of the survivors of poison gas attacks in Kurdistan-Iraq. All of them describe the intense smell of apples—or more precisely: the sweet smell of rotten apples that spread all over the place once the lethal poison-gas bombs touched the ground.

Visitors of today's Kurdistan-Iraq stumble over references to the apple as a memory symbol on many occasions. Apples decorated with cloves were traditionally used as a symbol of love and friendship in Kurdish communities; today, there is an additional tragic meaning to the symbol: the association with the lethal scent of chemical weapons. Visitors to the Ministry of Martyrs and Anfal Affairs in Erbil are given an apple studded with cloves with the inscription "a message of love and peace" as a gift. In the film "1001 Apples" by the late Kurdish director Taha Karimi (2013), survivors of the Anfal Campaign distribute apples to be decorated in remembrance of the victims. In the Zamwa Art Gallery in Sulaimania, a rocket shell with green apples gushing out of it like a waterfall symbolizes the Halabja attack. And at the 2015 ceremony for the victims of the chemical attack on Halabja at the Brandenburg Gate in Berlin, organized by Kurdish exile groups, young Kurdish women walked through the crowd of participants spraying air freshener with the scent of green apples to evoke the horror of the chemical weapons used.

2 Background

The Kurds often define themselves as the world's largest nation without a state. Indeed, after World War I and the collapse of the Ottoman Empire, their hopes for a Kurdish state were dashed and 30 million Kurds were scattered across five different nation states: Turkey, Syria, Iran, the former Soviet Union, and Iraq. Today, some 6 million Kurds live in Iraq alone and make up some 20% of Iraq's population. They inhabit a strategically important region of Iraq, rich in oil reserves and bordering on Iran and Turkey. There has been a conflict between the Kurds and the various Iraqi central governments at least since 1923.[2] The conflict with and the persecution of the Kurds intensified with the Baath Party's rise to power in 1968 and escalated when Saddam Hussein became president of Iraq in 1979. Saddam Hussein's regime spread a meticulous net of control, violence, and terror all over the Iraqi society, targeting Shia and Marsh Arab communities in the South, Kurds in the North and

[2]For the history of the Kurds in Iraq see McDowall (2004); van Bruinessen (1989).

political opponents of all ethnic and religious affiliations.[3] The Kurdish population in Iraq saw multiple waves of violence and persecution since the mid-1970s: Kurdish villages in border regions were evacuated and the population forcibly resettled; Kurds were expelled from oil-rich cities like Kirkuk and the regions repopulated by Arabs instead. Up to 1981 alone, 700,000 people had been forcibly displaced and relocated within the Kurdish areas (Vanly 1986, 163).

In 1980 Saddam Hussein led Iraq into a war with Iran that lasted eight years and left both countries devastated, with more than one million soldiers dead on both sides. Kurdish guerrilla fighters—the so-called *peshmerga*[4]—made use of the war situation, intensified their attacks against the Baath regime and entered into an alliance with Iran. The reprisals by the Baath regime reached genocidal proportions. In 1984, 8,000 men from the Kurdish Barzani tribe, the backbone of the Kurdish autonomy struggle during the 1960s and 1970s, were deported and killed—a prelude to the forthcoming episodes of mass killings in 1988 (Human Rights Watch 1993, 39–41). At the very end of the Iran-Iraq war, Saddam Hussein's regime threw its overblown military apparatus against the internal enemy: the Kurds.

3 The Anfal Campaign and the Poison Gas Attack Against Halabja in 1988[5]

In 1987 Saddam Hussein's cousin, General Ali Hassan al-Majid, was appointed General Secretary of the so called "High Office for Issues of the North" and assigned full military responsibility to—in al-Majid own words—"solve the Kurdish problem and slaughter the saboteurs" (Human Rights Watch 1993, 351). Al-Majid was the architect and commander of the so-called Anfal Campaign in 1988. Anfal is the title of the 8th sura of the Quran and served as a code word for a vast military operation against Kurdish rural areas where resistance fighters were active. The Anfal Campaign was planned long beforehand, openly announced, justified as punishment for the Kurds' cooperation with Iran, and meticulously documented by the regime. From February to September 1988 thousands of villages were razed to the ground. The population was rounded up; men between 15 and 60 and young women were separated from their families and deported to unknown destinations. For more than 15 years after Anfal, their fate remained unknown; the discovery of more than 300 mass graves after the fall of the Baath regime proved that these people fell victim to mass executions. Kurdish sources estimate the

[3]For a comprehensive analysis of the Baath regime's policy and repressive structure, see Farouk-Sluglett and Sluglett (1987); Al-Khalil (1989).

[4]Literally translated, the Kurdish term means "those who face death" and is used for Kurdish rebels and fighters, independently from their various party affiliations.

[5]A comprehensive documentation and analysis of the chemical attack against Halabja has been provided by Hiltermann (2007).

number of Anfal victims at 182,000.[6] As only a small number of the mass graves has been investigated, the individual fates of most of the victims are uncertain to this day.[7]

Elderly people and women with children were deported to detention camps and held there for months—many of them, especially children and the elderly, died. Those who survived were released in late autumn 1988 under what the regime called an "Amnesty" and forcibly resettled to so-called collective towns, where they continued to live under the control of the Iraqi Army (Human Rights Watch 1993, 306–311; Mlodoch 2014, 237–246).

In the course of the Anfal Campaign, the Baath regime used chemical weapons in dozens of locations (Human Rights Watch 1993, 22; Hiltermann 2007, 130); their use proved effective in making villagers flee in panic so that Iraqi soldiers could easily channel them towards concentration points, whence they were deported to be killed or detained (Hiltermann 2007, 130, 135).

Shortly after the beginning of the Anfal Campaign, on March 16, 1988, the Iraqi air force attacked Halabja, a Kurdish town of some 70,000 inhabitants located in the immediate proximity of the border with Iran, a reprisal for alleged joint cross-border operations of Iranian forces and Kurdish *peshmerga.* The town came under Iraqi attack days before by conventional weapons; therefore many of its inhabitants hid in provisional hand-dug underground air-raid shelters, basements or caves. These hideouts became deadly traps when the Iraqi Army threw tons of chemical agents on the town in the early evening of March 16.

The chemical attack lasted some 45 min. Survivors describe white-yellowish clouds that sank to the ground.[8] They describe the smell of apples[9]—some say it was rather a smell of garlic or banana, the sense of burning in eyes and on the skin, the inability to breathe.

An estimated 5,000 men, women, and children died a terrible death in Halabja. Many died immediately in the shelters or their houses; thousands ran out in panic to the streets and died there. Others tried to flee the town, but died minutes or hours later after "burning and blistering" and "coughing green vomit" (Human Rights

---

[6]This figure is based on the number of villages destroyed during the Anfal Campaign and the average village population and is generally used by Anfal survivors, Kurdish politicians, and local academics. Human Rights Watch estimates the number of victims as at least 50,000 and "possibly twice that number" (Human Rights Watch 1993, 20) after evaluating survivors' testimonies and the Baath regime's own documents regarding the Anfal Campaign. The responsible military commander of the Iraqi regime, Ali Hassan al-Majid, is reported as having confirmed "not more than 100,000 victims" (ibid., 345).

[7]Background, preparations and course of the Anfal Campaign have been meticulously documented by Human Rights Watch (1993; 1994) based on survivors' testimonies and the Baath-regime's own documents. For the long-term impact of Anfal on the survivors see Mlodoch (2014).

[8]1988: Thousands Die in Halabja Gas Attack. *BBC News*, March 16, 1988. http://news.bbc.co.uk/onthisday/hi/dates/stories/march/16/newsid_4304000/4304853.stm. Accessed October 30, 2015.

[9]Halabja: Survivors Talk About Horror of Attack, Continuing Ordeal. *ekurd daily*, March 15, 2008. http://ekurd.net/mismas/articles/misc2008/3/independentstate2078.htm. Accessed March 10, 2016.

Watch 1991). Survivors report victims hysterically laughing moments before their death (ibid.). Some 7,000 were severely injured—their eyes and skin burned.

Due to the presence of Iranian and Turkish journalists at the time, photos of the victims have gone around the world, giving stirring evidence of the destructive effects of chemical weapons: scores of corpses of men, women, and children in unnatural positions spread all over the streets of Halabja; dreadfully distorted faces of children captured at the moment of death; disoriented survivors, wandering among the corpses, crying out for their loved ones; images of injured survivors with burned eyes and skin peeling off from their faces, arms, and legs in big pieces.[10] These photos have been important as evidence, but are unbearable to look at both for the horror they document and for the viewer's sense of violation of intimacy of the victims at the moment of their agonizing death. One photo among the many has become especially well known: that of Omeri Khawer who throws himself upon his baby child in a desperate attempt to protect him at the moment of death. The image has become an icon in Kurdistan-Iraq; the scene has been reproduced as a diorama in the Halabja Central Memorial and in a statue erected in the centre of Halabja.[11]

Eyewitnesses who clearly identified Iraqi airplanes as the ones that dropped the chemical bombs and the Baath regime's own documents which were later evaluated by Human Rights Watch give evidence about the responsibility of the Baath regime (Human Rights Watch 1993; Hiltermann 2007). The commander of the poison gas attack against Halabja was once again Iraqi General Ali Hassan al-Majid. Ever since, Iraqi Kurds refer to him as "Chemical Ali."

The injuries of the victims, the testimonies and symptoms of the survivors of the Halabja attack as well as specimen of unexploded bombs analyzed in the immediate aftermath prove the use of highly concentrated mustard gas,[12] combined with at least one nerve agent, probably sarin (Hiltermann 2007, 199).[13]

---

[10]For some of the photos see the website of the Kurdistan Democratic Party, http://www.kdp.se/halabja.html. Accessed March 10, 2016.

[11]The original photo was taken by the Turkish photographer Ramazan Öztürk in the immediate aftermath of the chemical attack in 1988. For the original photo and its reproduction as a statue, see: http://www.hurriyetdailynews.com/halabja-monument-opens-in-the-hague.aspx?pageID=238&nID=65792&NewsCatID=359. Accessed March 10, 2016. For the diorama at the Halabja Monument, see photo by Adam Jones: www.flickr.com/photos/adam_jones/5640509079/in/photostream and www.hurriyetdailynews.com/halabja-monument-opens-in-the-hague.aspx?pageID=238&nID=65792&NewsCatID=359. Accessed March 10, 2016.

[12]See also the interview with the Austrian physician Dr. Freilinger who treated Halabja survivors in 1988: Kurdistan Regional Government Representation Austria. Halabja: Interview with Dr. Freilinger. *KRG AT Media,* March 14, 2014. http://www.gov.krd/a/d.aspx?l=12&a=51117. Accessed October 30, 2015.

[13]At the time Iraq possessed sarin, tabun, and VX (United Nations Monitoring, Verification and Inspection Commission 2006). However, as there were no medical or forensic investigations in the immediate aftermath of the Halabja attack, the process of finding evidence on the exact substances used in Halabja has proven difficult and controversial. See: Elisabeth Rosenthal. In Iraq Chemical Arms Trial, Scientists Face Many Burdens of Proof. *New York Times*, June, 19, 2006. http://www.nytimes.com/2006/06/19/world/europe/19iht-chem.2001719.html?_r=0&pagewanted=all.

At the time, there was no immediate international response to the poison gas attack on Halabja. The U.S. and West European governments then still stood firmly behind Saddam Hussein in his war against the Iranian Islamic Republic, which was regarded as the greatest danger for the West at the time. The international community remained silent.

Thus, the Anfal Campaign and the use of poison gas had fulfilled the Baath regime's aims: the Kurdish resistance was defeated, the Kurdish population frozen in shock and disbelief at the scale of the terror they had experienced and paralyzed by the prospect of seeing another poison gas attack. The Kurds felt forgotten by the world.

4 The Situation of Survivors of the Poison Gas Attacks

4.1 Haunting Memories, Enduring Grief

Survivors of the poison gas attack on Halabja have all gone through highly traumatic experiences. Their homes, their families, their social structures, their entire world was destroyed in a few minutes. They witnessed their children, partners, parents, siblings dying a terrible death right in front of them, while being themselves injured, immobilized, struggling with death and thus unable to help them. Many struggle with feelings of guilt for not having been able to protect their children or for not having been able to attend to their relatives during their last moments and adequately mourn their death thereafter. Their concepts of themselves and the world were deeply shattered.

Kamaran Haider was 11 years old then and survived hidden in a makeshift shelter. He stayed there for many hours. When he left the shelter, he found his father, mother, and siblings lying dead on the stairs. "I lost my feeling, all my feeling," he recounts. "I knew that my mum died. I knew that my brother died [...]. At that time, I didn't cry [...]. I didn't feel anything. No happiness, no stress. Well, I knew that people around me died, that's it."[14]

Iranian soldiers and the Iranian Red Crescent took hundreds of survivors to Iran for medical treatment after the attack. There was an indescribable chaos, panic, and fear of more attacks to come and of the gas that was still lingering around. Injured and uninjured alike were hectically loaded on trucks and brought outside the town and into Iran. Many survivors lost track of their family members.

(Footnote 13 continued)

Accessed March 10, 2016. Hiltermann (2007, 183–205) gives a detailed account on the controversial process of fixing the evidence.

[14]Alfred Joyner. Kurdish Genocide in Iraq: Survivors Tell Their Stories. *International Business Times*, January 18, 2013. http://www.ibtimes.co.uk/kurdish-genocide-in-iraq-survivors-tell-their-stories-2028. Accessed March 10, 2016.

Many of the severely injured survivors spent years in Iran going from one hospital to another for medical treatment. Others were taken abroad for treatment—to Austria, Germany, and other countries, unaware of what happened to their relatives. A number of children, babies at the time, could not be identified by Iranian authorities and were adopted and brought up by Iranian families. In 2011, the heartbreaking story of Ali Pour was covered by Kurdish and international media and a documentary film (Hidou 2011).[15] Ali grew up as the son of an affectionate Iranian family. At the age of 21 he learned that he was originally a baby survivor of the poison gas attack in Halabja. He went back to Halabja in search of his original family. Three families claimed and hoped for Ali to be their lost son. The evidence provided by a genetic test led to an outbreak of joy in one family and to a new emotional breakdown in two others. Ali, whose original name is Zimnako Mohammed Salih, went the Halabja graveyard and erased his name he found engraved on one of the tombstones.[16]

To this day memories and images of the attack are deeply impressed upon the survivors' minds and thoughts. They suffer from what can be clearly defined as traumatic symptoms: nightmares, anxiety, restlessness, depression, panic attacks, flash-backs. Ako Sirini's documentary film "There is Hope Behind the Tears" (2013), based on testimonies of survivors, shows the intensity and presence of pain and grief in survivors twenty five years after the event. The young man, a child of maybe ten years in 1988, was taken to Iran with his siblings after the attack and spent weeks hoping and waiting for his parents to join them. He describes the moment of reunion with his uncle who brought certainty that his parents were dead: "Imagine at that young age, I did not cry because of the presence of the other children. From the day I received that news, I behaved differently. To this day, I haven't cried as much as my heart ached for them" (Sirini 2013, 10:46–11:11). A woman in her seventies stated: "When I think, every moment is like death for me. This pain is not like a soul; once it leaves, it never comes back. The pain lies within you forever" (Sirini 2013, 17:38–17:49).

4.2 Survivors' Life Conditions Between 1988 and 2003

We know from psychological trauma research that for victims of extreme violence and trauma the ability to recover and reconstruct their lives largely depends on the life conditions they find themselves in after the traumatic experience.[17] Safety,

[15]See also: Halabja Gas Attack Survivor Reunited With Mother. *BBC News Online*, December 5, 2009. http://news.bbc.co.uk/2/hi/8397547.stm. Accessed March 10, 2016.

[16]Other similar cases have been covered by both Kurdish and international media, see e.g.: Osamah Golpy. Halabja Child To Be Reunited With Family 3 Decades Later. *Rudaw*, August 18, 2015. http://rudaw.net/english/kurdistan/180820154. Accessed March 9, 2015.

[17]Psychoanalyst Hans Keilson was the first to draw attention to the significance of post-trauma life conditions for coping with traumatic episodes in his post-war studies of child Holocaust survivors in the Netherlands. He developed the concept of sequential traumatization (Keilson 1979).

economic and political stability, social support, societal and political recognition of their experience, and punishment or at least accountability of the perpetrators help victims of extreme violence to recover and find a balance between the past and the present. On the other hand, ongoing conflict and violence, poverty, impunity of the perpetrators, and a lack of assistance prolong and aggravate their suffering and keep them frozen in the moment of their trauma.[18]

The life conditions of the Halabja survivors in the years following the attack were more than adverse to any kind of recovery. First of all, many survivors died in subsequent years from their injuries.[19] Numerous others suffered from the long-term impact of the poison gas, from skin and eye diseases, damage to the respiratory and neural systems, and various forms of cancer, infertility, miscarriages and congenital disorders. Even in the third generation, the rate of leukemia is high among children (Gosden et al. 2001). The soil in and around Halabja will remain contaminated for a long time to come (Ala'Aldeen 2005).

Immediately after the attack, the Baath regime had razed the town of Halabja to the ground; there was no possibility of return. Many survivors stayed in Iran; others continued to hide in the mountains. The majority of survivors were forcibly resettled by the Baath regime in a so called *mujamma* (collective town) cynically named "New Halabja" at a distance of some 70 km from the destroyed town. Like other urban settlements built for Anfal survivors, "New Halabja" was a camp-like structure with housing, schools, and medical facilities. Its population lived under military control and was forbidden to leave and, especially, to set foot in Halabja.[20]

In 1991 a U.S.-led military coalition attacked Iraq in reaction to the latter's invasion of Kuwait. The Kurds in Iraq—seeing finally a chance to defeat the dictator—followed the U.S. call to arms, but were let down and abandoned by the coalition after its armistice with Iraq. The Kurdish insurrection was subsequently brutally crushed by Iraqi forces. At the moment of defeat, two million Kurds fled in utmost panic to Iran and Turkey, leaving behind all their belongings. The images of their mass exodus went around the world in the Spring of 1991. The panic among the fleeing Kurds could only be explained by their fear of further poison gas attacks that had been deeply imprinted into the minds of the Kurdish people in 1988.

---

[18]The author refers to a socially and contextualized concept of trauma as developed in the work with Holocaust survivors, victims of torture and detention in Latin America and victims of political and gendered violence in various contexts. For an overview on the related concepts and debates see Mlodoch (2014, 29–66).

[19]Decades Later, Halabja Victims Still Dying of Wounds from Chemical Attacks. *Rudaw*, November 27, 2015. http://www.rudaw.net/english/kurdistan/271120153?keyword=Halabja&isArchive=True. Accessed March 10, 2016.

[20]Osamah Golpy. Tale of Two Cities: Halabja and New Halabja. *Rudaw,* March 24, 2015. http://rudaw.net/english/kurdistan/240320152. Accessed March 10, 2016.

In response to the mass exodus, Dutch and British troops set up safe havens for the refugees and the United Nations declared a no-fly zone over Iraq to prevent the Baath regime from carrying out airstrikes.[21] In the wake of these events the Kurds achieved provisional autonomy and a Kurdish government was democratically elected in 1992. Anfal survivors started to reconstruct their villages and Halabja poison gas attack survivors returned to their home town, from then on referred to by all Kurds as *Halabja shehid*—the "martyr's town of Halabja."

Despite these first steps toward their safety, the survivors would stay in precarious conditions for another twelve years. The Kurdish Regional Government had not been internationally recognized; the Kurdish region suffered from international sanctions against the whole of Iraq and an additional embargo from Baghdad imposed on the Kurds. There was no trustworthy agreement about Kurdish autonomy with the Iraqi regime, and the fear that the Baath-regime will come back was pervasive throughout those years. Iran and Turkey frequently intervened militarily in the unstable region and from the mid-1990s the two main Kurdish political parties, the Kurdistan Democratic Party (KDP) and the Patriotic Union of Kurdistan (PUK), engaged in a bloody internal struggle for power and resources that caused further violence and death in the region.[22]

During all those years until 2003, Halabja poison gas survivors lived in a city of ruins, in extreme poverty, facing multiple new waves of violent conflict, and fearing that the Baath regime will come back and the catastrophe will recur. Absorbed in a daily struggle for survival they had no possibility to rest, take care of themselves, reconstruct their town and lives and thus regain some trust, courage, and hope. Instead, anger and bitterness added to their injury and exacerbated their suffering. In the 1990s the people of Halabja were outraged about the lack of assistance by both the Kurdish parties and government and the international community. They denounced Western countries' previous complicity with the Iraqi Baath regime and its production of chemical weapons and urgently called upon the international community to engage in the reconstruction of Halabja. They felt betrayed and forgotten once again.[23]

4.3 Changes After 2003

The situation only changed in 2003 with the second U.S.-led invasion of Iraq. In the run-up to the invasion, U.S. president George W. Bush made frequent reference to Saddam Hussein's use of poison gas against his own population. After years of

[21]An overview on the developments and legal aspects of the international "humanitarian intervention" and the establishment of safe havens in Northern Iraq after the Kurdish mass exodus 1991 is given by Cook (1995).

[22]Winter (2002) provides a comprehensive analysis of the provisional situation in Kurdistan-Iraq during the 1990s.

[23]The author's own observations visiting Halabja in the 1990s and talking to survivors.

silencing, the chemical attack against Halabja now became an argument for legitimizing the invasion.[24] This time the military invasion led to the demise of the Baath regime and as such was enthusiastically welcomed and celebrated all over Kurdistan. For the first time after 1988, a sense of safety was restored to the Halabja survivors. The sanctioning of Kurdistan as an autonomous region in a federal state of Iraq by the Iraqi constitution of 2005 brought about a sense of political stability. The main perpetrators—Saddam Hussein and Ali Hassan al-Majid— were sentenced to death by the Iraqi High Tribunal and executed. Al-Majid received four death sentences for crimes against humanity, war crimes, and genocide against the Kurds and was hanged in 2010. Saddam Hussein was already executed on December 30, 2006 for the massacres against Shiites in Dujail before the Anfal and Halabja trials had come to an end.[25] Internationally, there has been a highly controversial debate about the legitimacy of the trials because of the strong U.S. role in the set up of the trials, the victor's justice involved, and the non-compliance with international law standards.[26] However, for the Anfal and Halabja survivors, these trials—the fact that survivors gave testimonies in a court of law facing the main perpetrators—were important milestones for restoring their sense of justice and satisfaction. Yet many survivors were disappointed that Saddam Hussein was not executed for Halabja and Anfal, as they wanted his death to be linked in the historical record to these crimes (Mlodoch 2014, 364–365).

After 2003 the Kurdish region saw a rapid process of economic development and modernization, which brought improvement to the life conditions of the Anfal and Halabja survivors as well. The Kurdistan Regional Government finally started to invest in the destroyed areas' infrastructure. Survivors' pensions were raised; they received grants for building houses and their children stipends for university or college education. Today, the survivors' economic situation has improved. Those who were children during the chemical attacks have meanwhile grown up, started their own families, finished their education, and became a source of pride for the entire survivors' community. Indeed there are strong collective structures and a sense of community among the survivors due to the shared experience. All these changes have at last created the possibility for the survivors to take some rest and engage in the reconstruction of their town and their social structures.

Mamosta Fakhradin, who saw two of his children die in his arms on March 16, 1988, is today a teacher at a primary school in Halabja and says that he regards each of his students as his own child (Hidou 2011). A young man in Ako Sirini's 2013 documentary who was a baby when he lost his parents in the gas attack talks about the day he took his degree at the Medical School. However, he said it was a sad day

[24]See, e.g., George W. Bush's speech at the 2003 Azores Summit in Portugal. President Bush: Monday "Moment of Truth" for World on Iraq. March 16, 2003. http://georgewbush-whitehouse.archives.gov/news/releases/2003/03/20030316-3.html. Accessed March 10, 2016.

[25]For the documentation of the trials see International Center for Transitional Justice (2006a, b).

[26]For an overview on the debate see Mlodoch (2014, 348–358).

for him as he imagined how proud he would have made his parents coming home with the university certificate. He is now practicing medicine in Halabja. The other young man in the film who lost his parents has become an artist. He gives art lectures to young people and says that he wants to bring color back to Halabja. They are examples of how survivors live with haunting memories and indelible images, but at the same time try to relate to the present and engage in reconstructing their lives.

Today Halabja is step by step turning into a lively town again. Streets and markets are crowded; the town has playgrounds, schools, a university, women centers, and cultural projects. The reconstruction of the town of Halabja gives a sense of triumph to survivors over the destructive impact of the poison gas. Yet the scars of the past and its representation are visible everywhere. Besides the huge graveyard for the victims with the endless-looking rows of 5,000 tombstones and the huge central memorial at the outskirts of the town, there are numerous smaller memorials, art pieces and wall inscriptions reminding of the poison gas attack all over the town.

However, there is still an intense feeling of rage and bitterness among Halabja survivors. They feel exploited by the Kurdish national discourse and political elite, who define the chemical attack against Halabja as a national trauma but fall short of addressing the survivors' claims and needs. At the occasion of an official remembrance ceremony at the Halabja anniversary in 2006, survivors turned against the attendant Kurdish politicians and their guests, demanding better services instead of high-profile ceremonies and ultimately set the central Halabja monument ablaze (Hiltermann 2008).[27]

Survivors are also bitter about the lack of international assistance and recognition. They strongly call upon international governments and parliaments to recognize the Anfal and Halabja attacks as genocidal and to take to justice the international companies which delivered supplies for the poison-gas production to Saddam Hussein's regime. Indeed, although UNSCOM inspections and the Baath regime's confiscated documents brought evidence about the implication of European and specifically West German companies in Iraq's poison gas production in the 1980s (Kelly 2013), there has not been to date any noteworthy legal prosecution of those responsible. Only one Dutch businessman, Frans van Anraat, whose company had delivered thousands of tons of chemical substances to the Iraqi regime in the 1980s, has been convicted to 17 years in prison for supporting war crimes by a District Court in The Hague in 2005, but was acquitted of the charge of supporting genocide (Oñate et al. 2007).

[27]See also Robert F. Worth. Kurds Destroy Monument in Rage at Leadership. *New York Times*, March 17, 2006. http://www.nytimes.com/2006/03/17/international/middleeast/17kurds.html?_r. Accessed March 10, 2016.

Fig. 1 On the anniversary of the poison gas attack, activists in Halabja protest against the chemical attack on Ghouta, Syria. March 16, 2015. WADI e.V. https://wadi-online.org/2016/03/16/halabja-day-2016-end-the-impunity/. Accessed June 26, 2017. Photo reproduced with permission from WADI e.V.

5 New Threats, New Fear, and Joint Initiatives of Poison Gas Survivors

After a period of stabilization, the Kurdish people in Iraq are currently facing new threats. In 2014 the terror militia ISIS took control of large parts of Central and Northern Iraq and committed horrendous massacres of especially the Yezidi Kurds. Once again Kurdish *peshmerga* are fighting and dying in combat, this time against ISIS. Once again the region is mired in conflict and violence, which stir up the traumatic memories of the Halabja and Anfal survivors. Over a million people who fled from ISIS terror in the provinces of Mossul and Central Iraq and another 250,000 Syrian Kurds who fled the war in their country currently seek refuge in Kurdistan-Iraq.[28]

Evidence suggests that in August 2013 chemical weapons were used against civilians by the Syrian government of Bashar al-Assad in the Ghouta area of Damascus (UN Mission to Syria 2013). Survivors from Ghouta and survivors from Halabja jointly founded an initiative named "Breathless."[29] On April 22, 2015, the

[28]United Nations High Commissioner for Refugees. Iraq 3RP Summary 2016: Regional Refugee and Resilience Plan. March 3, 2016. https://data2.unhcr.org/en/documents/details/44046. Accessed March 10, 2016.

[29]See https://www.facebook.com/Breathless-830986016974121/.

100th anniversary of the first use of poison gas by the German Army during World War I at Ypres, they came together in parallel commemoration activities in Kurdistan-Iraq and Syria and jointly called upon the international community to curb the use of chemical weapons in warfare, stating: "It takes one second to drop the bomb, but it takes decades to overcome its impact".[30] On the occasion of the anniversary of the chemical attack on Halabja on March 16, 2015, activists in Halabja commemorized the attack on their town in 1988 and at the same time protested against the chemical attacks in Ghouta. They held up signs asking: "After Halabja you said: Never again. After Ghouta you said: Never again. What will you say next time?"(Fig. 1).

References

Ala'Aldeen, Dlawer. 2005. Long term hazards of chemical weapon agents: analysis of soil samples from Kurdistan years after exposure to sulphur, mustard and nerve agents. DlawerNet. http://www.dlawer.net/?q=node/71. Accessed 30 October 2015.

Al-Khalil, Samir. 1989. *The republic of fear: the politics of modern Iraq*. London: Century Hutchinson.

Bruinessen, Martin van. 1989. *Agha, Sheikh und Staat. Politik und Gesellschaft Kurdistans*. Berlin: Edition Parabolis.

Cook, Helena. 1995. *The safe haven in Northern Iraq: International responsibility for Iraqi Kurdistan*. Essex, London: Human Rights Centre and Kurdistan Human Rights Project.

Farouk-Sluglett, Marion, and Peter Sluglett. 1987. *Iraq since 1958: From revolution to dictatorship*. London, New York: Routledge and Kegan Paul.

Gosden, Christine, Derek Gardener, and Mike Amitay. 2001. Effects of Genotoxic Weapons of Mass Destruction on Civilian Populations. In *Disarmament Forum*. Geneva: United Nations Institute for Disarmament Research.

Hiltermann, Joost. 2007. *A poisonous affair: America, Iraq, and the gassing of Halabja*. New York: Cambridge University Press.

Hiltermann, Joost. 2008. Halabja: The Politics of Memory. International Crisis Group. https://www.crisisgroup.org/middle-east-north-africa/gulf-and-arabian-peninsula/iraq/halabja-politics-memory. Accessed 10 March 2016.

Human Rights Watch. 1991. Whatever happened to the Iraqi Kurds? Human Rights Watch Report March 11. https://www.hrw.org/reports/1991/IRAQ913.htm. Accessed 10 March 2016.

Human Rights Watch. 1993. *Genocide in Iraq*. Human Rights Watch: The Anfal Campaign Against the Kurds. New York.

Human Rights Watch. 1994. *Bureaucracy of Repression*. Human Rights Watch: The Iraqi Government in its Own Words. New York.

International Center for Transitional Justice. 2006a. Dujail. Trial and Error. ICJT Briefing Paper, November 2006. https://www.ictj.org/sites/default/files/ICTJ-Iraq-Dujail-Trial-English-2006.pdf. Accessed 10 March 2016.

International Center for Transitional Justice. 2006b. The Anfal Trial and the Iraqi High Tribunal. Update Number 3: The Defence Phase and Closing Stages of the Anfal Trial. https://www.ictj.org/sites/default/files/ICTJ-Iraq-Anfal-Tribunal-2006-English2.pdf. Accessed 10 March 2016.

[30]Open Letter by Survivors of Chemical Attacks from Iraq, Syria, Iran and Kurdistan. "It Takes a Second to Drop a Bomb, But It Takes Decades to Overcome Its Impacts." *Breathless*. April 22, 2015. https://www.facebook.com/Breathless-830986016974121/. Accessed March 10, 2016.

Keilson, Hans. 1979. *Sequentielle Traumatisierung bei Kindern: Deskriptiv-klinische und quantifizierend-statistische follow-up Untersuchung zum Schicksal der jüdischen Kriegswaisen in den Niederlanden*. Stuttgart: Enke.
Kelly, Michael J. 2013. "Never Again"? German Chemical Corporation Complicity in the Kurdish Genocide. *Berkeley Journal of International Law* 31 (2): 348–391.
McDowall, David. 2004. *A modern history of the Kurds*. London: Palgrave Macmillan.
Mlodoch, Karin. 2014. *The limits of trauma discourse: Women Anfal survivors in Kurdistan-Iraq*. Berlin: Klaus Schwarz Verlag.
Oñate, Santiago, Brigitta Exterkate, Lisa Tabassi, and Erwin van der Borght. 2007. Lessons learned: Chemical traders convicted of war crimes. *Hague Justice Journal* 2 (1): 23–42.
United Nations Mission to Investigate Allegations of the Use of Chemical Weapons in the Syrian Arab Republic. 2013. Report on the Alleged Use of Chemical Weapons in the Ghouta Area of Damascus. United Nations, August, 21. https://undocs.org/A/67/997. Accessed 30 October 2015.
United Nations Monitoring, Verification and Inspection Commission (UNMOVIC). 2006. Summary of the Compendium of Iraq's Proscribed Weapons Programme in the Chemical, Biological and Missile Areas. United Nations. http://www.un.org/Depts/unmovic/new/documents/compendium_summary/s-2006-420-English.pdf. Accessed 30 October 2015.
Vanly, I. Cherif. 1986. *Kurdistan und die Kurden*, vol. 2: *Türkei und Irak—Fortsetzung*. Göttingen: progrom-Reihe Bedrohte Völker.
Winter, Bernhard. 2002. *Kurdistan-Irak: Untergehen im sicheren Hafen*. VAS: Studie über die Folgen einer humanitären Intervention. Frankfurt am Main.

Films

Hidou, Akram. 2011. *Halabja: The lost children*. Kurdistan-Iraq/Germany: Akram Hidou.
Karimi, Taha. 2013. *1001 Apples*. Kurdistan-Iraq: Ava Media/Taha Karimi.
Sirini, Ako. 2013. *There is hope behind the tears*. Kurdistan-Iraq: Ministry of Martyrs and Anfal Affairs.

The Use of Chemical Weapons in Syria: Implications and Consequences

Ralf Trapp

Abstract Chemical weapons are banned under customary international law, the 1925 Geneva Protocol and the 1997 Chemical Weapons Convention (CWC). The CWC today has achieved near universal adherence; a small number of states, however, remain outside its realm. Syria—until 2013 one of them—was long presumed to possess chemical weapons and in 2012 effectively admitted so. The Syrian civil war always carried the risk that one side or another would use these weapons. Reports to this end began to appear in 2012. In March 2013, following separate requests by Syria and several Western States, the UN Secretary-General began to investigate these allegations. Whilst the investigation team was in Damascus, a large-scale sarin attack was launched on Ghouta, killing hundreds of people. This incident and its subsequent confirmation by the UN team set in motion a series of unprecedented events leading to the elimination of Syria's chemical weapons stockpile under strict international control, supported by financial and in-kind assistance by more than 20 countries. But this multilateral effort did not end the use of toxic chemicals in Syria, and OPCW fact-finding missions have since confirmed several cases of chlorine attacks. Also, ISIS/Daesh reportedly has used chemical weapons including chlorine and mustard gas in Syria and Iraq. The paper concludes that it will be important to identify the perpetrators of these attacks and bring them to justice in order to protect the international norm against poison gas.

1 Introduction

On April 29, 1997, after two decades of negotiations in the Geneva Conference on Disarmament and another four years of work of the Preparatory Commission in The Hague, the Chemical Weapons Convention (CWC) entered into force. As of the

R. Trapp (✉)
International Disarmament Consultant, Chessenaz, France
e-mail: ralf.trapp@gmail.com

B. Friedrich et al. (eds.), *One Hundred Years of Chemical Warfare: Research, Deployment, Consequences*, DOI 10.1007/978-3-319-51664-6_19

summer of 2015, this global disarmament treaty, which aims at abolishing once and for all the threat of chemical warfare, had attracted 191 states as its parties.[1] Near universal adherence to this comprehensive ban combines with efforts of the States Parties to internalize its prohibitions and set them firmly into their domestic legislation, penal codes and administrative and enforcement systems. The chemical weapons arsenals and production facilities of the vast majority of countries are under the control of the Convention's treaty organization—the Organization for the Prohibition of Chemical Weapons (OPCW)—and measures to eliminate them are nearing completion. An effective verification system with on-site inspections has been established to verify declarations by the States Parties, ensure that all declared chemical weapons and production facilities are being destroyed, and that no new chemical weapons are produced. To many observers, it appeared that the menace of chemical warfare had become a matter of the past.

But whilst chemical weapons were successfully being taken out of the armory of the majority of countries that had acquired them in the past, a small number of countries remained outside the realm of the CWC.[2] Some of them are presumed to have chemical weapons in their arsenals. The dangers that emanate from the continuing presence of chemical weapons in military arsenals became apparent when Syria spiraled into civil war after the "Arab Spring".

2 Early Reports of Chemical Weapons Use in the Syrian Civil War

In July 2012, a spokesperson of the Syrian Foreign Ministry effectively admitted that Syria possessed chemical weapons; he went on to say that Syria would not use these weapons except against an external aggression (MacFarquhar and Schmitt 2012). Around the same time, reports appeared that claimed that terrorist groups were smuggling chemical weapons from Libya and other countries via Turkey into Syria. Beginning in late 2012, reports emerged on the Internet and in the media alleging the use of chemical weapons in Syria by government forces. At the same time, government sources in Syria claimed that terrorist groups were using improvised chemical weapons.

Whilst the Independent Commission of Inquiry into Syria of the UN Human Rights Council in February 2013 still concluded that there was no credible evidence for the use of chemical weapons by either side (UNGA 2013), allegations and counter-allegations of such uses were increasingly leveled against the Syrian government but also against certain factions of the military opposition, including elements of the Free Syrian Army and groups such as *Jabhat al Nusra*. Then, in

[1]For status updates, see http://www.opcw.org/about-opcw/member-states/status-of-participation/.

[2]As of October 2015, Israel—a signatory state—has yet to ratify its signature, and Egypt, South Sudan, and North Korea have yet to accede to the Convention.

March of 2013, the UN Secretary-General received a request from the government of Syria to investigate an alleged chemical weapons attack against government troops and civilians in Khan Al Asal near Aleppo.[3]

According to the Syrian report, 25 people had died and more than 110 civilians and soldiers been injured as the result of an attack with what was said to have been a chemical rocket. The Syrian government accused the opposition of being responsible for the chemical attack whilst an opposition military spokesperson blamed government troops for it. Subsequently, France and the United Kingdom reported allegations of the use of chemical weapons in several locations in Syria, including in Khan Al Asal and near Damascus, Aleppo, and Homs, near the Turkish border, and in other places, blaming the Syrian regime as the instigator (UNODA 2013, 2). They requested that the UN investigation be mandated to investigate all these alleged incidents. The United States, who had initially been cautious about confirming any chemical weapons use by either side, eventually also concluded that the government of Syria had used chemical weapons in several instances (ibid., 3–4).

3 The Secretary-General's Investigation

In response to these requests, the UN Secretary-General established an investigation mission under the leadership of Swedish scientist Åke Sellström, using a mechanism that had been devised during the Cold War as a tool to investigate alleged breaches of the 1925 Geneva Protocol (Ban 2013a). The mission was composed of a group of inspectors of the OPCW, medical experts of the WHO, and supported by UN staff in New York as well as Syria and the region. But Sellström's team was for several months unable to go into Syria, whilst in New York negotiations continued behind closed doors about which specific incidents and locations the team would be allowed to scrutinize.

All along, however, video footage and photographic images spread through social media and the Internet, showing what appeared to be further victims of poison gas attacks. There also were reports from journalists who had managed to gain access to activists in Syria or who had in fact spent time with one or another group of fighters, and who had collected information about the alleged chemical weapons attacks. Several States extended their own intelligence gathering operations in Syria. Russia, France, the United Kingdom, and the United States all acquired samples from different sources and locations of alleged gas attacks, and analyzed them for the presence of chemical agents. They concurred that their analysis showed that the nerve agent sarin had been used. But whilst Russia concluded that terrorist groups fighting for the opposition had used the nerve agent,

[3]For a chronology of the allegations and the steps taken by Secretary-General and his investigation team, see the Final Investigation Report (UNODA 2013).

Western countries concluded that it had been the Assad regime who was responsible for the gas attacks (Robinson 2013, 28–34; Gladstone 2013).

Any use of chemical weapons would, of course, constitute a war crime. Not only have chemical weapons been banned by the 1925 Geneva Protocol and the 1997 Chemical Weapons Convention, but there is also a common view that under customary international law, any use of a chemical weapon in armed conflict would constitute a prohibited act (ICRC 2005). Amongst the many atrocities committed in the Syrian civil war, and meticulously recorded by the Independent Commission of Inquiry into Syria of the Human Rights Council, the use of chemical weapons was specifically noted as an act that is prohibited in all circumstances under customary international humanitarian law, and a war crime under the Rome Statute (UNGA 2013).

But the situation remained unclear for some time. Sellström's investigation team could not get into Syria, and much of the information could not be independently verified. Videos and images posted on the Internet did not allow an unequivocal confirmation that sarin had in fact been used; the authenticity of the samples analyzed by the different countries could not be demonstrated independently; videos and photographs on the Internet appeared to show both Syrian army and opposition fighters preparing and firing what were described to be chemical rockets; and, unsurprisingly, the propaganda war surrounding the allegations and the debate about how the international community should respond increased in intensity, obscuring both the facts of the matter and the underlying intentions of the different actors involved. The stakes were high, framed by fears about possible military strikes on the one side and hopes for military intervention and support on the other.

After several months of negotiations between the UN and Syria about which particular locations the team would be allowed to access and investigate, the UN Secretary General's investigation mission was finally dispatched to Damascus in the middle of August 2013. Whilst the team was making final preparations for its fieldwork at its operations base in Damascus, in the night of August 21, 2013, a major gas attack was launched against civilian quarters in Ghouta, a suburb just outside of Damascus. Within hours, videos posted on the Internet showed large numbers of victims arriving in hospitals for treatment of what clearly were poison gas injuries, among them many women and children. The precise number of casualties has never been established and would be difficult if not impossible to verify—casualty figures ranged from 355 to more than 1500 people killed plus many more injured—but there was little doubt that a war gas had been used (France 2013; United States 2013).

What followed had no precedence in the history of chemical weapons disarmament. First, within days, despite the on-going civil war and in fact despite having been attacked by sniper fire on one occasion, the UN investigation team negotiated its way into the attacked sites and carried out as thorough an investigation as was possible under the conditions of the civil war. In the compressed time available, the team interviewed a significant number of victims of the attack, medically examined many of them and took blood samples for subsequent analysis. It also inspected several of the impact sites and conducted interviews with eye witnesses, assessed the remnants of the weapons used, estimated likely flight trajectories, and took samples from impact craters and from the weapons themselves (UN Mission 2013, p. 4 and appendices 5 and 6). Subsequent analysis of these biomedical and environmental samples undertaken by OPCW designated laboratories confirmed beyond doubt that sarin had been used, in significant amounts, in the attack on Ghouta.

The Sellström team subsequently also investigated other incidents of alleged chemical attacks in Syria and was able, with varying degrees of certainty, to confirm that sarin had been used in a number of cases (UNODA 2013, 19–21). The incidents for which the use of the nerve agent was confirmed included both scenarios where Syrian government forces were accused of having used chemical weapons and scenarios where the opposition had been accused of having done so. The UN team was not mandated, however, to establish which party had used the chemical weapon. In fact, such an attribution would have required a different type of investigation, and access rights as well as investigative tools reaching beyond what the team could bring to bear under the circumstances and the scope of the UN Secretary-General mechanism (for example, reference data with the precise analytical signatures of the chemical agent mixtures present in the Syrian CW stockpile and access to documents and individuals in the military command structures of the parties involved in the incidents).

4 The Elimination of the Syrian Chemical Weapons Program

As political pressure mounted after the confirmation of the use of sarin in Ghouta, and the risk of external military intervention increased, a second remarkable development took place: a last-minute agreement was reached in Geneva between the Russian Federation and the United States of America about a framework for the elimination of Syria's chemical weapons program (OPCW 2013a), combined with Syria's accession to the Chemical Weapons Convention on September 13, 2013 and its declaration that it would apply the Convention's norms and obligations immediately, even before becoming a State Party.

Within a mere two weeks, the OPCW Executive Council transferred this general framework agreement into detailed modalities for the disablement and destruction of the Syrian chemical weapons arsenal and production capacity (OPCW 2013b).[4] On the very same day, resolution 2118 (2013) of the UN Security Council

[4]Syria joined the CWC after the expiration of the ten-year destruction period prescribed under Article IV(6) of the CWC for the complete elimination of CW stockpiles (which expired April 29, 2007). In such a situation, the OPCW Executive Council under Article IV(8) of the CWC establishes the destruction deadlines and related verification measures for the newcomer State Party. This enabled the conversion of the bilateral US-Russian framework into a legally binding undertaking under the CWC without a formal amendment of the CWC.

harnessed this decision and added the strength of the UN Charter to this endeavor.[5] Within days, OPCW inspectors began inspecting the Syrian chemical weapons facilities to confirm weapons and equipment inventories, advise Syria on how to apply the requirements of the CWC, and start preparations for the destruction and removal operations envisaged. In October, a Joint Mission of the United Nations and the OPCW was established to implement and supervise the chemical weapons disarmament of Syria, led by Sigrid Kaag.

Syria declared a total of 41 CW stockpile and production facilities at 23 locations (OPCW 2013c). 18 of them were CW production facilities, 12 were storage facilities, and there were 8 mobile filling units and 3 other CW-related facilities. The declared stockpile amounted to 1000 tons of category-1 chemicals (these are chemicals listed in Schedule 1 of the CWC and included the sarin key precursor DF, a key precursor for VX, and mustard gas). There also were some 290 tons of other chemicals that formed part of the CW stockpile, 1230 items of unfilled munitions, and 2 cylinders which the Syrian government claimed did not belong to it but might contain chemical agent.

By the beginning of November, the initial inspections of the Syrian chemical weapons stocks and production facilities had been completed (OPCW 2013c). Some sites could only be inspected by authenticated video links but these turned out to be empty. The vast majority of the declared stockpile and production locations were inspected by OPCW inspectors, and their inventories of weapons, materials, and specialized production and filling equipment verified and secured. The stationary and mobile production and mixing units were functionally disabled: they were physically damaged so that they could no longer be used for their intended purpose. That was important because the deployment system that the Syrian army had developed for its main chemical weapon—the nerve agent sarin—depended on mixing two precursor materials (DF and iso-Propanol) and an acid scavenger (hexamethylene tetramine) to form sarin shortly before the weapons were deployed and used. Once the mixing and filling equipment had been rendered inoperable under the watchful eyes of the inspectors, that deployment system could no longer be employed and the Syrian army had lost its means of delivering sarin effectively, even though it still had access to the precursor and other chemicals it would need to make sarin.

The preparations for the transportation of most of the chemical materials of Syria's stockpile out of the country began immediately after the initial inspection campaign was completed, supported by an Operations Planning Group established by the OPCW to coordinate the support of Member States and the Joint Mission (OPCW 2013d) for the removal and destruction operation. This removal and the decision to destroy the chemical weapons materials outside of Syrian territory were two of several modifications to the rules of the CWC that became necessary under the special circumstances of Syria (Trapp 2014a). The sequence of this operation

[5]This resolution also removed any possible doubts about the legal power of some of the elements of the Executive Council decision, given the effect of Article 103 of the UN Charter.

followed a simple logic: to prevent any further use of these chemical weapons, one had to immediately disable the system for delivering the chemical agents, then remove as soon as possible all chemical weapons materials that would have required dedicated destruction facilities—which did not exist—from Syrian soil, destroy the remaining material (Isopropanol) in Syria, and then destroy the materials removed from Syria as soon as practicable. At the same time, these measures had to prevent any new production of chemical weapons, and get the weapons, materials and specialized equipment away from anyone who might want to lay their hands on them to use them.

This approach involved adapting some of the legal and regulatory provisions of the Chemical Weapons Convention and its verification procedures to the special situation in Syria (Krutzsch et al. 2014). That in itself was a novelty in the way the OPCW implemented the CWC. States Parties are usually reluctant to change treaty provisions, in particular when it comes to arms control and security agreements. But special circumstances warranted special measures and the political will to tweak the provisions of the treaty and make them work under these unusual conditions did prevail over the usual caution states exercise in treaty interpretation.

Equally important, the elimination of Syria's chemical weapons program under the conditions of an on-going civil war required a high degree of international cooperation and support for what turned into a complex multilateral overland and maritime operation. The removal operation was supported by countries as diverse as Denmark, Norway, the UK, Germany, China, Russia, and the USA, coordinated with the help of the Joint OPCW-UN Mission and special task forces set up for the different aspects of the operation (OPCW 2014a). The Mediterranean Seaport of Latakia was selected as the exit point for removing the weapons materials from Syria. The chemical weapons materials were packed in ISO transport containers and moved by road to Latakia in several shipments, beginning in January 2014. The last shipment arrived in Latakia in June 2014, some six months later than originally planned; but given the circumstances, this was a remarkable achievement.

Upon arrival of each shipment at the port of Latakia, the containers were loaded onto transport ships provided by Denmark and Norway. These transport ships did not permanently dock in Latakia but remained at sea or in Cyprus most of the time, protected by a multilateral contingent of navy ships from a number of countries including China and Russia. Once the removal operation was completed (on June 23, 2014), the transport ships sailed under escort to the Italian container port of Gioia Tauro where the most dangerous materials (sulfur mustard and DF) were reloaded onto the US maritime vessel Cape Ray. The remaining materials were shipped to facilities in the US, the UK and Finland. The effluents from the destruction aboard the Cape Ray would eventually be shipped to facilities in Germany and Finland for final treatment.

5 Chemical Weapons Destruction at Sea

A critical step was the destruction of the sarin precursor DF and the mustard gas. Well before Syria's accession to the CWC, at the beginning of 2013, the United States had commissioned the development of a field-deployable CW destruction facility that could be airlifted to wherever it was to be deployed, and after a 10-day set-up period could operate independently of any local supplies and destroy chemical agents and precursors by chemical hydrolysis (CBARR 2013). The system was called the "Field Deployable Hydrolysis System" or FDHS. The design capacity of the system would allow it to destroy, under normal operating conditions, the entire Syrian stockpile in approximately 30 days.

The original plan had been to find a country in the neighborhood of Syria that would be willing to host the mobile destruction facility (or to offer one of its destruction facilities if it had one) to destroy the Syrian chemical materials. But a combination of regulatory, political, and technical factors worked against this plan, and no country could be found to take on the destruction task or host the US facility. As time progressed it became clear that a more unconventional approach was needed to eliminate Syria's chemical weapons materials in the agreed timeframe.

In November 2013, it became clear that the only remaining option to eliminate the Syrian CW stockpile outside of Syria would be to destroy the most dangerous chemical materials at sea (OPCW 2013e). The US began to reconfigure its field-deployable hydrolysis facility so it could be set up inside a Navy Reserve container ship, the Cape Ray. A chemical operation that would under normal conditions have been set up on land on an area the size of several football fields now had to be stacked into the three dimensions of a container ship.

The destruction of chemical weapons on a floating platform is not in itself a new concept, but the technical difficulties that needed to be overcome for operating a chemical destruction plant at high sea were nevertheless significant. Here are some of the issues that needed to be considered and resolved before destruction operations could begin in earnest (Trapp 2014b):

- Safety: the need to protect sailors, workers, and OPCW inspectors on board as well as the environment, in case of any spills or accidents, was paramount. Many of the procedures for operating the destruction plant and for responding to incidents needed to be adapted to the compact and at the same time three-dimensional work environment on board the Cape Ray. It also required very careful planning of how the different types of containers had to be loaded, to determine which containers went onto which deck and place within a deck in order to minimize the need to move containers with highly toxic chemicals between or across decks, and to facilitate the way in which the containers' content could be piped to bring the toxic/precursor chemicals and the reaction water and caustics needed for the hydrolysis to the plant, and for back-loading the reaction masses into empty containers. The operation was to be essentially self-contained, that is to say, all chemicals needed for the plant operation were on board; no additional materials should be brought on board whilst the facility was in operation; and no reaction masses were to be taken off the ship until the destruction operation was completed.

- The operational environment: normally, a chemical plant operation requires a stable platform to work on, but a ship at sea moves with and within the environment; it was important to establish under which sea condition the facility could be operated safely, and under which it needed to be shut down. Also, the pace with which the facility was being operated needed to be adjusted to the environmental conditions, and operations had to be phased in gradually to make sure the operators had full control over the process and equipment. A significant amount of sea training with the facility operating in simulation mode was undertaken before actual destruction operations could begin.
- Environmental protection: the need to prevent any leaks and accidental releases of the precursor materials and agents as well as of the reaction masses into the environment. This was a particular concern given the vulnerability of the maritime environment in the Mediterranean Sea, and quickly became a political issue in many of the countries with coastlines in the Mediterranean Sea (Walker et al. 2014; Üzümcü 2014).
- Verifiability: despite the unusual environment, the destruction operation had to be undertaken in a way that was consistent with the requirements of the CWC, in particular with regard to the systematic monitoring of all destruction operations by OPCW inspectors to ensure full accountability and to confirm that all chemical weapons materials that had been removed from Syria were in fact destroyed.

Despite these complexities, the destruction at sea proceeded without incidents. It began in late June 2014, and on August 18, 2014, Secretary of State John Kerry confirmed that 100% of DF (methylphosphonyl difluoride), the precursor for sarin nerve agent, as well as 20 tons of mustard gas had been destroyed aboard the Cape Ray (Kerry 2014). The reaction masses were shipped to facilities in Germany and Finland for final disposal by incineration.

The importance of this collective effort to eliminate the Syrian chemical weapons arsenal cannot be overstated. It was a complex multilateral operation implemented by the OPCW, verified by its inspectors, and supported by financial and in-kind contributions from 21 different countries and the European Union. A total of 10 countries supported the operation with a range of assets and practical support measures, ranging from the delivery of transportation containers to the provision of security and naval support, making available port facilities, and undertaking actual destruction operations of the declared chemical agents and precursors as well as of other chemicals and the reaction masses of the primary destruction operation on board the Cape Ray.

This complex and truly collaborative effort took a stockpile of extremely dangerous and effective chemical weapons out of the context of a horrendous civil war that has in the meantime spread well beyond Syria's borders and become a regional insurgency with global ramifications. Work on completing the destruction of the

Syrian CW production facilities has yet to be completed but is well under way. The OPCW will continue to verify that the production of chemical weapons in Syria will not be resumed.

6 New Incidents of Chemical Weapons Uses

But chemical weapons disarmament cannot remove all toxic chemicals from a country. Many materials that are in daily use in society are poisonous and could be used in improvised chemical weapons. We have today confirmation that improvised chemical weapons in the form of chlorine-filled barrel bombs are being used in Syria (OPCW 2014b). The OPCW has undertaken several fact-finding missions and has on a number of occasions confirmed with a high degree of confidence that chlorine has been employed as a means of war fighting in Syria (OPCW 2015). There is also no doubt that "Daesh" and groups associated with it have shown an interest in acquiring chemical weapons, and reports—yet to be independently verified—suggest that Daesh may have used mustard gas against Kurdish fighters in Iraq (Associated Press 2015).

The OPCW's Executive Council as well as the UN Security Council have condemned the use of chlorine in the Syrian armed conflict, and further investigations are being conducted by the OPCW Fact-Finding Mission and the OPCW-UN Joint Investigation Mechanism to establish what actually happened. Does this imply that, as a consequence of the recent uses of toxic chemicals in the Syrian conflict, the threshold against the use of toxic chemicals in armed conflict has in fact been lowered? Statements by political leaders and international bodies including the Security Council and the OPCW seem to point in the opposite direction; to use Ban Ki Moon's words: "the use of chemical weapons by any side under any circumstances would constitute an outrageous crime with dire consequences, and a crime against humanity" (Ban 2013b). But to prevent a lowering of the threshold for the use of chemical weapons, it will be important to bring those responsible for the use of chemical weapons in Syria to justice.

7 Attribution and Accountability

Issues of attribution and accountability were not taken up by the investigations under the UN Secretary-General Mechanism (UN-SGM). Investigating culpability would have required a type of investigation different from what the SGM is: a science-based fact finding mission. This is not to say that science cannot help in identifying the responsible individuals who ordered the use of sarin, but a fact-finding mission conducted at the sites of the alleged chemical weapons use and hosted by the Syrian government differs in certain respects from a criminal investigation.

There are, however, other international mechanisms such as the Human Rights Council's Independent Commission of Inquiry for Syria or the International Criminal Court (ICC), which could be used to investigate issues related to attribution and accountability. Whilst the Syria Commission has been actively collecting and securing evidence with regard to alleged human rights violations, by all parties involved in the Syrian conflict, the ICC has not been activated. Any use of chemical weapons in armed conflict would fall under its jurisdiction under the Rome Statutes as amended in Cartagena; but as Syria is not a member of the ICC, this would have required an express authorization by the Security Council.

However, the Security Council did not make use of this mechanism but instead decided to establish a separate, dedicated investigation mechanism: a Joint Investigation Mechanism of the UN Secretary-General in close coordination with the OPCW (UN 2015). It is too early to comment on how effective this mechanism will be, but past experience strongly suggests that it will only yield results if the Security Council remains united in its support of the work of the Joint Investigation Mission, and can avoid the politicization of issues related to its mandate, scope and the nature of the investigation.

8 Conclusions

Many lessons have been learnt and are still being learnt in the process of eliminating the chemical weapons program of Syria. Some issues that are being pursued already by the United Nations, the OPCW, the WHO, and other relevant actors include:

- Further strengthen the authority and operational capacity of the UN Secretary-General Mechanism to investigate allegations of the use of chemical and biological weapons
- Organize effective interagency cooperation to implement complex and demanding operations in dangerous and potentially hostile environments
- Appreciate the role that disarmament and arms control can play (but also their limitations) in extreme circumstances including during armed conflict
- Preserve and strengthen the ways in which effective multilateral collaboration can be orchestrated despite existing disagreements on a number of key policy and security issues.

References

Associated Press. 2015. German intelligence confirms Isis used mustard gas in Iraq, says news report. *The Guardian*, September 8. http://www.theguardian.com/world/2015/sep/08/german-intelligence-confirms-isis-used-mustard-gas-in-iraq-says-news-report. Accessed 27 Oct 2015.

Ban Ki-moon. 2013a. Secretary-general's press encounter on Syrian government request. New York, March 21. http://www.un.org/sg/offthecuff/index.asp?nid=2745. Accessed 27 Oct 2015.
Ban Ki-moon. 2013b. Secretary-General's remarks to the third review conference of the states parties to the Chemical Weapons Convention. *The Hague*, April 8, http://www.un.org/sg/statements/index.asp?nid=6722 . Accessed 27 Oct 2015.
Chemical, Biological, Application and Risk Reduction Business Unit (CBARR). 2013. The field deployable hydrolysis system. *CBARR News* 1.8 Special Edition. http://www.ecbc.army.mil/cbarr/newsletter/2013/CBARR_August2013.pdf. Accessed 29 Sept 2016.
France. 2013. Syria/Syrian Chemical Programme—National executive summary of declassified intelligence. Paris, September 3. http://www.diplomatie.gouv.fr/en/IMG/pdf/Syrian_Chemical_Programme.pdf . Accessed 27 Oct 2015.
Gladstone, Rick. 2013. Russia says study suggests Syria rebels used sarin. *The New York Times*, July 9. http://www.nytimes.com/2013/07/10/world/middleeast/russia-says-study-suggests-syria-rebels-used-sarin.html. Accessed 27 Oct 2015.
ICRC. 2005. Customary law. *International Review of the Red Cross* 87.857, Rule 74.
Kerry, John. 2014. *Milestone in Eliminating Syria's Chemical Weapons Programme.* Press Statement, August 18. www.state.gov. Accessed 27 Oct 2015.
Krutzsch, Walter, Eric Myjer, and Ralf Trapp. 2014. Part Six. Annex to the commentary: Issues raised by the accession of Syria to the chemical weapons convention. *The chemical weapons convention: A commentary,* eds. Walter Krutzsch, Eric Myjer, and Ralf Trapp, 689–701. Oxford: Oxford University Press.
MacFarquhar, Neil, and Eric Schmitt. 2012. Syria threatens chemical attack on foreign forces. *The New York Times,* July 23. http://www.nytimes.com/2012/07/24/world/middleeast/chemical-weapons-wont-be-used-in-rebellion-syria-says.html. Accessed 27 Oct 2015.
OPCW 2013a. Russian Federation and the United States of America. *Framework for Elimination of Syrian Chemical Weapons.* Executive Council document EC-M-33/Nat.1, September 17.
OPCW 2013b. *Destruction of Syrian Chemical Weapons*. Executive Council Decision EC-M-33/Dec.1, September 27. https://www.opcw.org/?id=1902. Accessed 27 Oct 2015.
OPCW 2013c. *Progress in the Elimination of the Syrian Chemical Weapons Programme*. Note by the Director-General, Executive Council document EC-M-34/DG.01, October 25.
OPCW 2013d. *Progress in the Elimination of the Syrian Chemical Weapons Programme*. Note by the Director-General, Executive Council document EC-M-35/DG.01, November 25.
OPCW 2013e. *Progress in the Elimination of the Syrian Chemical Weapons Programme*. Note by the Director-General, Executive Council document EC-M-36/DG.01, December 2.
OPCW 2014a. *Progress in the Elimination of the Syrian Chemical Weapons Programme*. Note by the Director-General, Executive Council document EC-M-44/DG.02, August 25.
OPCW 2014b. *OPCW Fact Finding Mission: "Compelling Confirmation" That Chlorine Gas Used as Weapon in Syria.* Press Release, September 10. https://www.opcw.org/news/article/opcw-fact-finding-mission-compelling-confirmation-that-chlorine-gas-used-as-weapon-in-syria/. Accessed 27 Oct 2015.
OPCW 2015. *Allegations of Toxic Chemical Attacks in Syria.* Press Release, March 25. https://www.opcw.org/news/article/allegations-of-toxic-chemical-attacks-in-syria/. Accessed 27 Oct 2015.
Robinson, Julian Perry. 2013. *Alleged Use of Chemical Weapons in Syria*. Harvard Sussex Project Occasional Paper 4. http://www.sussex.ac.uk/Units/spru/hsp/occasional%20papers/HSPOP_4.pdf. Accessed 29 Sept 2016.
Trapp, Ralf. 2014a. Elimination of the chemical weapons stockpile of Syria. *Journal of Conflict and Security Law* 19 (1): 7–23.
Trapp, Ralf. 2014b. *The Elimination of State Chemical Weapons Programmes: Past Experiences and Current Challenges.* Paper presented at the 4th International Conference on Industrial and Hazardous Waste Management, September 2–4. Chania (Crete), Greece.
UNGA. 2013. United Nations General Assembly Document A/HRC/23/58, July 18, 2013. Paragraph 136. https://documents-dds-ny.un.org/doc/UNDOC/GEN/G13/156/20/PDF/G1315620.pdf?OpenElement. Accessed 27 Oct 2015.

UN Mission. 2013. United Nations Mission to investigate allegations of the use of chemical weapons in the Syrian Arab Republic. Report on the alleged use of chemical weapons in the Ghouta Area of Damascus on 21 August 2013. http://www.un.org/zh/focus/northafrica/cwinvestigation.pdf. Accessed 27 Oct 2015.

UNODA. 2013. United Nations Office of Disarmament Affairs. Final Report of the United Nations Mission to investigate allegations of the use of chemical weapons in the Syrian Arab Republic. https://unoda-web.s3.amazonaws.com/wp-content/uploads/2013/12/report.pdf. Accessed 27 Oct 2015.

UN. 2015. Security Council Unanimously Adopts Resolution 2235 (2015), establishing mechanism to identify perpetrators using chemical weapons in Syria. Security Council meetings coverage. http://www.un.org/press/en/2015/sc12001.doc.htm. Accessed 27 Oct 2015.

United States, 2013. U.S. Government Assessment of the Syrian Government's Use of chemical weapons on August 21, 2013. https://www.whitehouse.gov/the-press-office/2013/08/30/government-assessment-syrian-government-s-use-chemical-weapons-august-21. Accessed 27 Oct 2015.

Üzümcü, Ahmet. 2014. Letter of OPCW Director-General to the Pancretan Commission Against the Destruction of Syrian Chemical Weapons in the Enclosed Sea of the Mediterranean Sea. July 29, 2014. https://www.opcw.org/fileadmin/OPCW/ODG/uzumcu/DG_Letter_Pancretan_Commission.pdf. Accessed 27 Oct 2015.

Walker, Paul, et al. 2014. Open Letter to Secretaries John Kerry and Chuck Hagel: public outreach and stakeholder involvement in destruction of Syrian chemical weapons. http://www.the-trench.org/open-letter/. Accessed 27 Oct 2015.

Part IV
Commemoration Ceremony

A Century of Chemical Warfare: Building a World Free of Chemical Weapons

Paul F. Walker

Abstract The first major use of chemical weapons in warfare was on April 22, 1915, when Germany attacked Allied forces along the Ypres Salient in Belgium in World War I. Since that historic attack a century ago, dozens of countries have researched, developed, tested, and deployed still more deadly chemical weapons. These inhumane and indiscriminate weapons of mass destruction were again used in 1924 by Spain against Morocco, by Italy against Libya and Ethiopia in the 1920s and 1930s, and by Japan against China in World War II (Robinson 1971). More recently they were deployed by Iraq against Iran and Iraq's Kurdish population in the 1980s, and from 2012 to the present in the Syrian civil war. The 1993 Chemical Weapons Convention (CWC) in 2016 includes 192 countries, 98% of the world's population, with only four countries—Egypt, Israel, North Korea, and South Sudan—still missing. And of the 72,525 metric tons of chemical agents declared to date in eight possessor states, over 66,000 metric tons—92%—have been safely destroyed in the last 25 years. This is a historic achievement in global disarmament and peace-building and needs to continue until we rid the world of all chemical weapons, prevent their re-emergence, and promote peaceful uses of chemistry.

1 Introduction

Chemicals have been used as weapons for centuries, primarily in poison arrows and darts and in targeted assassination attempts. But just a century ago, on April 22, 1915, a chemical, in this case chlorine, was used on a massive scale in major warfare. The advance of the German 4th Army against Ypres, Belgium, in November 1914 had been stalemated for months by British, French, Belgian, Canadian, Algerian, Senegalese, and other Allied forces which were dug into trenches along the Ypres Salient in World War I. At 5 o'clock in the afternoon of April 22, when the wind had finally turned to blow from the northeast, German

P.F. Walker (✉)
Environmental Security and Sustainability, Green Cross International, Geneva, Switzerland
e-mail: pwalker@globalgreen.org

B. Friedrich et al. (eds.), *One Hundred Years of Chemical Warfare: Research, Deployment, Consequences*, DOI 10.1007/978-3-319-51664-6_20

troops opened valves on 5,700 canisters of pressurized liquid chlorine and let the green-yellowish toxic industrial chemical mix with air, turn gaseous, and waft across the Allied trenches.

The Allied forces, primarily Canadian, British and French, had two bad choices—either remain in their trenches and choke on the chlorine gas, or jump above the heavier-than-air toxic cloud and be machine-gunned by the German troops. This impossible choice caused thousands to be killed that day as the Germans advanced on the town of Ypres from the northeast.

Over the next several years of major warfare, over 190,000 tons of chemical agents, much more deadly than chlorine, were produced and used by Germany, France, Britain, the United States, and others in WWI. As a result, some 90,000 troops were killed and another million or more injured with chemical weapons in the war.[1] A famous photo of Allied troops, walking one behind the other with their arms on the shoulders of the soldier in front of them and their eyes bandaged, illustrates the horrible injuries of the widespread use of mustard agent in WWI, and the international outcry thereafter against the cruel and inhumane use of chemicals in warfare.

On April 22, 2015, a very moving nightly memorial ceremony took place at the Menin Gate Memorial to the Missing in Ypres, Belgium, dedicated to those British soldiers who took part and died in the Ypres Salient and who remain missing—still over 54,000 names engraved on the gate. This historic ceremony has been ongoing since 1928 except for the years during World War II when Ypres was occupied by Germany, and recently surpassed its 30,000th evening ceremony.

2 The Geneva Protocol

On June 17, 1925, just a few years after the end of World War I, the Protocol for the Prohibition of the Use in War of Asphyxiating, Poisonous or Other Gases, and of Bacteriological Methods of Warfare, otherwise known as the Geneva Protocol, was opened for signature. This historic international treaty stated:

> Whereas the use in war of asphyxiating, poisonous or other gases, and of all analogous liquids, materials or devices, has been justly condemned by the general opinion of the civilized world; and Whereas the prohibition of such use has been declared in Treaties to which the majority of Powers of the world are Parties; and To the end that this prohibition shall be universally accepted as a part of International Law, binding alike the conscience and the practice of nations.[2]

[1]For a fuller account of estimated casualties in World War I, see https://en.wikipedia.org/wiki/Chemical_weapons_in_World_War_I. We will never know exact numbers of deaths and injuries from chemical agents in WWI, but Michael Duffy in *Weapons of War—Poison Gas*, estimates 90,198 deaths and 1,140,655 injuries from use of chemicals, see http://www.firstworldwar.com/weaponry/gas.htm.

[2]See the United Nations Office of Disarmament Affairs Treaties Database, accessed at http://www.un.org/disarmament/treaties/t/1925.html.

36 countries, including a number of chemical weapons states and victims of WWI chemicals—Austria-Hungary, Belgium, Canada, France, Germany, Italy, Russia, the United Kingdom, and the United States—signed the protocol that day, but many of the early signatories took decades to ratify and join the chemical ban. The Geneva Protocol banned the use of both chemical and biological weapons, but unfortunately did not limit research, development, testing, evaluation, and stockpiling of these weapons. It entered into force on February 8, 1928, but only 25 countries, including France, Germany, the Soviet Union, and the UK, had ratified the treaty by 1930; other chemical weapons powers such as the United States did not ratify it until many decades later—1975 in the case of the US. And of 137 States Parties today, 46 or one-third have ratified or acceded with reservations. The US and other reservations stated that the country reserved the right to retaliate in kind if attacked by chemical and/or biological weapons. US President Franklin Roosevelt stated specifically in 1943 that the "use of such [gas] weapons has been outlawed by the general opinion of civilized mankind."[3]

3 Chemical Weapons Convention

Fortunately the broad condemnation of chemical weapons after World War I appeared to impact the non-use of them for over half a century thereafter. Although many countries researched, developed, and produced thousands of tons of chemical agents and launch systems before, during, and after World War II, few countries deployed them in any major WWI-type attacks. Britain used adamsite against Russian troops in 1919 and possibly against Iraq in the 1920s; Bolsheviks used gas in the peasant Tambov Rebellion of 1920 in Russia; Spain used chemicals against Moroccan RiF tribesmen in the 1920s; and Italy used mustard agent against Libya in 1930 and in Ethiopia in 1936. Japan had shipped thousands of chemical weapons with their troops to China in WWII, but left them behind after minimal use when they retreated back to Japan. Both Italy and Germany produced large chemical weapon stockpiles in WWII, along with Russia, the United States, and Britain, but never used them in Europe except for a 1939 reported accidental attack by Germany on Warsaw with mustard agent. And Egypt used chemical weapons, although ineffectively, in Yemen in 1963.[4]

In 1968 Sweden was successful in placing both biological and chemical weapons on the agenda of the Eighteen Nation Disarmament Conference (ENDC) convening in Geneva, Switzerland and co-chaired by the Soviet Union and the United States. A year later the United Kingdom introduced a draft convention banning biological weapons which eventually resulted in the Biological and Toxin

---

[3]See https://www.state.gov/t/isn/4784.htm for the Roosevelt quote and a brief history of the Geneva Protocol. Also http://disarmament.un.org/treaties/t/1925.

[4]See https://en.wikipedia.org/wiki/Chemical_weapons_in_World_War_I.

Weapons Convention of 1972.[5] Article IX of the BTWC (or more commonly, BWC) was a step towards a ban on chemical weapons by stating:

> Each State Party to this Convention affirms the recognized objective of effective prohibition of Chemical Weapons and, to this end, undertakes to continue negotiations in good faith with a view to reaching early agreement on effective measures for the prohibition of their development, production and stockpiling and for their Destruction, and on appropriate measures concerning equipment and means of delivery specifically designed for the production or use of chemical agents for weapons purposes.[6]

The BWC entered into force in 1975, effectively banning all development, production, stockpiling, and use of biological weapons although it did not include any verification or inspection mechanisms. This was the first time such an international treaty banned a whole class of weapons of mass destruction and it thereby opened the door to follow-on negotiations on a similar chemical weapons ban. In 2016 the BWC has 175 States Parties and eight signatories, but 13 countries still remain outside its regime.[7]

The issue of chemical weapons remained part of ongoing discussions at the United Nations in Geneva and a number of countries tabled various drafts throughout the next decade. Both the United States and the Soviet Union recognized the need to limit, perhaps even to eliminate, their large and dangerous stockpiles of chemical weapons and initiated a bilateral working group thereon. In 1978 the Geneva group of 40 countries established an ad hoc working group on a chemical weapons treaty, and an increased focus thereon was launched with a US draft convention in 1984 (Kenyon and Feakes 2007, especially Chap. 1). The use of chemical weapons by Iraq against Iran in the 1980s, the first major use of chemical agents in warfare since WWI, also increased pressure on diplomats to negotiate a treaty; it is estimated that 20,000 Iranians died, and another 80,000 were injured, from Iraqi chemical attacks throughout the 1980's Iran-Iraq War (Kenyon and Feakes 2007, 9–10).[8]

The Geneva Conference on Disarmament (CD), newly renamed in 1980, began ongoing negotiations on a chemical weapons ban, and intensified bilateral discussions between the US and the Soviet Union, now with a new and forward-looking president, Mikhail Gorbachev, led to a bilateral agreement to reduce US and Soviet CW stockpiles to 20% of their then-current inventories and to reduce them down to 5,000 metric tons each by 2002. This agreement, called the "Wyoming

[5]For the text of the BWC, see http://www.unog.ch/80256EE600585943/(httpPages)/04FBBDD6315AC720C1257180004B1B2F? OpenDocument.

[6]https://www.state.gov/t/isn/4718.htm English.pdf.

[7]http://www.unog.ch/80256EE600585943/(httpPages)/7BE6CBBEA0477B52C12571860035FD5C? OpenDocument.

[8]See https://en.wikipedia.org/wiki/Chemical_weapons_in_World_War_I.

Agreement," was signed in 1990 but never entered into force. It however allowed both the US and the Soviet Union to move forward unilaterally and reciprocally with the destruction of their existing large CW stockpiles.[9]

Another chemical weapons attack, again by Iraq under Saddam Hussein's leadership, took place in 1988 against Kurdish citizens in northern Iraq, including the town of Halabja, killing several thousand Iraqi citizens and shocking the world. This also propelled the negotiations forward in Geneva on a multilateral treaty banning these inhumane and indiscriminate weapons.[10]

In 1990 the US began unilaterally operating its prototype incinerator for the destruction of chemical weapons on Johnston Atoll in the middle of the Pacific Ocean. The US had secretly moved forward-deployed chemical stockpiles from both Germany and Okinawa to this remote atoll over a decade earlier and had accumulated 1,842 metric tons of mustard agent there.

After a series of diplomatic breakthroughs including agreement on intrusive on-site inspections, sharing of peaceful chemistry training and technology, and national declarations, agreement on an international convention was reached in Geneva on September 2, 1992. The draft text was transmitted to the United Nations and opened for signature in Paris on January 13, 1993.[11] Two-thirds of the world—over 130 countries—signed the Chemical Weapons Convention (CWC) in its first two days, but it didn't enter into force until April 29, 1997, 180 days after 65 countries had ratified it. After a long political battle in Washington, D.C., the US ratified the CWC on April 25, 1997, just in time to become a full-fledged member before formal entry into force. Russia also faced a difficult political fight over ratification and finally ratified the treaty on November 5, 1997.

Until 2016 192 countries have joined the Chemical Weapons Convention, with both Myanmar/Burma and Angola having joined in 2015. This leaves just four countries—Egypt, Israel, North Korea, and South Sudan—outside the Convention, although Israel signed the CWC in 1993. Other regions including both Taiwan and Palestine are not members.[12] This makes the CWC the most universal arms control and disarmament treaty so far, and the largest multinational organization outside of the United Nations.[13]

[9]See http://www.acq.osd.mil/tc/treaties/bda/text.htm.

[10]For more history, see: https://en.wikipedia.org/wiki/Chemical_warfare; https://en.wikipedia.org/wiki/Chemical_weapons_in_the_Rif_War.

[11]For a full text of the Convention, see https://www.opcw.org/chemical-weapons-convention/preamble/.

[12]For a full list of members and dates of signature, ratification, accession, and national entry into force, see OPCW, Note by the Technical Secretariat: Status of Participation in the Chemical Weapons Convention as at 17 October 2015, S/1315/2015, October 19, 2015, at https://www.opcw.org/fileadmin/OPCW/S_series/2015/en/s-1315-2015_e_.pdf.

[13]See www.opcw.org for more detail on verified chemical weapons destruction.

4 Chemical Weapons Destruction

Article IV of the Chemical Weapons Convention requires all States Parties which have declared chemical weapons stockpiles to begin destroying them "not later than two years after this Convention enters into force for it [the State Party]," and to "finish not later than 10 years after entry into force of this Convention." This means that each country possessing chemical weapons should complete its destruction program by April 29, 2007, a decade after the CWC's 1997 entry into force. There is also a provision for a maximum five-year extension of this deadline until 2012.

Eight countries have declared existing chemical weapons stockpiles to date, with both Russia and the US accounting for about 95% of the total declared tonnage. Table 1 shows rounded estimates of declared stockpiles, although the figures do not add to the estimated total. While most figures are within a few tons of variance, both the Indian and South Korea figures remain rough estimates due to the lack of accurate and transparent numbers from these two possessor states.

Russia: The largest declared stockpile is in Russia which declared seven stockpile sites in five oblasts (states) and one republic when it joined the CWC in December 1997. Table 2 shows the declared tonnage—almost 40,000 metric tons of both blister and nerve agents. The two most worrisome sites in the 1990s were the easternmost stockpile, Shchuch'ye, and one of two sites in the Udmurt Republic, Kizner; both of these sites housed weaponized nerve agents in millions of artillery shells which would fit in a briefcase or gym bag, thereby increasing the risk of theft or diversion. They also housed much larger missile warheads with multiple mini-munitions filled with nerve agents. The major concern was that these shells could very easily disappear from either site, and the Shchuch'ye stockpile was located very close to the new Kazakhstan border in Central Asia. The other five declared Russian CW stockpiles all contained blister and nerve agents in bulk storage—large barrels, containers, and railway cars which would be very difficult to steal but were still vulnerable to terrorist attack.

In July 1994 the US and Russian governments organized a high-level, on-site inspection of the Shchuch'ye chemical weapons stockpile, part of the bilateral effort to promote transparency and confidence-building measures between the two major CW possessor states. The US delegation included Dr. Harold Smith, a senior advisor to the US Secretary of Defense, and two US representatives, Glen Browder from Alabama (who had a CW stockpile, Anniston, in his congressional district) and John Spratt from South Carolina, both on the House Armed Services Committee.[14] It also included General Robert Orton, Program Manager of Chemical Demilitarization (PMCD) for the US Army. The delegation was hosted by Russian Army General S.V. Petrov, head of the Russian chemical corps.

The chemical weapons stockpile at Shchuch'ye was massive, housing over 2 million nerve agent artillery shells and almost 1,000 short- and medium-range

[14]The author was fortunate to participate in this on-site inspection in 1994 while he was a Professional Staff Member of the Armed Services Committee in the US House of Representatives.

Table 1 Declared chemical weapons stockpiles[a]

Russian Federation	39,965 metric tons
United States	28,577 metric tons
India	1,056 metric tons (est.)
South Korea	605 metric tons (est.)
Libya	26 metric tons
Albania	16 metric tons
Iraq	n.a
Syria	1,308 metric tons
Total	72,525 metric tons (est.)

[a]See www.opcw.org as well as historical documents for national estimates of CW stockpiles. There are several reasons for the variance in figures, including the 1997 CWC entry into force which came seven years after the US had initiated its CW destruction process in 1990. The US had incinerated 1,436 metric tons at Johnston Atoll and Tooele prior to CWC EiF. The South Korean and Indian figures are estimates. The Iraqi numbers are unknown

Table 2 Declared Russian chemical weapons stockpiles[a]

Gorny, Saratov Oblast	1,142 metric tons
Kambarka, Udmurt Republic	6,349 metric tons
Shchuch'ye, Kurgan Oblast	5,457 metric tons
Kizner, Udmurt Republic	5,745 metric tons
Maradykovsky, Kirov Oblast	6,890 metric tons
Leonidovka, Penza Oblast	6,885 metric tons
Pochep, Bryansk Oblast	7,498 metric tons
TOTAL	39,965 metric tons[b]

[a]These figures are taken from a 2004 presentation of Viktor Ivanovich Kholstov, the Deputy Chief of the Federal Agency for Industry at the time, at a Green Cross annual national dialogue in Moscow, Russia

[b]The total may not add up due to rounding

missile warheads in old, decrepit, above-ground warehouses of corrugated metal and wood, with large barn-like doors locked with bicycle padlocks. The shells were stored in wine rack-type shelving as far as the eye could see, while the warheads were on railroad dollies, all very battlefield-ready. While this bilateral US-Russian inspection was a major historic step forward in destroying both US and Russian CW stockpiles, the visit documented the lack of any comprehensive inventory of chemical weapons, and also illustrated the disturbing lack of high security at the site. It was clear that the Shchuch'ye stockpile was highly vulnerable to theft, diversion, and proliferation, right in the middle of a region prone to growing terrorism and Islamic jihadism.

Dr. Smith, representing the US Department of Defense, offered his Russian counterparts a turnkey incinerator, similar to what had been operating since 1990 on Johnston Atoll in the Pacific Ocean as the first prototype US destruction facility for chemical weapons. This would be fully funded by the US Cooperative Threat Reduction (CTR) Program which Congress had initiated two years earlier under the bipartisan leadership of Senators Sam Nunn, a Democrat from Georgia, and Richard Lugar, a Republican from Indiana, amongst others (Walker 2016b). Russian General Petrov, along with the Chairman of the Russian Military Committee in the Duma in Moscow, rejected the offer stating that incineration was opposed by the environmental and public health communities in Russia, was too expensive and high-maintenance, and too risky to build. Petrov and his colleagues offered, however, to establish a Joint Evaluation and Research Program (JREP) with US colleagues to study what Russian technologies might be available and/or developed to safely destroy Russia's enormous stockpiles.

Russia eventually chose neutralization, a wet chemistry process in which the weapons and tanks are drained of their liquid chemical agents which are in turn mixed with hot water and a caustic reagent such as sodium hydroxide. This chemical mixing destroys most of the toxicity of the agent but produces 10 times the volume of liquid waste which must in turn be treated in a secondary process. The neutralization process is preferred by many experts because it can contain and manage all emissions, gaseous, solid, and liquid, while incineration pumps large amounts of potentially toxic gaseous emissions out the smokestack. For Shchuch'ye, Russia chose bituminization for its secondary treatment process, mixing the toxic liquid waste with asphalt to solidify it for long-term, retrievable storage in barrels and bunkers. Russia first sought to use this asphalt to pave roads and parking lots, but it was discovered that the liquid waste was slightly carcinogenic, thereby precluding any possible reuse.

Russia's chemical weapons destruction process began in December 2002 at Gorny in the Saratov Oblast, where Germany had helped Russia build a prototype neutralization facility for the lewisite stockpile. The facility was built in larger scale at Kambarka in the Udmurt Republic for lewisite neutralization as well, also with the support of Germany. In 2016 Russia has been successful at eliminating six of its seven chemical weapons stockpiles, neutralizing 92%—about 37,000 metric tons—of its declared stockpile, with its remaining 3,000 metric tons still in process of neutralization at Kizner, the site in the Udmurt Republic similar to the Shchuch'ye stockpile.

The Organization for the Prohibition of Chemical Weapons (OPCW) in The Hague, the multilateral group which oversees implementation of the Chemical Weapons Convention, approved a five-year extension for Russia's CW destruction program until 2012, but Russia missed this final legally binding deadline, along with the US and Libya, and has now projected that it will complete its program in September 2020.[15]

---

[15]See OPCW 2016c, paragraph 1.15. See also the decision of the OPCW Conference of States Parties in December 2011: Decision: Final Extended Deadline of 29 April 2012, C-16/DEC.11, December 1, 2011, https://www.opcw.org/fileadmin/OPCW/CSP/C-16/en/c16dec11_e_.pdf.

Table 3 Declared US chemical weapons stockpiles[a]

Johnston Atoll, Pacific Ocean	1,842 metric tons
Tooele, Utah	12,353 metric tons
Edgewood, Maryland	1,471 metric tons
Anniston, Alabama	2,045 metric tons
Umatilla, Oregon	3,374 metric tons
Pine Bluff, Arkansas	3,494 metric tons
Newport, Indiana	1,152 metric tons
Blue Grass, Kentucky	475 metric tons
Pueblo, Colorado	2,369 metric tons
Total	28,577 metric tons[b]

[a]These figures are taken from US Chemical Materials Agency (CMA) reports, formerly the Program Manager for Chemical Demilitarization (PMCD), and now the Program Executive Office, Assembled Chemical Weapons Alternatives (PEOACWA) in Aberdeen, Maryland; see http://www.peoacwa.army.mil/

[b]The total may not add up due to rounding of tonnage

United States: The second largest chemical weapons possessor country is the US, which declared 28,577 metric tons (31,501 US tons) at nine sites in the early 1990s. The US National Defense Authorization Act for 1986 directed the Secretary of Defense to carry out the safe destruction of the US chemical weapons stockpile.[16] The early destruction plan for the US was to build three centralized incinerator facilities on Johnston Atoll; Tooele, Utah; and in the south (likely Anniston, Alabama or Pine Bluff, Arkansas), to which the nine stockpiles would be shipped by train and truck and burned. The US Congress, upon hearing of this plan, banned transportation of these old and leaking chemical weapons and storage tanks, necessitating the US Army to build destruction facilities at all nine sites (Table 3).[17]

The next challenge for the US CW destruction program was to address public concerns over incineration. The US Army was very reluctant to introduce alternative technologies for destruction due to the potential for additional costs, schedule delays, and technical complexities. Very early in the program, however, it became clear that there would be strong and vocal opposition to incineration; a national grassroots group, the Chemical Weapons Working Group, was formed in Kentucky in 1991, and state public health and environmental regulators also began to raise concerns.[18]

In order to help overcome this dilemma, which was mounting between the US Army and state officials and incinerator opponents, the Assembled Chemical Weapons Assessment (ACWA) Program was created by Congress in the 1997

[16]National Defense Authorization Act (NDAA) for Fiscal Year 1986. Public Law 99–145.

[17]National Defense Authorization Act (NDAA) for Fiscal Year 1995. Public Law 103–337.

[18]For the Chemical Weapons Working Group, see http://www.kyenvironmentalfoundation.org/cwwg-history-and-accomplishments.html.

appropriations process to establish "a pilot program to identify and demonstrate at least two alternatives to incineration for the destruction of assembled chemical weapons".[19] The ACWA Program also established a national dialogue of stakeholders, including federal and state regulators, grassroots activists, state governors' representatives, the US Army, and interested engineering firms to discuss options such as neutralization for safe and timely destruction of chemical agents, explosives, and rocket propellant. It also established Citizens' Advisory Commissions (CACs) and public outreach offices at all chemical weapons stockpile sites in order to build trust, confidence, and transparency in the process.

By the early 2000s, the US had completed the destruction of the CW stockpile on Johnston Atoll, and had begun operating several other incinerators. In 2016 the US has safely destroyed 90%—about 25,700 metric tons—of its declared stockpile and closed seven of its nine destruction facilities. The US built and operated five incinerators and two neutralization facilities, and has projected to complete operations by 2020 at Pueblo, Colorado and by 2023 at Blue Grass, Kentucky. The Pueblo, Colorado facility to neutralize 2,369 metric tons of mustard agent started initial operations in September 2016, and the last facility at Blue Grass, Kentucky should open in the next two years. Similar to the Russian program noted earlier, the US missed its last OPCW deadline in 2012 and continues to brief the OPCW on progress towards the 2023 completion date for stockpile destruction.[20]

Albania: When Albania joined the Chemical Weapons Convention in 1994 it joined as a non-possessor state and therefore did not declare any chemical weapons. In the early 2000s, it discovered suspicious barrels in a small garage in the mountains outside of Tirana and called in OPCW experts to determine if the barrels contained chemical agents or not. It turned out to be about sixteen metric tons of mustard agent, apparently imported from China under the former government, and a program was established under the G-8 Global Partnership to help Albania secure the site, design a destruction plan, and safely destroy the agent (OPCW Conference of States Parties 2004).

In late 2006 and early 2007 the German engineering firm Eisenmann built a small incinerator which was moved to the remote site in the Albanian mountains and began to burn the barrels. Unfortunately, the volatility of the mustard agent was underestimated and the first barrel burned a hole in the bottom of the furnace and also burned out the afterburner. These repairs took six weeks or more and caused Albania to be the first CWC possessor state to miss a legally binding deadline—April 29, 2007—for completing its CW destruction. Ironically, Albania could have asked the OPCW for an extension to this deadline, but had not, assuming the destruction process would go smoothly. By July 2007, Albania's sixteen tons of

[19]OCAA–Omnibus Consolidated Appropriations Act. 1997. Public Law 104–208; Strom Thurmond National Defense Authorization Act (NDAA) for Fiscal Year 1999. Public Law 105–261.

[20]See the US press release of September 7, 2016, http://www.peoacwa.army.mil/2016/09/07/first-chemical-weapons-processed-today-in-pueblo-chemical-agent-destruction-pilot-plant/.

declared chemical weapons had been safely incinerated, but the remaining toxic waste still is sitting on the site.

South Korea: The Republic of Korea declared a stockpile of about 605 metric tons of binary chemical weapons when it joined the CWC in 1997, and it apparently safely destroyed this stockpile by 2008 (OPCW 2005).[21] But very little is known publicly about this stockpile or destruction process due to the high degree of secrecy demanded by South Korea concerning its arsenal. There is speculation about why South Korea is so secretive about this, including theories that it is sensitive due to the existing large CW stockpile in North Korea; it may not want its citizens to know where, when, and how it destroyed the stockpile due to environmental and public health concerns; or perhaps it is politically sensitive because its stockpile very closely resembled the newest US binary stockpile, showing that the arsenal (and/or the technology for producing it) was probably transferred from the US before the CWC entered into force for either country. South Korea, by the way, does not even allow its name to be included as a declared possessor state at the OPCW, so it is always referred to as "A State Party" (OPCW Conference of the States Parties 1999)."[22]

India: Another somewhat secretive declared possessor state is India, which declared a stockpile of about 1,056 metric tons of mustard agent when it joined the CWC in 1997. This stockpile was incinerated by March 16, 2009 under OPCW verification inspectors, but there is little known about its location or arsenal specifications. While India is transparent that it is a declared possessor state, it refuses to provide any details about its chemical weapons program or destruction process (OPCW Conference of the States Parties 1999; OPCW 2005).[23]

Libya: The Libyan Arab Republic joined the CWC in 2004 after its former leader, Colonel Muammar Gaddafi, agreed publicly to destroy his weapons of mass destruction. Libya declared about 23 metric tons of bulk mustard agent, which was subsequently neutralized and verified by the OPCW in 2010–2013, although the destruction process was interrupted by the civil war in 2011. After the 2011 death of Colonel Gaddafi, the new Libyan government declared another secret stockpile of weaponized mustard agent, about three metric tons, and this was verifiably destroyed by 2015. This was the first known time that a CWC State Party intentionally misled the OPCW by hiding a CW stockpile (OPCW Conference of the States Parties 2005; OPCW 2015).

In 2016 a major effort was made by the OPCW and several States Parties to remove several hundred tons of precursor chemicals from Libya and neutralize the

---

[21]Note that the report states that "A State Party" had "destroyed 302.716 metric tonnes, or approximately 50%, of its Category 1 chemical weapons" (OPCW 2005). This is one of the very few times that South Korea's declared stockpile is noted in tonnage.

[22]Discussion was of "four States Parties—India, the Russian Federation, the United States of America and one other," https://www.opcw.org/fileadmin/OPCW/CSP/C-IV/en/C-IV_5-EN.pdf, p. 8.

[23]Note that India's stockpile is described as "45.14%" destroyed at "476.545 metric tonnes," indicating that its total stockpile is 1,056 metric tons (OPCW 2005).

toxic materials in Germany. This was catalyzed by an early 2016 request from Libya to the OPCW Executive Council to help destroy these remaining chemicals before non-state actors were able to capture them (OPCW 2016b).

Iraq: Iraq was another latecomer to the Chemical Weapons Convention, acceding to the CWC in 2009.[24] It also declared itself a CW possessor state with two large bunkers at Al Muthanna near Falluja containing unknown quantities of chemical agents and related equipment. These bunkers had been bombed in the 1991 Gulf War by the US and reportedly still contained a large unexploded aerial bomb, but were sealed with concrete by United Nations inspectors in the mid-1990s. While Iraq is obligated to destroy these old chemical agents from the 1980s CW program of Saddam Hussein, there has been considerable discussion at the OPCW about how best to evaluate the risks involved and to begin a destruction process. These bunkers were reportedly taken over in 2014 by ISIS but have now been retaken by Iraqi forces.[25] This has led to concern that ISIS could have gained access to the bunkers and/or taken related laboratory equipment which was located nearby to analyze the bunker contents (Cirincione and Walker 2014). Iraq is considering filling the bunkers with concrete to eliminate any further threats of proliferation, although the CWC expressly forbids any burial or dumping of chemical weapons as a means of irreversible destruction.[26]

Syria: Syria joined the Chemical Weapons Convention in September 2013, just a month after the sarin nerve agent attack on Ghouta, a suburb of Damascus, in which about 1,400 people died (Human Rights Watch 2013). Under threat of attack from the United States, and with considerable pressure from Russia as well, Syrian President Bashar al-Assad declared 1,308 metric tons of chemical weapons and was presented with a very ambitious timeline to eliminate them in the midst of his ongoing civil war.[27]

The Chemical Weapons Convention prohibits the removal of declared chemical weapons stockpiles out of a country, but an exception was made in this case, given the high risks involved with establishing a safe and secure destruction facility in

[24]See the opening statement of OPCW Director-General Rogelio Pfirter welcoming Iraq as the 186th State Party, February 20, 2009, before the OPCW Executive Council. https://www.opcw.org/news/article/opening-statement-by-the-director-general-to-the-executive-council-at-its-fifty-fifth-session/.

[25]See CIA fact sheet, "Al Muthanna Chemical Weapons Complex," https://www.cia.gov/library/reports/general-reports-1/iraq_wmd_2004/chap5_annxB.html; see also http://www.globalsecurity.org/wmd/world/iraq/muthanna.htm.

[26]Article IV, paragraph 10, of the CWC states: "Each State Party, during transportation, sampling, storage and destruction of chemical weapons, shall assign the highest priority to ensuring the safety of people and to protecting the environment." Part IV(A), paragraph 13, of the CWC Verification Annex states: "the following processes may not be used: dumping in any body of water, land burial or open pit burning." See https://www.opcw.org/chemical-weapons-convention/. See also http://www.nytimes.com/interactive/2014/10/14/world/middleeast/iraqs-plan-to-entomb-remnant-chemical-weapons-in-bunker-complex.html?_r=0.

[27]See the special section on "Syria and the OPCW," https://www.opcw.org/special-sections/syria/. See also the Green Cross blog postings on Syria, 2014–2016, http://www.gcint.org/?s=Syria.

Syria. The removal of this tonnage, most of it precursor chemicals in storage barrels, but also including about 20 metric tons of mustard agent, took place from January until June 23, 2014, out of the Syrian port of Latakia. The chemicals were loaded on board two freighters, the *Ark Futura* from Denmark and the *Taiko* from Norway. The *Ark Futura* departed for the southwest Italian port of Gioia Tauro where it transferred about 600 metric tons of chemicals to an American Merchant Marine ship, the *Cape Ray*, which had been outfitted with two semi-mobile neutralization units, "field deployable hydrolysis systems."[28] The reason for this ship-to-ship transfer in Italy was that Syrian President Assad refused entry of any US ships into Syrian waters.

Of the 1,308 metric tons of chemical agents and precursor chemicals removed from Syria, 600 metric tons were neutralized on board the *Cape Ray* in the Mediterranean, without any serious incidents, and the resultant hydrolyzed liquid was delivered by the *Cape Ray* to Germany and Finland for second-stage incineration. The *Ark Futura* delivered the remainder of its tonnage, about 150 metric tons, to the United Kingdom, where it was incinerated at two sites. The *Taiko* delivered its chemical cargo to Finland and to Port Arthur, Texas, in the United States, where its tonnage was incinerated (Walker 2014).

By mid-October 2014, about 98% of the Syrian chemicals were fully destroyed; the final 2% took another fourteen months in the US due to technical challenges with corroding tanks and was completed in January 2016 (OPCW 2016a).[29] In the end, the Syrian chemical destruction operation was judged a great success, thanks to the ten or more countries which participated in the naval convoy and destruction operations, to the OPCW and United Nations which jointly managed the enormous logistics, and to the two dozen or more countries which contributed financial resources to the OPCW totaling over 50 million euros to fund inspections and operations.

The Syrian operation was not, however, without a few major challenges. One of the largest was the lack of transparency in the effort which helped to catalyze large citizen protests throughout the Mediterranean, including politicians, environmental activists, the fishing industry, and the tourist industry, all of whom worried that any at-sea neutralization operation could impact the environment and public health. Efforts by non-profit environmental groups such as Green Cross International to facilitate dialogues and proactive outreach in and around the Mediterranean were rebuffed by the OPCW, the United Nations, and the US which argued that the tight schedule just did not allow for more democratic consensus-building.[30] Public concern created large demonstrations in Greece, Crete, Italy, Turkey, and

[28]See the US Army Edgewood Chemical Biological Center description of the Field Deployable Hydrolysis System, http://www.ecbc.army.mil/about/posters/2015/D13.pdf.

[29]See also http://archive.defense.gov/home/features/2014/0114_caperay/.

[30]This point was underlined in several meetings and calls by this author with UN, OPCW, and US State and Defense Department officials in 2013 and 2014. See also the letter from OPCW Director-General Ahmet Uzumcu to the Pancretan Commission, July 29, 2014, https://www.opcw.org/fileadmin/OPCW/ODG/uzumcu/DG_Letter_Pancretan_Commission.pdf.

elsewhere. In the end, the demilitarization operations went well, but a lesson was learned that any such future operation must include proactive outreach, dialogue, and information-sharing as a central feature and best practice in order to build more broad support and to preclude such public opposition.[31]

The Syrian process, however, still continues as both chlorine and mustard are being used in Syria as well as in Iraq, and the latest report from the United Nations-OPCW Joint Investigative Mechanism (JIM) confirms that the Syrian government used chlorine at least twice, while ISIS used mustard at least once.[32] Also, the OPCW Director-General stated at the OPCW Executive Council meeting on July 12, 2016 that Syria's declaration to the OPCW of its chemical weapons program and activities still remains replete with "gaps, inconsistencies, and discrepancies."[33]

5 Other Chemical Weapons Challenges

Buried Chemical Weapons: The Chemical Weapons Convention also takes note of "old" and "abandoned" chemical weapons, recognizing the fact that many chemical weapons stockpiles were dumped at sea or buried on land long before the Convention entered into force. We know, for example, that there were many other countries with chemical weapons stockpiles than the eight which have officially declared them under the CWC.[34]

The United States has been the most transparent on this issue, publicly identifying 224 suspected burial sites at 96 locations in 38 states, the Virgin Islands, and

[31]See, for example, an NGO letter to US Secretary of State John Kerry and US Secretary of Defense Chuck Hagel proposing public dialogues in Rome, Athens, Istanbul, and elsewhere, as well as video uplinks from the *Cape Ray* to build confidence in the at-sea neutralization operations; there was no response to these written suggestions. The NGO letter of February 2, 2014, can be found on the Green Cross International website, http://www.gcint.org/public-awareness-over-syria-chemical-weapons-destruction-needed-amid-mediterranean-region-concerns/. The Greek NGO Archipelagos Institute of Marine Conservation also organized a joint protest letter signed by many NGOs in Europe in 2014; see the Green Cross blog posting at http://www.gcint.org/page/6/?s=Syria.

[32]Joint Investigative Mechanism, "Third Report of the Organization for the Prohibition of Chemical Weapons-United Nations Joint Investigative Mechanism," August 24, 2016, http://www.un.org/ga/search/view_doc.asp?symbol=s/2016/738. The JIM concluded that two incidents of chlorine use in Talmenes, April 21, 2014, and in Sarmen, March 16, 2015, were caused by the Syrian government, and one incident with mustard use in Marea, August 21, 2015, was caused by ISIS.

[33]See the statement by the US Ambassador to the OPCW, Kenneth D. Ward, https://www.opcw.org/fileadmin/OPCW/EC/82/en/merged.pdf.

[34]See Article IV, paragraph 17, of the CWC which states that destruction requirements shall not "apply to chemical weapons buried on its territory before 1 January 1977 and which remain buried, or which had been dumped at sea before 1 January 1985." See also Part IV(B) of the CWC Verification Annex which defines old and abandoned CW.

in Washington, D.C. Most of these sites have not yet been surveyed and investigated, but one very large dump site, which the US Army Corps of Engineers has focused on since 1993, is a region of downtown Washington, D.C. called Spring Valley. With over 1,000 expensive private homes and the campus of the American University, this area was a testing and dumping ground during and after World War I, given that the university housed the US research and development laboratory for chemical weapons during the war. In 1993 buried chemical weapons were discovered during a private construction effort, and the ongoing survey, excavation, and remediation project has now taken over two decades and hundreds of millions of dollars.[35]

Many other countries, including most of Europe, Japan, Russia, China, and Australia have buried chemical weapons on their territories, primarily from the two World Wars in the last century, and these dangerous weapons are typically discovered in land excavation and development projects. Germany has been carefully evaluating and destroying both chemical and conventional weapons for years at its site in northern Germany, GEKA-Munster, just south of Hamburg.[36] Belgium has a weapons destruction site at Poelkapelle, not far from Ypres, which was also a famous WWI battlefield in West Flanders (De Bisschop et al. 2006). The largest current buried CW excavation project is in China, where Japan left hundreds of thousands of chemical weapons after World War II at dozens of sites.[37]

Sea-Dumped Chemical Weapons: After World War II most of the warring parties had large stockpiles of chemical weapons which were dumped at sea. This was seen at the time as the safest and most efficient means to rid the world of these dangerous stockpiles, but the practice has now left over 300,000 metric tons dumped in all oceans and seas of the world between 1946 and 1965.[38] There are an estimated 29,000 metric tons dumped in US coastal waters, another 40,000 metric tons in the relatively shallow Baltic Sea, 21,000 metric tons off the coasts of Australia, and 6,600 metric tons around Japan.[39]

While sea-dumped chemical weapons do not necessarily pose a serious terrorist or proliferation threat, they are now corroding and releasing their toxic agents into nearby waters, posing potential health and food-chain threats. They have also been appearing in fishermen's nets, injuring or killing dozens of fishermen around the world, and have washed up on beaches in the Baltic, Mediterranean, and Black Seas as well as in the Caribbean and elsewhere. While the CWC does not deal directly with sea-dumped CW, it does recognize that any such weapons which are raised

[35]For the US survey of buried chemical weapons, see US Army Program Manager for Chemical Demilitarization, *Survey and Analysis Report, Second Edition*, Aberdeen, Maryland, December 1996. For the Spring Valley ongoing cleanup and remediation effort, see http://www.nab.usace.army.mil/Home/SpringValley.aspx.

[36]http://www.geka-munster.de/index.php?id=2.

[37]http://www.china.org.cn/english/2003/Dec/81536.htm.

[38]http://www.nonproliferation.org/chemical-weapon-munitions-dumped-at-sea/.

[39]http://www.helcom.fi/baltic-sea-trends/hazardous-substances/sea-dumped-chemical-munitions/.

from the ocean floor must be declared to the OPCW and destroyed under international verification.

6 Other Lessons Learned in Chemical Weapons Destruction

Cost Escalation: When the United States began unilaterally destroying its large chemical weapons stockpile in 1990, the cost estimate for its destruction program was about $2 billion. After over 25 years of CW destruction operations at nine declared stockpile sites, this estimate is now well over $40 billion and still growing (US GAO 1991; Freeman and Alikhan 2013).[40] Each of the nine demilitarization sites, both incineration and neutralization facilities, will average well over $4–5 billion each, many times original project estimates.

The Russian chemical weapons destruction program was initially estimated at $3–4 billion in the mid-late 1990s, but its current cost is well over $10 billion (Green Cross Russia et al. 2008). These US and Russian cost escalations have been caused by the complexity of destruction operations, by the need to address environmental and public health risks and protections, and by the need for transparency and public involvement. But the unpredictable growth in costs has also extended schedules, stretched federal budgets, required $2–3 billion in support for Russia from the G-8 Global Partnership, and much foreign support for the Albanian, Libyan, and Syrian CW destruction operations. None of the eight declared CW possessor countries has been able to meet all of its legally binding destruction deadlines under the CWC, necessitating official requests for extensions and finally acknowledging that at least three countries—Libya, Russia, and the US—have missed the final 2012 deadline, and one other, Albania, missed its final 2007 deadline.

Technology Development: High temperature furnaces were deemed the technology of choice early on in CW destruction programs, seen by thermal engineers as the most mature, most cost-effective, and most manageable and safe. However, many questions were raised in both the US and Russian programs about the safety of the incinerators, even with new, high-tech, and effective scrubbers, and about the manageability and scientific understanding of toxic emissions. Both the Clean Air Act and the Clean Water Act in the US pose certain limits to toxic gaseous and liquid emissions, complicating the monitoring of emissions.

As noted earlier, the US finally determined that it would introduce alternative technologies, primarily neutralization, at four US CW stockpile sites, while Russia refused to use incineration as its first-stage destruction technology. Other secondary technologies were developed under the Assembled Chemical Weapons Assessment (ACWA) Program including bioremediation, super-critical water oxidation

[40]Also https://en.wikipedia.org/wiki/United_States_chemical_weapons_program.

(SCWO), high steam treatments, plasma arc, and other systems. Closed detonation and treatment systems have also been developed, primarily for old and abandoned chemical weapons, to preclude the need to use harmful open-burn and open-detonation (OB/OD) practices.[41]

The abolition of chemical weapons has thereby spawned a whole range of new treatment systems for high toxic waste, which is being found useful and relevant for many dangerous waste management programs.

Emergency Preparedness: Most communities around chemical weapons stockpile sites in the US, Russia, and elsewhere felt totally ill-prepared for such major destruction programs and demanded much more effective warning and evacuation planning. In the US, communities became very involved in planning efforts, with local schools outfitted with sealed, air-conditioned facilities to house all faculty and students in an emergency. Gas masks and emergency "shelter-in-place" kits were widely distributed to communities, and early warning sirens and radios were given to each household. All of this planning went hand in hand with local outreach and information efforts and Citizens' Advisory Commissions (CACs) at each site.[42]

Russia agreed to very similar emergency preparedness planning in its first decade or more of chemical weapons destruction efforts, including establishing CACs and local outreach offices managed by Green Cross Russia. Russia also actively participated in community study visits between the US and Russia to share best practices in the 1990s and early 2000s (Green Cross Russia et al. 2007). Unfortunately, much of this outreach effort came to an end in Russia as facilities began active operations in the early 2000s and Moscow wanted to maintain more control over public information. Green Cross Russia, which had operated public information and outreach offices at all Russian chemical weapons stockpile sites since the mid-late 1990s, was forced to close all offices at the request of federal authorities and the pullout of all Global Partnership countries by 2010.[43]

Albania, India, Libya, and South Korea no doubt had some minimum emergency and evacuation planning in place for their operations and workers, but not much is known about these efforts due to the secrecy of the projects, especially in India and South Korea.[44]

---

[41]http://www.peoacwa.army.mil/. This author was very involved in the creation and management of the ACWA Program in the 1990s and 2000s, and was a member of the ACWA National Dialogue process to build consensus on acceptable and effective technologies.

[42]https://www.peoacwa.army.mil/about-peo-acwa/acwa-public_involvment/.

[43]See the websites of Green Cross Russia, http://www.green-cross.ru/programms/legacy/, and Green Cross International, www.gcint.org.

[44]No public information is available for these programs except what is on the OPCW website, www.opcw.org. At the 13th Conference of States Parties in The Hague, the OPCW Director-General, referring to South Korea, stated: "On 10 July 2008, A State Party became the second possessor State, after Albania, to eliminate its entire chemical weapons stockpile. This notable achievement deserves to be commended and represents yet another important milestone in the process towards complete chemical disarmament." https://www.opcw.org/fileadmin/OPCW/CSP/C-13/en/c13dg09_en.pdf. And at the 14th Conference of States Parties in 2009, the Director-General stated: "India became, on 16 March 2009, the third State Party, after Albania and

Community Involvement: The chemical weapons abolition program has clearly shown the value of involving local communities and stakeholders. Some countries are reluctant to engage in public processes which can be contentious, costly, and time-consuming, but the Russian and US programs have documented the value of engaging interested and relevant segments of local and regional populations.

Public engagement is particularly important when local populations are at risk of environmental and public health impacts, and need to be involved in order to determine best practices, including choice of technology, emergency planning, risk assessments, and risk mitigation actions.[45] If a community is not involved, a program manager runs the risk of public demonstrations and lawsuits which can also complicate and stall, perhaps even kill, a project.[46]

Transparency: A critical part of any public involvement effort is to provide as much information and be as transparent as possible within limits of national security. Unfortunately, transparency is sometimes a victim of exaggerated national security concerns or overly ambitious schedules, leaving no time for any public discourse. This was shown in the Syrian chemical weapons destruction program, and has also been noted in both the Indian and South Korean programs.[47]

But even in the US CW destruction program there have been moments when transparency was intentionally blocked; in 2006 and 2007, when the US was trying to meet its interim 45% destruction deadline under the CWC, the US Army secretly devised a program to ship all the neutralized nerve agent by truck from Newport, Indiana to Port Arthur, Texas, against all prior agreements with the local community to treat the secondary waste on site with super-critical water oxidation. This catalyzed Indiana lawsuits and protests in Port Arthur, a poor African-American community sensitive to environmental justice issues (see Middleton 2007).

In Russia, the very first effort to secretly build a centralized facility at Chapayevsk met with thousands of protesters who caused the Russian military to cancel plans.[48] The OPCW in The Hague has also sought to increase transparency

(Footnote 44 continued)

A State Party, to complete the destruction of all its chemical weapons stockpiles. I have commended India, and I do so again today, for the exemplary commitment it has shown to fulfilling its obligations under the Convention, and I think it deserves the recognition of us all." https://www.opcw.org/fileadmin/OPCW/CSP/C-14/en/c14dg13_en.pdf.

[45]For a recent discussion of chemical safety and security, see http://www.gcint.org/green-cross-co-hosts-washington-dc-discussion-chemical-safety-security/.

[46]For more on the US outreach program, see https://www.peoacwa.army.mil/about-peo-acwa/acwa-public_involvment/. For the Russian outreach efforts, see annual national dialogue reports from Green Cross Russia.

[47]The OPCW, United Nations, and the US Defense Department finally organized an "Open House" on board the *Cape Ray* which had been docked at the US naval base in Rota, Spain for a month or more, waiting for the removal of chemicals from Syria, but this was organized with less than a week's notice to non-governmental organizations and with no travel support, so was limited to only a few media representatives.

[48]See, for example, a summary of Russian chemical weapons and their destruction at Federation of American Scientists, https://fas.org/nuke/guide/russia/cbw/cw.htm.

and involve non-governmental stakeholders and experts in the last decade; representatives of the CWC Coalition, a global network of non-governmental experts, was given permission, for example, to address the plenary sessions at annual OPCW meetings for the first time in 2013.[49]

7 Conclusions

The elimination of a whole class of weapons of mass destruction—in this case, chemical weapons—has been a long time in coming to the current success which we have seen in recent years. The use of chemicals in warfare has been widely condemned for well over a century, but it took the horrors of World War I, and the more recent indiscriminate use of chemicals in Iraq in the 1980s, to strengthen the taboo of chemical weapons use embodied in the 1925 Geneva Protocol. The 1993 Chemical Weapons Convention, bolstered by the bilateral disarmament commitments of the two largest possessor states—Russia and the United States—further solidified the specific plans for abolition of these dangerous and inhumane weapons.

But to see this process through so that we not only rid the world of all chemical weapons, but also never allow them to re-emerge in any capacity as we now see in Iraq again and in Syria, the following concluding remarks remain very important:

- The world must acknowledge that chemical agents are no longer viable military weapons, and have become "taboo," morally reprehensible, and a dangerous and costly burden for all countries.
- All possessor states must complete safe elimination of chemical weapons stockpiles in the near term—Iraq, Libya, Russia, and the United States.
- All non-member states must join the CWC—Egypt, Israel, North Korea, and South Sudan. The membership of Palestine and Taiwan must be resolved.
- The ongoing use of chlorine, a dual-use industrial chemical, in barrel bombs in Syria must stop, along with the most recent use of mustard by ISIS in Iraq. The OPCW and the United Nations must continue their investigations of chemical use by States Parties and by non-state actors in Iraq and Syria and hold those guilty parties accountable.
- All CWC States Parties must fulfill their national obligations under the Convention, including accurate annual trade reporting, implementation of a National Authority, and criminalization of nefarious chemical use.
- Protection of the environment, public health, and worker safety in weapons demilitarization activities is an absolute necessity, trumping deadlines and budget limits.

---

[49]For the NGO presentations at the OPCW, see the 2014 19th annual Conference of States Parties, https://www.opcw.org/documents-reports/conference-states-parties/nineteenth-session/national-statements/.

- Transparency, stakeholder involvement, public dialogue, and consensus-building are essential to program success.
- All States Parties must support—financially and politically—the OPCW to continue both stockpile and industrial inspections, to prevent the reemergence of chemical weapons, to promote chemical safety and security, and to remain a strong and accountable implementer of the Chemical Weapons Convention (Walker 2016a).
- The CWC is an excellent model for non-discriminatory and verified abolition of a whole class of weapons of mass destruction, with implications for a future ban on nuclear weapons.

Building a world free of chemical weapons is a historic achievement which has taken over a century to realize, including the loss of tens of thousands of soldiers and civilians in global and regional wars and, most recently, in terrorist acts, but this major step forward will hopefully serve to accomplish similar arms control and disarmament steps in other areas, including nuclear and biological weapons, and will certainly serve to build a more safe, secure, and peaceful world.

References

Cirincione, Joe, and Paul Walker. 2014. Is ISIS Using chemical weapons? Defense One. http://www.defenseone.com/ideas/2014/10/isis-using-chemical-weapons/96425/?oref=search_Paul Walker.

De Bisschop, Herbert C., Christel Meert, Bart R. Smedts, and Christiaan P. Pernell. 2006. Personal monitoring of old chemical weapons' dismantling operations at Poelkapelle (Belgium). In *ecological risks associated with the destruction of chemical weapons,* eds. Vladimir M. Kolodkin, and Wolfgang Ruck, 137–152. Springer.

Freeman, Ben, and Faris Alikhan. 2013. How much will it cost to destroy Syria's chemical weapons? *The Atlantic*, September 16. http://www.theatlantic.com/international/archive/2013/09/how-much-will-it-cost-to-destroy-syrias-chemical-weapons/279715/.

Green Cross Russia, Green Cross Switzerland, and Global Green USA. 2007. In *Proceedings of the ninth Russian national dialogue on chemical weapons nonproliferation and destruction.* Moscow, Russia.

Russia, Green Cross, Green Cross Switzerland, and U.S.A. Global Green. 2008. *Tenth Russian national dialogue on chemical weapons nonproliferation and destruction*. Russia: Moscow.

Human Rights Watch. 2013. Attacks on Ghouta: Analysis of alleged use of chemical weapons in Syria. https://www.hrw.org/report/2013/09/10/attacks-ghouta/analysis-alleged-use-chemical-weapons-syria.

Kenyon, Ian R., and Daniel Feakes. 2007. *The creation of the Organisation for the Prohibition of Chemical Weapons: A case study in the birth of an intergovernmental organisation*. The Hague: TMC Asser Press.

Middleton, Rusty. 2007. A lot of nerve: The army's deadly VX waste is burning in Port Arthur. *The Texas Observer*, September 7. https://www.texasobserver.org/2573-a-lot-of-nerve-the-armys-deadly-vx-waste-is-burning-in-port-arthur/.

OPCW. 2005. Report of the OPCW: On the implementation of the Convention on the Prohibition of the Development, Production, Stockpiling and Use of Chemical Weapons and their Destruction in 2004. C-10/4, especially, 6–7. https://www.opcw.org/fileadmin/OPCW/CSP/C-10/en/C-10_4.pdf.

OPCW. 2008. Report of the OPCW: On the implementation of the Convention on the Prohibition of the Development, Production, Stockpiling and Use of Chemical Weapons and their Destruction in 2007. C-13/4, especially, 3–4. https://www.opcw.org/fileadmin/OPCW/CSP/C-13/en/c1304_en.pdf.

OPCW. 2015. Draft Report of the OPCW: On the implementation of the Convention on the Prohibition of the Development, Production, Stockpiling and Use of Chemical Weapons and on their Destruction. EC-79/5. C-20/CRP.1, especially, 1, 4–5. https://www.opcw.org/fileadmin/OPCW/EC/79/en/c20crp01_ec7905_e_.pdf.

OPCW. 2016a. Destruction of Syrian chemical weapons completed. https://www.opcw.org/news/article/destruction-of-syrian-chemical-weapons-completed/.

OPCW. 2016b. Libya's remaining chemical weapon precursors arrive safely and securely at German facility for destruction. Press Release, September 8. https://www.opcw.org/news/article/libyas-remaining-chemical-weapon-precursors-arrive-safely-and-securely-at-german-facility-for-destruction/.

OPCW. 2016c. Draft Report of the OPCW: On the implementation of the Convention on the Prohibition of the Development, Production, Stockpiling and Use of Chemical Weapons and on their Destruction in 2015. EC-82/4. C-21/CRP.1, especially, 4–6. https://www.opcw.org/fileadmin/OPCW/EC/83/en/ec8204_c21crp01_e_.pdf.

OPCW Conference of the States Parties, Fourth Session. 1999. Report of the organisation on the implementation of the convention (1 January–31 December 1998). C-IV/5, especially, 8–10. https://www.opcw.org/fileadmin/OPCW/CSP/C-IV/en/C-IV_5-EN.pdf.

OPCW Conference of the States Parties, Ninth Session. 2004. Decision: Request by Albania for extensions of the intermediate deadlines for the destruction of its category 1 chemical weapons stockpiles. C-9/DEC.8. https://www.opcw.org/fileadmin/OPCW/CSP/C-9/en/C-9_DEC.8-EN.pdf.

OPCW Conference of the States Parties, Tenth Session. 2005. Opening statement by the director-general to the conference of the states parties at its tenth session. C-10/DG.11. https://www.opcw.org/fileadmin/OPCW/CSP/C-10/en/C-10_DG.11.pdf.

Robinson, Julian Perry. 1971. *The problem of chemical and biological warfare: The rise of CB weapons. A study of the historical, technical, military, legal and political aspects of CBW, and possible disarmament measures*, ed. Stockholm International Peace Research Institute (SIPRI), Vol. 1. Stockholm: Almqvist & Wiksell and New York: Humanities Press.

US Government Accountability Office (GAO). 1991. Chemical weapons: Stockpile destruction cost growth and schedule slippages are likely to continue. http://www.gao.gov/products/145681.

Walker, Paul F. 2014. Syrian chemical weapons destruction: Taking stock and looking ahead. *Arms Control Today*, December. http://www.armscontrol.org/ACT/2014_12/Features/Syrian-Chemical-Weapons-Destruction-Taking-Stock-And-Looking-Ahead.

Walker, Paul F. 2016a. Reading the riot act: Toward a more comprehensive ban on chemical weapons. *Nonproliferation Review* 22 (3–4): 517–525. doi:10.1080/10736700.2016.1161881.

Walker, Paul F. 2016b. Cooperative threat reduction in the former Soviet states: Legislative history, implementation, and lessons learned. *Nonproliferation Review* 23 (1–2): 115–129. doi:10.1080/10736700.2016.1178442.

Statement by HE Ghislain D'hoop, Ambassador of the Kingdom of Belgium

Ghislain D'hoop

In 1929, survivors of the French 418th infantry regiment erected a monument in Zuidschote, just to the North of Ypres. It was dedicated to the memory of their comrades killed or maimed by the chlorine gas attack at 5 pm on the 22nd of April, exactly one hundred years ago. The main victims of the attacks on that day and two days later were French, Moroccan, Algerian and other colonial forces, as well as 2,000 Canadian soldiers and more than 1,000 men of the Belgian Grenadier Guards and Carabiniers. Together with the British and Commonwealth forces, all were engaged in what later became known as the Second Battle of Ypres.

The 1929 French monument no longer exists. It was destroyed in 1942, during the Second World War. In its place, on the same spot, now stands a cross of reconciliation erected by the French and Belgian authorities. A moving Canadian memorial, the Brooding Soldier, stands somewhat further along the former front-line. And in Ottawa, paintings by William Patrick Roberts and Richard Jack also evoke the first chemical attack in the Ypres Salient. The Council Room at OPCW's headquarters in The Hague is called the Ypres Room. And famously, of course, the gas attacks are given strong poetical meaning in Wilfred Owen's poem, *Dulce Et Decorum Est*.

Commemoration, reconciliation, peace, international cooperation: they are key principles of humanity, embedded in our minds and in our hearts, and they must always be rekindled and never forgotten. Monuments in stone or bronze help us to achieve this difficult but necessary undertaking. So do paintings, poems and music since they touch our soul, help us to comprehend the unfathomable and make us better human beings. There is a special role to play for diplomats and scientists, in making sure that the world fully understands the horrors of chemical warfare and unites in condemning its manufacturing, stockpiling and use.

G. D'hoop (✉)
Belgian Embassy, Berlin, Germany
e-mail: ghislain.dhoop@diplobel.fed.de

B. Friedrich et al. (eds.), *One Hundred Years of Chemical Warfare: Research, Deployment, Consequences*, DOI 10.1007/978-3-319-51664-6_21

I therefore congratulate you on this initiative to take stock of historical and scientific research on the use of chemical weapons. Today, it is fitting that we commemorate all the men and women, soldiers and civilians alike, who fell victim to the atrocious, horribly painful and debilitating effects of chlorine gas, of phosgene and of mustard gas, all used in the trenches of Flanders.

If the attack one hundred years ago teaches us anything, it is not only the particularly horrifying and perverse impact of chemical weapons upon the human body but also the extreme fear and psychological damage they cause. Indeed, the mere possibility of their use instils primeval fear and feelings of horror into any decent human being. Because of these effects, chemical weapons are truly a deterrent and can easily be construed as an instrument of mass annihilation.

A hundred years later, much has been achieved in our global reaction to chemical weapons. We abhor those who use them or threaten to do so, we are working together to destroy the remaining stockpiles, and we set up an international organisation, the OPCW, which won the Nobel Peace Prize for its unrelenting efforts toward a universal ban on chemical weapons.

But as the horrific events in Syria show, our vigilance must never abate. Sarin gas was used in a Damascus suburb in August 2013, and chlorine, the same chemical released on the Ypres front one hundred years ago, was used against innocent civilians.

Belgium knows what it means to remain vigilant. Especially the Flemish farmers ploughing their lands on what a hundred years ago was an international battlefield know the importance of careful handling of unexploded ammunition. Every spring, the land yields a harvest of warfare when around 200 tonnes of such explosive ordnance are found. Some 5% of the ammunition found is of a chemical nature. These shells have to be disposed of through specialized techniques. The Explosive Ordnance Disposal Company of the Belgian army works all year round to eliminate these dangerous relics of World War I.

Because Belgium has suffered so severely under the impact of chemical attacks, we—its government, researchers, NGOs, and industry—are forerunners in international efforts to guarantee that the ban on the production, storage, and use of chemical weapons is effectively implemented worldwide.

Chemical weapons continue to be produced and even deployed. For this reason, it is crucial that their devastating impact continues to be widely recognized and understood. The current volume contributes magnificently to this noble undertaking. I am optimistic that the strong message conveyed by this joint scientific endeavor, a message of understanding through research and the power of international cooperation, will be heard and that in the coming years will also be acted upon.

Fritz Haber and His Institute

Gerhard Ertl

Dulce et decorum est pro patria mori
(It is sweet and honorable to die for one's fatherland).

This verse by Horace not only served during the First World War to stimulate the enthusiasm for the war. It is also the title of a poem by Wilfred Owen, which ends with the cry "the old lie!" Owen himself was killed as a young British volunteer just one week before the end of the war.

It was not only young men who were enthusiastic about the war in those days but also a large fraction of the university teachers and scientists. Among them was Fritz Haber with his maxim: "In peace for mankind, in war for the fatherland,"—a philosophy that he adopted also for his institute, the recently founded Kaiser-Wilhelm-Institut für Physikalische Chemie und Elektrochemie, to which he had been appointed first director in 1911.

Two years earlier Haber had made his most important scientific discovery while still professor at the Technical University of Karlsruhe. While today, climate and energy are the greatest problems for mankind for which solutions are expected to come from science, in those days it was the need for food. As a consequence of progress in technology and medicine, the world population was growing so rapidly during the nineteenth century that, unless crops could be increased considerably by the development of artificial fertilizers, there was a great danger of famine. In a worldwide competition it was Fritz Haber who succeeded in 1909 in his laboratory to produce ammonia from nitrogen (from air) and hydrogen, which then could be readily converted into nitrogen fertilizers ("bread from air"). This reaction was then transferred within a surprisingly short period into a large-scale industrial process, now known as the Haber-Bosch process, by Carl Bosch and Alwin Mittasch at the BASF company, which started production in 1913. At present about 140 million tons of ammonia are produced this way every year, of which 90% are converted

G. Ertl (✉)
Department of Physical Chemistry, Fritz-Haber-Institut of the Max Planck Society in Berlin, Berlin, Germany
e-mail: ertl@fhi-berlin.mpg.de

B. Friedrich et al. (eds.), *One Hundred Years of Chemical Warfare: Research, Deployment, Consequences*, DOI 10.1007/978-3-319-51664-6_22

into fertilizers. According to a realistic estimate, about one third of today's world population would starve if this source were not available.

Alfred Nobel decided in his testament that the prize named after him should be awarded to the person whose research had brought the greatest benefit to mankind. Under this viewpoint no one deserves this prize more than Fritz Haber, who received it immediately after the end of the First World War for the year 1918, despite international protests because of his role during the war. The Janus face of science becomes manifest in this context. Historians agree that without the Haber-Bosch process, the war would very likely have been over within less than a year, since the import of nitrates necessary for the production of explosives was blocked. This problem could be circumvented by the "saltpeter promise" ("*Salpeterversprechen*") of the German chemical industry, which was building large plants for the production of ammonia to be converted into explosives.

The general enthusiasm at the outbreak of the war also seized Fritz Haber, who became a volunteer like many other intellectuals. In a letter to his Swedish colleague S. Arrhenius he explained his reasons as follows:

> ... now we see it as our ethical duty, to take down our enemies with the use of all our strength and bring them to a peace that will make the return of such a war impossible for generations and give a solid foundation for the peaceful development of western Europe.

What an illusion, one can only say!

Haber followed this maxim not only personally, but also immediately converted the complete research program of his institute to military purposes. At the beginning the work concentrated on the development of alternatives for existing explosives, as for example the replacement of toluene, which could be used for the production of TNT. A severe accident happened in the laboratory on December 17, 1914, when a gifted young scientist, Otto Sackur, was killed and Gerhard Just, Haber's co-worker for many years, lost his right hand.

From the beginning of 1915 the work concentrated on problems of chemical warfare by poison gas, which was partly personally supervised by Haber at the front. On April 22, 1915 favorable wind conditions allowed the valves of containers filled with pressurized chlorine gas to be opened, thus causing the death or injury of many enemy soldiers. This military success prompted the emperor to promote Haber to the rank of *Hauptmann* (captain). In a state dominated by the military, this was an extraordinarily high distinction for him as a Jew.

A particularly tragic event has to be mentioned in this context. Haber's wife Clara Immerwahr, one of the very few female chemists at that time, lived in a broken marriage and also disagreed with the military activities of her husband. After a severe quarrel she shot herself with her husband's weapon during the night of May 2, 1915 in the garden of their home on the grounds of the institute. She was found dying by their 13-year-old son. Nevertheless, this did not prevent Haber from going to the front again the next day.

From then on the institute was the German center for research on poison gas and Haber became head of the chemistry department of the Ministry of War. His institute was transformed into a large-scale institution with 10 departments and up to about 150 scientists and engineers, for whom barracks were built and other rooms hired. The effect of poison gas became characterized by the Haber constant, that is, the product of concentration and duration of impact.

After the war, Haber went temporarily into hiding, and even grew a beard. The military activities were discontinued; later, Haber was even awarded the Nobel Prize. The institute rapidly returned to its normal research activities and experienced its "golden years" with world-class scientists such as Karl-Friedrich Bonhoeffer, James Franck, Herbert Freundlich and Michael Polanyi. Haber remained a patriot and tried to help his country with the immense reparations by starting a project in which gold was to be extracted from seawater, which, however, failed because of erroneous data analysis. His contribution to the foundation of an organization that was later to become the German Science Foundation (DFG) and the establishment of strong scientific contacts with Japan are two further examples for his outstanding organizational abilities.

After the seizure of power by the Nazis, Haber was forced by law to fire his Jewish collaborators. (He himself was exempted from this sanction because of his active participation in the war.) As a protest, he resigned from his position as director of the institute and died a broken man shortly afterwards (1934). In his testament he asked for a common gravesite together with his first wife Clara, which was then established in Basel.

Under the new government, fundamental research was rapidly abandoned and the projects were again concerned with military problems (however, no longer with poison gas), for which the institute was even honored with the title "*NS-Musterbetrieb.*"

There was another war during which many millions had to sacrifice their lives. In memory of this terror, Benjamin Britten composed his "War Requiem," for which he used poems by the above-mentioned Winfred Owen. And again we may quote a text from an ancient poet concerning the Thermopyles, which reads in the words of Schiller: "*Wanderer kommst du nach Sparta, verkündige dorten, du habest uns liegen gesehen, wie es das Gesetz befahl*" (Wanderer, if you come to Sparta, report thither that you have seen us lying here, as the law commanded). A shocking story by Heinrich Böll reports of a heavily injured young man who recognizes that he is lying in his former class room where he can still read the words on the blackboard: "*Wanderer kommst du nach Spa....*"

Today, we look back with satisfaction and gratitude on 70 years without war in Germany, during which the institute also continuously developed. In 1953, at the proposal of its then acting director Max von Laue, it was named after its first director in order to honor a great scientist and remember the injustice he suffered. Four years ago, the institute celebrated its centenary. During this period, among others, seven Nobel Prize winners worked at the institute. This glory is, however, tarnished by some dark stains, which have imbued later generations with sorrow and shame.

The manufacturer's authorised representative in the EU is Springer Nature Customer Service Centre GmbH, Europaplatz 3, 69115 Heidelberg, Germany. If you have any concerns regarding our products, please contact ProductSafety@springernature.com

Printed and bound by CPI Group (UK) Ltd, Croydon, CR0 4YY
01/10/2026
02225851-0015